GEORGE S. LONG
Timber Statesman

George S. Long

GEORGE S. LONG

Timber Statesman

CHARLES E. TWINING

University of Washington Press

Seattle and London

For George S. Long, Jr.

Library of Congress Cataloging-in-Publication Data
George S. Long, timber statesman / Charles E. Twining.
    p.   cm.
Includes bibliographical references and index.
ISBN 0-295-97322-6 (alk. paper)
    1. Long, George S., 1853–1930.   2. Lumbermen—Washington (State)—
Biography.   3. Lumber trade—Washington (State)—History.
4. Forest products industry—Washington (State)—History.
I. Title.
HD9760.L65T85   1994
338.1'7498'092—dc20   93-48388
[B]   CIP

Contents

Preface

SCHOLARS HABITUALLY TRY TO JUSTIFY THEIR EFFORTS, and the handiest means of doing so is to claim that their work fills a void. Nearly twenty years ago I did just that in explaining the need for *Downriver,* a biography of Wisconsin lumberman Orrin Ingram, asserting, "Trappers have trapped, miners have mined, farmers have farmed, and seemingly each time that lonesome cowpoke climbed aboard Old Paint to chase another longhorn, a new folk hero was born. . . . But through it all, the lumbermen continued to chop and saw away in relative obscurity. Perhaps Paul Bunyan was simply too big." Whether that contention was true or not in 1975, it would be far less true today: the field of forest history has grown more popular in the intervening years.

I was reminded of this when recently I read a thoughtful piece by David H. Stratton, "Oh, Nova Albion, Ye Were Young So Long," in the Winter 1991/92 issue of *Columbia,* the Washington State Historical Society's fine quarterly. Professor Stratton was lamenting the fact that Pacific Northwest history seemed to get such short shrift within the professional circle of so-called western historians. He has a point, as many who attended the 1991 annual confab of the Western Historical Association concluded after reviewing the program. Clearly, the Pacific Northwest was not favored ground. As a possible explanation, Stratton suggests that Pacific Northwest historians are themselves largely to blame, in not having "done a good enough job of researching, writing, and publishing the region's history." With that admonition in mind, we shall get on with the task.

George S. Long merits our attention because his impact

was so great, both as the manager of his organization and as a regional lumberman. He came to Tacoma early in 1900 to supervise the 900,000 acres the newly organized Weyerhaeuser Timber Company had purchased from the Northern Pacific Railway Company. He would remain in place for nearly three decades. Long obviously built from strength, but he also built wisely. And as one looks at the Weyerhaeuser Company operations in the Pacific Northwest from today's vantage, one immediately sees the influence of George Long. For the most part, the operating units are what he envisioned and, in many cases, established. In addition, the manner in which business is done still reflects to a considerable degree his style. Finally, and perhaps of greatest importance, Long was the one to whom others in the industry looked for leadership, for he saw beyond the moment into a future of growing trees.

There are doubtless some who will look back upon the likes of George Long and accuse him of being a crony and paternalistic. There is truth in this. As for cronyism, businessmen of his day did wheel and deal on a personal level, often reaching agreements with a handshake and an oral commitment. Long would not have minded being called a "good ol' boy." As a matter of fact, he would have taken such reference as a compliment. He valued his colleagues; they may have been competitors but many were also his closest friends. Regarding paternalism, it is true that he cared deeply about his associates and employees. And those with more recent corporate experience may be willing to admit that now and then a little paternalism might not be such a bad thing after all.

It is good to be reminded of the context in which Long operated. He worked before the advent of chain saws and extensive use of logging trucks, before unions were effectively organized, and considerably before environmentalism became the ethos of the land. But, given a bridge across time, he and today's environmentalists would most likely have been empathic. Their interests were much the same, even if their experience and objectives were quite different. It is hard for us to appreciate that throughout the course of Long's lifetime, the practice of forestry was a primitive exercise at best. On the last day of 1929, within a year of his death, Long addressed that disappointing fact in a letter to Austin Cary of the Forest Service. "Forestry matters out here [in the Pacific Northwest]," he acknowledged, "have progressed a certain amount since you used to be in the region. But there is still little actual management of properties in a systematic manner with a view to any operation growing timber for its own continuing needs." That there had been any progress at all was the result, to a very considerable extent, of the efforts of George S. Long. He was able to maintain an optimistic outlook, foreseeing a day in which those who followed would achieve what he and his more visionary colleague could only imagine.

Going from the woods to the office, readers who also happen to be managers will marvel at the breadth and extent of Long's responsibilities. It is amazing. One might argue that he should have delegated more. Still, it is difficult to dispute his success. In part, he worked long and hard because that is what he most enjoyed; it was his reason for being. Another striking feature is that George Long was such a perfect extension of Frederick Weyerhaeuser in nearly every respect. Frederick observed on more than one occasion that his only bad trades were those he had failed to make. Long felt much the same, seeing opportunities where others saw only risk and likely failure. It is ironic in a way. Frederick had seven children, four of whom were sons. Although each son had talents of one sort or another, each was also far more conservative than his father. The reason for this can be easily explained: the father had started with next to nothing and his accomplishments must have surprised even himself. In any case, he hadn't succeeded by being conservative. The sons, however, must have appreciated from the beginning that their ultimate sin would be to lose what Frederick had built. So they had every reason to be conservative, and the wonder is that they took as many chances as they did. This was particularly true of the youngest, Frederick Edward, who proved a staunch supporter of George Long and his commitment to expansion.

At this point it seems only fair to forewarn the unwary reader. I wish it were otherwise, but the truth is that those most likely to be drawn to these pages will already be somewhat familiar with the organization, the industry, or the region, and curious to know more. As one colleague wryly observed, "This is business biography of an industrial strength." So be it. Others will assert that the writer could have organized the book so as to make the reading a bit easier, specifically that we might have dealt with subjects one by one and adhered less to chronology. True enough. There are numberless right ways to write a book. But it is also true that part of George Long's genius was his capacity to juggle a variety of issues simultaneously. Most of us are "task oriented," and would prefer to handle problems in sequence. Few of us enjoy such a luxury, however, for life demands otherwise.

Throughout most of Long's tenure, the Weyerhaeuser Timber Company was primarily in the business of managing lands rather than manufacturing lumber. But under his direction, by the late 1920s the company had become a major producer, with large modern mills operating in western Washington at Everett, Snoqualmie Falls, Longview, and at Klamath Falls, Oregon. Solid wood products were the only commodity. There had been discussion about pulp production, but no serious thought given to further diversification.

One final reminder of context seems appropriate. Timberland aside—and by the end of 1923 Weyerhaeuser owned approximately 1.8 million acres—the company was a small, close-knit, and, in many respects, quaint organization. That same year, 1923, the entire Tacoma headquarters office staff totaled 26. And in May 1930, at George Long's last annual meeting, there were 168 shareholders altogether. Few of them attended the meeting, thereby passing up their only opportunity to learn in any detail about the health of their investment.

Several years ago, I was strolling across the University of Washington campus with the late Professor Emeritus Vernon Carstensen and one of his colleagues, headed for lunch at the Faculty Club. At a fork in the walkway, the colleague suggested one path, the shorter. Vernon vigorously dissented, preferring the longer but leveler route. Ever the historian, he shared the wisdom of James J. Hill: "Never give up altitude easily, boys." Now I don't know if the Empire Builder ever spoke those words. No matter; he practiced them.

I later thought it curious that Vernon had cited James J., because without the efforts of these two, this book would never have happened. As a student of Vernon's, when he was professing at the University of Wisconsin, I was pushed into Andrew

Clark's historical geography course. One of Clark's requirements was to study a subject of our own choosing which significantly involved both history and geography. For a reason since forgotten, I picked nineteenth-century logging in the Chippewa Valley of Wisconsin, where Frederick Weyerhaeuser was the predominant operator. And it was James J. Hill who, at the turn of the century, convinced Weyerhaeuser that his future as a lumberman would be in the Pacific Northwest. And, to complete the circle, it was Frederick who agreed to appoint George S. Long general manager of those Pacific Northwest properties acquired at Hill's urging.

In the fall of 1980, I commenced work on a biography of Phil Weyerhaeuser, a grandson of Frederick and chief operating officer of the Weyerhaeuser Timber Company from 1933 to 1956. During research on this project, I was introduced to George S. Long, Jr. He had served for many years as the Timber Company's secretary, and I valued him initially as the keyhole to Phil's office. We soon became good friends, and out of that friendship developed a mutual interest in a story about his father.

George, Jr., was a wonderful man, a keen observer who was filled with memories. In his younger days, he happily shadowed his father. Later on, and a bit sadly, he was never quite able to escape his father's shadow. But if ever there was a devoted son, it was George, Jr., and the dedication of this book could be to no other. I can only hope that he would generally approve.

Few companies are so well documented as Weyerhaeuser. The reason for this is obvious. While Weyerhaeuser was never strictly a family organization, individual family members have always played key roles in the business. They, partly in recognition of their forebear Frederick's accomplishments, have sought to preserve the record. Indeed, Frederick's son, the aforementioned F. E., titled his own historical effort *The Record*.

Geography also explains the relative completeness of the company's documentation. George S. Long worked out of a Tacoma office from 1900 to 1929, and though he became a power unto himself, he nonetheless reported to absentee owners, most of whom lived in Minnesota. Their correspondence documents the developments in great detail, affording us an intimate, dynamic accounting of not just what they did, but why they did it. One can't help wondering how the business historians of this generation will fare. They will, of course, have an abundance of paper with which to work, but will it mean as much? Today, decisions are made over the phone or face-to-face. Even if managers want to cooperate, their memories of those decisive moments inevitably fade, or worse, change.

Recognizing the value of the written record, George H. Weyerhaeuser, when he was CEO of the company, approved organization of an official archives. The result is historically invaluable and truly a source of corporate pride. The 100th anniversary of Weyerhaeuser is on the horizon, and this patron hopes that the archival treasure will continue to receive the care it deserves.

As with any project of this scope, there are a great many people who provided assistance and encouragement along the way. I have already referred to the key figure in this regard, George Long, Jr. When he died in February 1984, the wind went out of my sails. I still miss him. I also wish to thank his son, George, and daughter, Carol Feige, especially for their patience. There are several at the Weyerhaeuser Company who have helped, beginning with George Weyerhaeuser. Also my friends at the Weyerhaeuser Company Archives—Donnie Crespo, Pauline Larsen, and Megan Moholt—have assisted in countless ways and always in good spirit. Many others have done likewise, notably staff members of the Weyerhaeuser Research Room of Minnesota History Center, and Tom White, curator of the James Jerome Hill Reference Library in St. Paul.

I must also note the contributions of Pete Steen and members of the Forest History Society staff, particularly editorial associate Nancy Margolis, who labored long and hard over the manuscript. Lita Tarver and Leila Charbonneau of the University of Washington Press offered similar assistance, providing their editorial skills with patience and sensitivity.

Finally, I would be terribly remiss if I failed to acknowledge the help of my wife, Dianne Vars. A musician by profession, her love of notes has not dulled her ear for prose. If there remain errors in syntax or punctuation, I assure you that these are minor compared to what would have been the case without her efforts. Most important, I thank her for keeping the faith.

May 1, 1993   CHARLES E. TWINING
                                                  Federal Way, Washington

GEORGE S. LONG
Timber Statesman

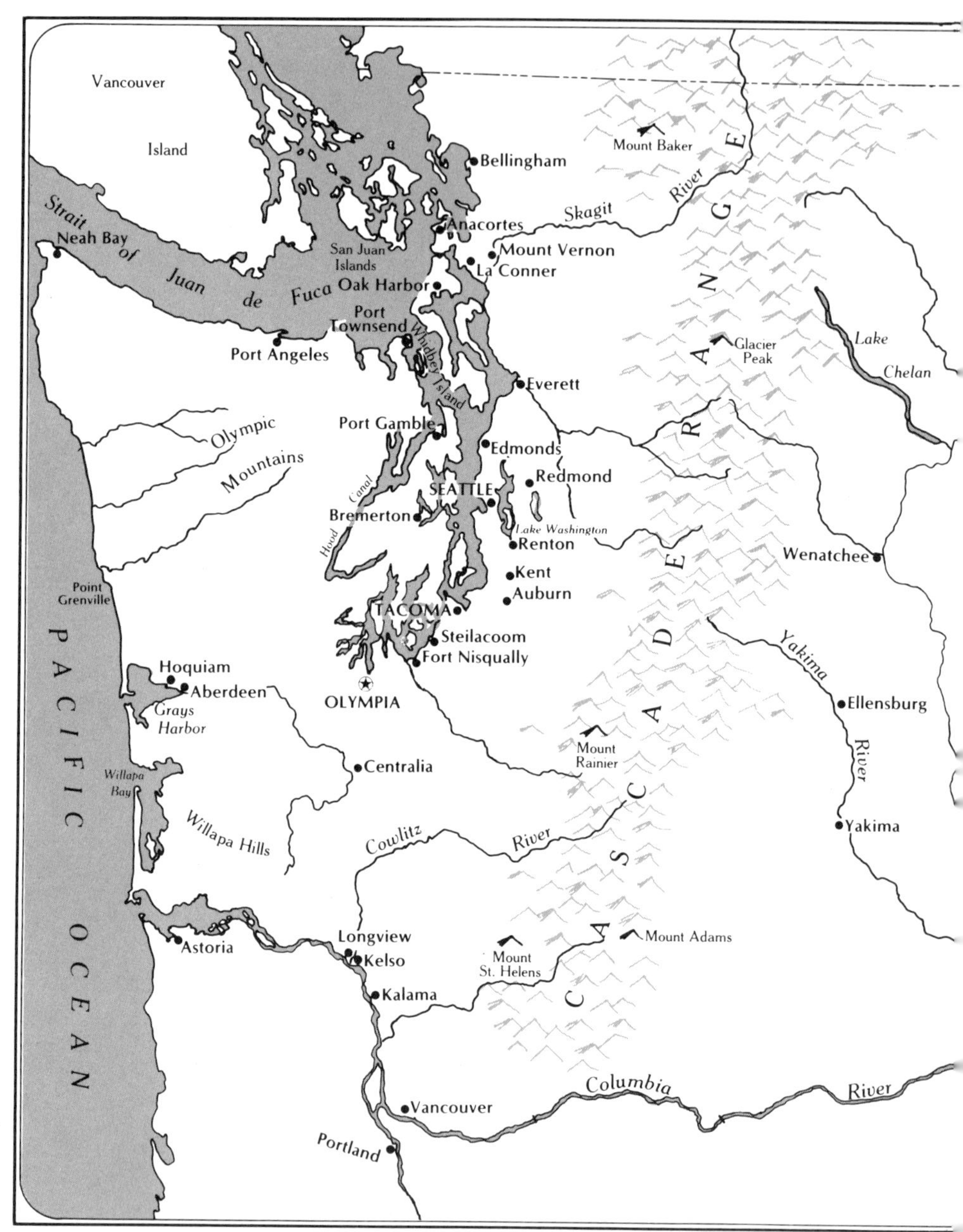

Western Washington.

ONE

Hoosier Beginnings

ALTHOUGH FREDERICK EDWARD WEYERHAEUSER WAS trying his best to stay awake, he kept nodding off, for the room was warm, the hour late, and the conversation tiresome. F.E., Yale '97, had been working hard lately, intent on helping his father Frederick, but he was finding the job difficult. He still knew little about the lumber business of the Upper Mississippi Valley; his father's affairs were incredibly complex; and, while sixty-five-year-old Frederick relished the company of his youngest son, he wasn't at all anxious to pass along business burdens. Not then, and not for some years to come.

It was the summer of 1899, and the four of them—Frederick and wife Sarah, F.E., and neighbor James J. Hill—were sitting in the parlor of Weyerhaeuser's home at 266 Summit Avenue in St. Paul, Minnesota. As usual, Hill dominated the conversation, and the subject was all too familiar. He was a rough-looking sort, stocky, with a full salt-and-pepper beard, and blind in one eye. Indeed, his appearance belied his intelligence. No American knew railroading better than James J. Hill. His Great Northern was testimony to that. At the moment, however, he seemed far more interested in another transcontinental line, the Northern Pacific. The Northern Pacific had fallen victim to the panic of 1893, and was just beginning to show promise of recovery. Hill, for a variety of reasons, had sought to direct if not control the NP's fortunes, and he saw in Frederick a key ingredient in the recipe for success.

Frederick's sagacity was already legendary. Indeed, that was part of the problem: he and his contemporaries had nearly exhausted the timberlands of Wisconsin and Minnesota. If they

3

were to stay in the business, they would have to move their operations elsewhere, to the South or the Pacific Northwest. While Hill may not have understood lumbering in detail, he understood enough to know that it all started with trees. And he knew where there were lots of trees.

Thanks to 1864 federal land grants, the Northern Pacific possessed vast tracts of timber in the Pacific Northwest. Hill saw a wonderful opportunity to fit the pieces together: his friend Weyerhaeuser needed timber and the Northern Pacific needed cash immediately and business—especially east-bound freight—in the long term.

But it wasn't all that simple. Frederick had visited those western parts on a couple of occasions, and while he was impressed with the size of the trees and extent of the forest, the risks were clearly large. The principal species, Douglas-fir, was unfamiliar. The terrain in many areas was difficult, and the logging would be expensive. And where was the marketplace? Perhaps in the end, Weyerhaeuser's faith in Hill's railroading abilities tipped the scales. In any case, negotiations went on for months, with the breakthrough apparently occurring a few days before Christmas. On December 19, Hill sent a coded telegram from New York City to his son James N. in St. Paul: "Thwarting [tell] Weyerhaeuser toothed [that] pensiveness [parties] here faction [expect] tripped [to] additionally [accept] humdrum [his] oscillate [offer] caleb [without] [conditions] implorer [if] tower [they] cannot duchy [do] bleeds [better]."

The final agreement, however, wasn't reached for another couple of weeks. The *St. Paul Pioneer Press* of January 3, 1900, announced the details: ". . . for the purchase of 1,000,000 acres of pine lands in Washington, the Weyerhaeuser syndicate yesterday paid over to the Northern Pacific $6,500,000." The reported figures were close to being correct. The actual terms were 900,000 acres at $6 an acre, or $5,400,000, of which $3 million was due immediately, with the balance to be paid in eight semiannual installments at 5 percent interest. The public had become accustomed to Weyerhaeuser's ways, but this greatly exceeded anything in the past, and was nearly beyond belief. The territory was larger than the state of Rhode Island, and some doubtless wondered whether private enterprise had finally gone too far. The figures merely hinted at the extent of the involvement. When George Long arrived on the scene to take charge, he began to understand. He thought he had done big things previously, but in 1900, "I comprehended that this was an empire instead of a bailiwick."

Frederick had orchestrated the arrangements in the usual way, convincing as many friends and associates as possible to make the effort a joint one. That was the part of dealing which he most enjoyed, and, given his record, he was easy to believe. Besides, the larger the pie, the larger the slices. Now the problem awaiting resolution was who would be selected to do the managing. Frederick had no one in mind. His four sons each had responsibilities of their own: John at Lake Nebagamon, Wisconsin; Charlie with the Pine Tree Company in Little Falls, Minnesota; Rudolph watching over the companies at Cloquet, Minnesota; and young Fred working in the St. Paul office. More important, Frederick would never have wanted any of his children to live in so distant an outpost as Tacoma, Washington.

Thus it was that one of Frederick's associates, Sumner McKnight, nominated a former employee, George S. Long. George was a hard fellow to forget, tall and spare, with a fair gift of a nose, receding hair and chin, and an Adam's apple of note. His accoutrements were a cigar and a celluloid collar. He was seldom seen without either. Although he looked something like a caricature, he was for real. Contemporaries occasionally described him as Lincolnesque. Many would have assumed him to be a schoolteacher or possibly a preacher, had it not been for the cigar. Few if any would have guessed he was a lumberman. But he was, and as such he had firmly established a reputation for intelligence, integrity, loyalty, and hard work. Though Frederick had no firsthand knowledge of Long's abilities, he decided to accept McKnight's recommendation—a decision that would prove nearly as wise as the purchase itself.

George Smith Long was born in Marion County, Indiana, near Indianapolis, on December 3, 1853. He was the seventh of twelve children born to Isaac and Sarah Long, eight of whom lived to adulthood. Isaac Long was a merchant, and in the fall of 1860 he moved his large family to Danville, county seat of Hendricks County, some fifteen miles west of Indianapolis. There he ran a little clothing store, one of several small businesses around the perimeter of the town square. The family lived in an adjacent house, providing the ideal opportunity for young George to observe whatever might be happening.

George Long later wondered whether all small towns were like Danville, recalling that it seemed to him that he "knew everybody" and that he and his school chums enjoyed doing "all manner of stunts together." One of the most pleasant memories involved the swimming hole, about a half mile from town and bordered by "the proverbial sycamore trees." Years later when James Whitcomb Riley published his poem, "The Old Swimmin'-Hole," Long was transported back through time to hot summer days.

> Oh! the old swimmin'-hole! whare the crick so still and deep
> Looked like a baby-river that was laying half asleep,
> And the gurgle of the worter round the drift jest below
> Sounded like the laugh of something we onc't ust to know.

Long was convinced that Riley must have had Danville's swimming hole in mind when he wrote those words, and eventually he had the chance to ask the poet himself. Riley replied that his inspiration had been Greenfield, not Danville, but it didn't much matter; the two seemed one and the same.

In Danville George was also introduced to another pastime, one that would remain a love throughout his life— baseball. As a grown man, he could call a halt to the day's activities, stopping off at any field to watch the game in progress. It didn't seem to matter what was being interrupted or if he was paying for a car and chauffeur, there he would sit, happily unaware of everything except the field of play. Baseball was one of the adult George Long's very few indulgences.

It was also in Danville that George first tested the free enterprise system. Alexander Chambers ran a little combination gristmill and sawmill, alternating power between the two, allowing three days a week to each activity. It occurred to George that there was money to be made on days when farmers brought their grain to the mill; the farmers had time to kill, perhaps a little cash, and maybe an appetite too. With the cooperation of their mothers, who provided bread and butter, pies, and cakes, George and some friends went into business.

But customers were too few, and finally one old farmer told them why: "You can't run a store unless you keep tobacco." So the boys invested fifteen cents, "for which we got a plug of tobacco and three Pittsburgh stogies.

The initial enthusiasm of the partners soon began to wane—largely because they had to "neglect the old swimming hole." Finding that irksome, they decided to go out of business. Eliminating the pie and cake inventory was no problem, but the tobacco was another matter. The boys eventually went off to the woods, "sat down on a log, and each member of the firm lit a stogie," with the predictable results. An older and wiser George looked back at the experience many years later and smiled, concluding that his momentary nausea had served a good purpose: "I did not tackle another cigar until after I had reached man's estate, so it was rather a profitable investment after all." The cure was not complete, however, for when George Long reached man's estate, a cigar became his trademark.

Early in 1865, Isaac Long purchased an "old worn out sawmill," and, chores aside, the first real work George ever did was to help his father dismantle it. Isaac then shipped the boiler, the engine, and a few other usable parts up into Tipton County, some four miles from Sharpsville and fifty miles northeast of Danville. There he reassembled the thing, becoming the owner-manager of a little sash sawmill that, on a good day, could produce upwards of five thousand feet of lumber.

The family subsequently joined Isaac in Tipton County, and George "was around this little mill more or less all the time," helping his father as best he could. For the family, however, the situation in Tipton County proved to be disappointing and worrisome. Malaria was a constant concern, and to escape its threat the family moved to Indianapolis in 1866. Isaac stayed behind, trying to make a success of his sawmill. He would ship about a carload a month of low-grade lumber down to Indianapolis to be piled in the vacant lot adjacent to their little home. As the oldest son available to help with the heavy work, George assumed the responsibility at that end, "selling it to carpenters and other users of lumber in small quantities." George could not have known it, but in 1865 he had already found his life's work. He was twelve years old.

Eventually Isaac moved the little sawmill to an Indianapolis location, where he attempted to develop a specialty trade in walnut. He failed. George later recalled the circumstances matter-of-factly: "He met with adverse business conditions and . . . he was down and out."

Although George had completed only two years of high school, he sadly put aside his formal education to meet his family responsibilities. But jobs were scarce, and the best he could do was a clerk's position in a local real estate office. For four years (1869–73) he worked for the firm of George W. Alexander, after which things "flattened out completely" and the company went out of business. Long, by this time "thoroughly disgusted" with real estate, hired on as a lumber tallyman with Holt & Buggee, a Boston company that operated a large wholesale yard in Indianapolis. But when within a few months Holt & Buggee also declared bankruptcy, Long found himself once more unemployed.

Though Long was at this point hardly what one could call a lumberman, events seemed to be pushing him in that direction. Henry C. Long—no relation—was a prominent Indianapolis hardwood lumber dealer in need of help, busily extending his operations into adjoining states. The firm specialized in walnut, cherry, and oak, and its plans for expansion included not only sales but also new sources of supply. Long started with the company at a salary of $50 a month, a fair sum. The benefit of the association with H.C., however, far exceeded his salary: "I think I learned more under his tutelage than from any man with whom I have ever been connected."

Men and lumber and work were not the only objects of George Long's attention. During high school he had made the acquaintance of Carrie Robinson, the daughter of Colonel Edward J. Robinson, commander of the Indiana 137th Volunteer Regiment at war's end. Carrie, born on September 30, 1854, was a year younger than George. He first noticed her because she was pretty, but they soon discovered common interests of lasting importance. Carrie was an outstanding student, like George fascinated with language and literature. This was a serious young lady, and she and George were serious about each other. We cannot know what early understandings they may have reached. We do know that they were henceforth devoted solely to each other. Marriage, however, would be long delayed. George Long was not going to subject his family-to-be to those privations that had been so painfully a part of his own experience—not if he could help it.

Meanwhile, Long's employment with H. C. Long was taking him farther and farther afield. Lumberman H. C. Long was confronting a problem all too common to the industry: he was running out of timber. Shortages were particularly acute in black walnut, a species nearly exhausted in Indiana for commercial purposes. Since demand for walnut continued to increase, H.C. was forever seeking new sources of supply. The procedure was simple enough. H.C. purchased the timber and dispatched an agent to supervise the logging and then the manufacture of lumber with a portable sawmill. In this capacity George Long spent two and a half years in Iowa, first on the banks of the Skunk River between Keokuk and Fort Madison, and later at other locations, primarily along the Des Moines River. In time Long himself sought out timberland opportunities, managing the responsibility from beginning to end. And following the Iowa campaign, it was on to Missouri, from Gallatin to Albany to St. Joseph, the work varying only in detail.

His next assignment was different, taking him into the Indian territory, "at a time when it was rather wild and woolly, it being somewhat of a rendezvous for escaped criminals." After a couple of years Long's tour was ended by malaria, which, he later noted, came very near to "being my undoing." He returned home sick and weak, vowing never again to work in unhealthy climes. But H. C. Long had other ideas. He knew where the walnut was and that was where he would operate, malaria or no.

Specifically, he had in mind a major effort on the banks of the Kentucky River, and he assumed that George Long would be the on-site manager. He tried to convince George that Kentucky was "a much healthier country than Indiana," but without success. "I was stubborn about the matter," Long later acknowledged, "and would not even go down there to look it over, in spite of the fact that he offered me a handsome increase in salary."

Maybe it was just time for a change. In any case, the association came to an end, with both Longs sincerely feeling the loss. Thanks to the opportunities that H.C. had afforded,

George knew more about a profession than most his age. He had experience in the hardwood lumber business, and, more important, he had been in contact with all types of people in all kinds of situations. It wasn't education in the formal sense, but it proved invaluable.

TWO

North to the Chippeway

GEORGE LONG WAS THIRTY-ONE IN 1884, THE YEAR HE
ended his association with Henry C. Long. Although there is no
supporting evidence, it seems unlikely that his decision to resign
was made without some assurance of another opportunity.
George noted only his eagerness "to live in a country where
they did not have malaria." In the course of his search he had
written to an acquaintance in the employ of the North Western
Lumber Company, a fairly large operation headquartered in Eau
Claire, Wisconsin.

Like many lumber manufacturers in that region, North
Western rafted its product out of the Chippewa and down the
Mississippi to strategic sites. From there the lumber was
shipped by rail, supplying the needs of those who were spilling
out onto the prairies and the plains of the trans-Mississippi
West. The North Western Lumber Company's Hannibal, Missouri, yard was one of many such raft-to-rail transfer points.

By the mid-1880s, however, fundamental changes were
occurring within the industry. One of these involved the increasing competition between white pine and southern yellow pine.
Although yellow pine had difficulty competing in terms of quality, it enjoyed a price advantage; and farmers—building sheds,
cribs, and shelters—thought first of costs. Also, some lumbermen were questioning the economic sense of subjecting high
quality white pine to rafting. These upriver producers were
beginning to ship their lumber entirely by rail, from sawmill to
wholesalers and retailers.

So it was with the North Western Lumber Company in
1884. The owners were phasing out their wholesale yard in

Hannibal, replacing it with one located near the mill site, at a little town called Porters Mills, about five miles downstream from Eau Claire. And it was to Porters Mills that George Long went, hiring on as the new yard's shipping clerk. His starting salary was $75 per month.

This was a transitional period for the firm, and in the first year of Long's employ only about fifteen million feet of lumber was piled at Porters Mills, the balance rafted downriver as before. But the changeover was soon complete, and in 1887 the entire cut went to the new yard. Long's responsibilities grew along with Porters Mills. After only three months, he was promoted to yard foreman. It all seemed new to him, as well it should have. The people were different, their manners and style, and there was little about the manufacture and sale of white pine lumber that was familiar. His hardwood experience, however, did provide an important advantage when it came to grading, as he knew "more about the manipulation of lumber and the grading of it than was characteristic [of those in] the pine mills." The concern for proper grading would prove to be central to his success and a subject that he forever held dear. It was the test of institutional performance and integrity, and the point at which money was made or lost.

The decade of the 1880s was generally prosperous for Lake States lumbermen, and the North Western ownership sought to make the most of the good times. Delos R. Moon was the president, residing in Eau Claire; Vice-President J. T. Barber watched over the Hannibal interest until the dissolution of that branch; and Sumner T. McKnight was the secretary-treasurer, dividing his time between Milwaukee and Minneapolis. The problem they faced, ever increasing in seriousness, was obtaining enough logs. Competition for the remaining pine was keen. What some had presumed to be an inexhaustible resource clearly was not.

The North Western Lumber Company constructed a new mill at Stanley, Wisconsin, a few miles east of Eau Claire, and another at Sterling, after which the firm purchased two Eau Claire mills formerly owned by the Weyerhaeuser interests. And finally a controlling interest was bought in a company near Hurley, the Montreal River Lumber Company. George Long had been promoted to sales manager, responsible for the sale of all of the lumber produced, but by now he was much more than strictly a manager of sales.

Coincident with the company's expansion, North Western's ownership was reducing its involvement with the day-to-day management, and it was George Long who quietly picked up the reins. McKnight, like many others, had found southern California—specifically Santa Barbara—to his liking, especially in the winter months. More and more he contented himself with secondhand reports such as one from Long in late January 1887. "I have just received a nice long letter from Geo.," he reported to partner Moon, "which done me lots of good."

But McKnight and his partners had not retired. Not yet—not quite. They and many other Lake States lumbermen were finding all too few opportunities for new investment in familiar territory, and some had begun to consider distant forests. The most prominent of these was Frederick Weyerhaeuser, already something of a legend. Moon and his associates readily acknowledged the remarkable abilities of their friendly competitor. No one was Weyerhaeuser's equal when it came to seeing far into lumbering's future, and they realized they could do no better than to attach themselves to Weyerhaeuser's coattails, if given the chance, and hang on.

But until such a chance presented itself, the partners were determined to do everything possible to take advantage of the present favorable conditions. As McKnight told Barber: "If you cut 225,000 feet of logs per day, you and Jim Brown [the Eau Claire mill superintendent] may draw on me this fall for a good suit and I won't be stingy about the quality either and I don't believe Geo. Long's back will trouble him very much in selling if he don't get married."

The reference to Long's marital status may well have been in jest. He was nearly thirty-five years old and must have seemed every bit the confirmed bachelor. He shared a small house with his brother James, then twenty-two, and while they both enjoyed the company of others, their circle of friends was mostly limited to male companions and two sisters, Rhoda Belle and Susie, who had come to Eau Claire to keep house for their brothers.

Through the years Carrie had continued to wait in Indianapolis, patience personified. The lengthy betrothal finally ended on June 6, 1889, when George and Carrie married in Bedford, Indiana. There was little in the way of a honeymoon, the newlyweds hastening to Eau Claire, where the lumber awaited.

George Long managed to sell what the North Western mill produced in 1887, but the selling did not come easily. Indeed, too often it was done at little or no profit. Competition was always severe, the industry chronically producing more than the markets could bear. There were countless manufacturers, many of whom were compelled to continue operations with little regard for demand and price, a small return being better than none at all. These circumstances seemed a given. Thus salesmen such as George Long were locked in a never-ending struggle, trying to outmaneuver or outwit one another, seldom considering what might be done to bring some order to their chaotic world.

Doubtless George Long's success as a salesman had much to do with his increasing knowledge of the territory coupled with his appreciation for differences in lumber grades and species. But as is the case with all exceptional sales personnel, it was his personality that opened doors and kept them ajar. No one who met George Long would forget him, and almost without exception the remembrances were pleasurable. If ever one walked among his colleagues in absolute fellowship, it was George Long. His customers were his friends, or soon would be, and they would subsequently delight in repeating stories they heard first from him. Usually these stories related in some way or another to the subject at hand, making a point clearly and effectively. Depending on the audience, they could border on the risqué, or at least were earthy. Even if they failed to illustrate a point, they amused and helped to cement relationships. They were Long's way of putting his customers at ease, for he possessed that rare quality of always thinking first of how the other person felt. Aside from the storytelling, dealing with Long was an exercise in square dealing. If George Long said it, it was either true or he believed it was true.

Long's talent for relating to people was innate. If the word "charisma" had been in general use then, it would have made people think of him. George Long had charisma, and he knew it. One realizes that immediately from the photographs: There he stands, somewhat apart, presenting his profile to the camera. In picture after picture, he dominates the scene.

Because Long carried the title of sales manager, some assumed—then and later—that his experience was limited to sales. Not so. He carefully watched over the entire process, from the stump through the mill into the piles. But he spent more time in the sorting yard than anywhere else, for that was where those numberless decisions were made, each one critical in terms of profits and losses. He seemed never to tire of watching those who worked away at the sorting and marking, and after hours he would inspect the piles, accumulating questions to ask the mill superintendents.

That Long's responsibilities were broadening was not something planned or even ordered. The company's operations were expanding at a time when its owners were lessening their personal management, in part because they were growing old, but also because they had complete confidence in their manager.

One of the results of Long's increasing workload was that he had to spend considerable time on the road. He enjoyed dealing directly with company personnel and with those whose business he sought, but it wasn't easy to be absent when so much was happening at home. Long and Carrie had become the parents of two daughters: Margaret, born on March 15, 1891, and Helen on October 30, 1893. When a son, George, Jr., was born on October 30, 1895, the family circle was complete. But duty called and, for better or worse, George Long seldom let family get in the way of business.

Things had gone well for Long in Eau Claire, with regard to both family and work. As he looked ahead, however, uncertainties abounded. Delos Moon had died and McKnight, his successor as president, was elderly. More important, North Western's supply of available lumber was limited. By the end of 1899 only the mill at Stanley, Wisconsin, had any possibility of extended operation. The company, like most similar organizations in the region, had liquidated its resource. If there was to be a future in lumber, it would have to be elsewhere.

George Long hated to see things coming to an end, but he had to find another opportunity, even if it meant leaving Eau Claire. What he found was a position with a Madison, Wisconsin, lumber dealership, Brittingham & Hixon. It was difficult informing those at the North Western Lumber Company of his decision, and it was also difficult for those receiving the news. On Christmas Day 1899, Sumner McKnight sat down and penned a note to his longtime employee and good friend: "The event of your severing your connection with our Company is very much regretted. Our relations both social and in a business way during the whole time that we [have] been associated together (seventeen years) have been most agreeable and pleasant. If there has been any dissatisfaction or feeling other than friendship during all these years I do not know it. Now as our business relations are about to terminate, our Company wish to acknowledge their appreciation of your ability, your loyalty, your integrity, and your constant energy in our behalf. We will miss your judgment and personality in our councils." Events would shortly prove that McKnight meant every word of it.

At that moment, other lumbermen were making plans of their own, on a very large scale. For the same reasons that George Long had decided to move on, these lumbermen, including the North Western Lumber Company owners, were beginning to investigate new opportunities. Leading the search was Frederick Weyerhaeuser.

THREE

West with Weyerhaeuser

IN THE SPRING OF 1899, THE EVENT THAT WOULD change the course of the U.S. lumber industry was yet only an idea. Frederick Weyerhaeuser and his associates had begun to acquaint themselves with opportunities in the Pacific Northwest during the mid-1880s, but the territory was unfamiliar, and they proceeded with caution and conservatism. Others took the lead. In 1888 four midwestern partners purchased 80,000 acres from the Northern Pacific Railroad, committing themselves to constructing a mill at Tacoma's Commencement Bay. Their partnership, the St. Paul & Tacoma Lumber Company, caught the attention of many, not the least of whom was Weyerhaeuser.

The panic of 1893 brought a halt to most plans for new investment. But recovery gradually occurred and, encouraged by the Klondike Gold Rush, development resumed in the Pacific Northwest. As the economy improved, the lumbermen dusted off their old dreams of western expansion. In 1897 Weyerhaeuser joined with Lindsay & Phelps, a Davenport, Iowa, firm, and purchased some Skagit Valley timberland in the northern Puget Sound area. The Sound Timber Company was subsequently organized to manage these holdings, with Weyerhaeuser & Denkmann, the original Rock Island, Illinois, partnership, purchasing half of the Sound Timber Company's $600,000 capitalization. Then in 1898 Weyerhaeuser and several colleagues became partners in what was known as the Coast Lumber Company. This proved to be a short-lived endeavor, but at the time they hoped that by means of the Coast Company they could stabilize the price of shingles while taking advantage of low spring freight rates over the Great Northern.

By the spring of 1899, Frederick Weyerhaeuser had known William H. Phipps for several years, having negotiated with him at various times in the purchase of Chippewa Valley lands when Phipps was an agent with the Chicago, St. Paul, Minneapolis, and Omaha Railroad. Both Phipps and James J. Hill were St. Paul residents, and in fact Hill's mansion was next door to the Weyerhaeuser home on fashionable Summit Avenue. In truth, they were closer as neighbors than as friends. Still it was easy to get together and chat about matters of mutual interest, of which there were always plenty. Phipps was now land commissioner for the Northern Pacific Railway Company, and Hill, as earlier noted, was extending his influence in that organization.

It would have been interesting to listen in. No doubt the discussions were to the point; there was little reason to be indirect. Also, neither Hill nor Weyerhaeuser was much accustomed to wasting time. It appears that Phipps was the prime mover, the essential negotiator, even if the deal was one whose time had come, or nearly so. Weyerhaeuser was eager to make a major purchase of Northern Pacific timberland in western Washington. And Hill and Phipps were anxious to sell. The only questions seemed to be how much land and at what price.

Initially Weyerhaeuser made an offer of $5 per acre for a million acres, which, as Phipps wrote to C. S. Mellen, Northern Pacific president, "I rejected because [it was] too low." Phipps was convinced that Weyerhaeuser would pay as much as $6 an acre, but even that seemed low, since he believed their worth, "taken as an entirety," to be from $7 to $7.50.

Weyerhaeuser and Phipps met again in August and, although differences remained, it was apparent that considerable progress had been made. Phipps reported as much to Mellen: "Mr. Weyerhaeuser is a very successful and wealthy lumberman and, should he purchase, I understand he will associate with him other lumbermen of strong financial standing." In short, this was a very special opportunity, one that Northern Pacific would not likely encounter again. Mellen's response encouraged the negotiation, but stipulated that "all of the timber cut and lumber manufactured shall be shipped by Northern Pacific lines to destination, rates being in all cases equal to those by any other line."

That freight shipment clause immediately became the

focal point of contention: Weyerhaeuser was unwilling to place his future in the hands of a single railroad, and the Northern Pacific officials were unwilling to do anything that might benefit their competitors, specifically the Great Northern and the Canadian Pacific. A compromise was eventually reached in December, when Weyerhaeuser insisted that the transportation clause be limited to a period of fifteen years. Under those terms, from Weyerhaeuser's vantage the "compromise" amounted to little if anything since he had no plans to undertake manufacturing any time soon.

The new Weyerhaeuser Timber Company initially offered sixty shares of $100,000 each, totaling $6 million. The original investors in the company were old friends and competitors, reasonably comfortable in relationships that had evolved over the years. Weyerhaeuser & Denkmann agreed to subscribe $1.8 million, and Laird Norton & Company, $1.2 million. Sumner T. McKnight, Orrin H. Ingram, and Robert L. McCormick each subscribed $350,000, and the two Lamb brothers of Clinton, Iowa, $300,000. Nine other incorporators contributed at varying lesser amounts. The original investors considered this simply one of a long series of Weyerhaeuser-inspired investment opportunities. In this instance, those who chose to participate did so more out of faith than conviction. None could have foreseen where their decisions would take them.

When the agreement became public on January 3, 1900, interest naturally focused on the obvious, the purchase of 900,000 acres for the sum of $5.4 million. Those figures were sufficiently impressive. What was probably overlooked by casual readers was mention of options for additional purchases. The 1864 land grant to the Northern Pacific had been colossal, amounting to 25,600 acres per mile across the territories, and 12,800 acres per mile through the states; of course at that time everything to the west of Minnesota was territorial. In short, Weyerhaeuser and associates had purchased but a fraction of NP's total holdings, and both parties were agreeable to trades in the years ahead. And these options would be exercised, but only after there was a better understanding as to the particulars. That would be a central responsibility for the Timber Company's general manager.

Years later George Long would recall that he was "not

especially pleased" with his employment with the Madison firm of Brittingham & Hixon, although he had received "a very attractive offer and [was] given a great deal of freedom in the management and development of the company." T. E. Brittingham had been in charge in recent years, but for reasons of health and an interest in travel, he was anxious to take leave of his business responsibilities. If Long had any misgivings about the agreement, they were not shared by Brittingham, who wasted no time in making reservations for a round-the-world excursion.

But three weeks after he commenced work in Madison, Long received an urgent telegram from his former employer, Sumner McKnight. McKnight outlined the organization of the Weyerhaeuser Timber Company, noting that he had recommended Long to be "its western representative." Was Long interested? It was an awkward situation. To request an immediate release simply wasn't Long's idea of fair dealing. And yet, the offer from McKnight wasn't just any old opportunity. Deciding he had nothing to lose, Long shared the telegram with Brittingham, who was disturbed over it and asked what Long proposed to do.

What Long proposed to do, of course, was to abide by the terms of his contract. At the same time, however, Brittingham had to know that Long found himself in this predicament not by design. It was a dilemma.

McKnight, in the meantime, refused to take no for an answer, telephoning to inquire whether things might not change given a little time. On January 22, 1900, he wrote a letter to that effect: "As is usually the case, when the prospects are that we cannot get you, we want you all the worse, and we were quite disappointed in receiving your dispatch and letter on the subject. In talking with Mr. Weyerhaeuser, I asked him if it would be possible to put a man there temporarily and let you take the place later—say in two or three months. He said it might be that we would have to do that if we could not get the right man for the place. But it is quite necessary that we have a permanent man there as soon as possible." McKnight again expressed the hope that Long would find some way to accept, adding, "It is generally conceded by all with whom I have talked that you are the right man." In closing, he urged Long to come up to St. Paul to discuss the matter with Weyerhaeuser. "I have no doubt," he added, "that when you see Mr. Weyerhaeuser it will be all satisfactory." McKnight wanted to make certain that his friend understood just what an opportunity this was, "such a rare chance for a good thing that I hate to see you miss it."

As fate would have it, on that same day Long received a telegram that solved his dilemma and changed his life. While traveling in the South, Brittingham wired the following to his new manager: "You can go away within thirty days or less. Owing to your short stay and our loss you [are] to waive salary. Treat [this] strictly confidential. Fix up Sturgeon Bay at once." A letter followed, Brittingham promising to be home within ten days.

Long doubtless fixed up the Sturgeon Bay problems, but his mind was elsewhere: "I am rather idle and restless today," he wrote Carrie on February 6, "and inasmuch as my salary is not marching along, wish I was out of it, and on with the new." Some of his new associates had visited him on the previous day, before heading to Tacoma for a meeting of the Weyerhaeuser trustees. They "expressed regret at my not being able to go with them . . . and I felt very much like going anyhow but of course did not."

Brittingham showed up on February 7, promising Long that he could leave within the week. And in fact, Long was on his way on February 11, after convincing Carrie that it was foolish for them to attempt to go "all in a bunch without knowing where we go, and how long it would be before we got our goods and became settled." In addition, there was uncertainty about it all. It was as close to a jump into the dark as George Long would ever make. What did he really know about timberland? And what could he know of western Washington, Douglas-fir, and the problems associated with managing nearly a million acres? As he wrote in his letter to Carrie on February 7: "Who knows (judging by the past), I may get a job in California after being in Washington a few weeks, and may not want to move to Tacoma."

On February 9, the Weyerhaeuser Timber Company trustees met for the first time at the grand Tacoma Hotel, where Ingram offered the key resolution, enthusiastically seconded by

McKnight: "That George S. Long be and he is hereby appointed agent of the Weyerhaeuser Timber Company in the State of Washington, for the term of one year at a salary of Five Thousand Dollars per year, with authority to employ and discharge subordinates, agents and employees of this corporation, and fix their compensation, to make contracts for the purchase and sale of real and personal property, to accept, sign or endorse checks, drafts or bills of exchange, all subject to the approval of the president of this corporation." None, not even McKnight, could have realized the wisdom of their decision.

The great adventure began at the Eau Claire station on Sunday evening, February 11. George wrote to Carrie and the children on Tuesday, as the train began its climb into the foothills of the Montana Rockies, "in the midst of snow, and commencing to get some mountain scenery." Everything was new and promising, but the future was by no means assured. As he put it, "I have a growing feeling that we will like the change, if I can fill the position, and it is a big one, but we will count no chickens yet."

Tacoma's welcome was less than gracious, what with temperatures below freezing and snow on the ground. Long arrived at noon on Wednesday, February 14, greeted by Timber Company trustees R. L. McCormick and F. S. Bell. He was given no chance to wander around Tacoma, as plans were already in place to introduce him to the region at large, including Portland, Grays Harbor, Olympia, and Everett. Long listened respectfully to what others had to say and repeated much of what he learned to Carrie: "People say Seattle for business and Tacoma for residence, so we will come to the right town—am inclined to think the 400 here are gay and chic—hear about clubs galore, golf links and so on, but of course we do not belong in that class."

As for the natural setting, Long's initial reaction was more like that of a midwestern farmer than a lumberman. It seemed such a big country, everything from the mountains to the trees almost beyond belief and certainly beyond description. "West of the mountains," he wrote his family, "the all pervading forest seems to surround and hem you in—there is an absence of farm and village, which is rather dismal," immediately adding for Carrie's benefit, "but of course this is not noticeable at Tacoma."

The tour ended on the evening of February 17. By then Tacoma's weather had returned to normal, with temperatures balmy and "a soft, springlike smell in the air which is quite delightful." Long had yet to view Mount Tacoma (Rainier), although he was convinced that "it is a real thing" to which he was prepared "to bow down and admire whenever the clouds will let me see it."

But Long had more pressing concerns than sightseeing: they involved down-to-earth decisions concerning a home for his family, an office location for the Weyerhaeuser Timber Company, and the personnel to staff it. Long's first staff move was hiring a secretary "and plat maker," Monemia Evans, a woman who had come to Portland from the Chippewa Logging Company. Monemia Evans was therefore known to those who mattered, the Chippewa Logging Company being the 1881 Weyerhaeuser creation, "the pool" that managed the distribution of logs for the valley's operators.

As for locating the office, there was only one logical place, the headquarters building of the Northern Pacific. Not only was it situated in the center of the commercial district, but it also housed the railroad's land office with the indispensable maps. Thus in early March, Long wrote to Monemia Evans in Portland: "We will be ready any time after this week to occupy our new offices . . . Rooms 19 & 20." It was time to go to work. (The building still stands, recently renovated and today an office complex at 621 Pacific Avenue, but the wing that housed the Timber Company office is no more.)

In the meantime, Frederick Weyerhaeuser waited impatiently for word from the West. Although he had confidence in George Long, it was a confidence based largely on the opinions of those he respected, and Weyerhaeuser was not one to let matters take entirely their own course. He appreciated being reassured that they had indeed made the best possible choice. Captain W. R. Bourne of the Shell Lake Lumber Company, another of those small Wisconsin operations in which Weyerhaeuser held a partial interest, reported on a recent discussion that he knew would be of interest to Long. Weyerhaeuser and Bourne had been on a woods tour and "the matter of your engagement came up." Bourne was asked his opinion. "I told him," or so he wrote, "I did not know of a man in my whole lumber experience that I felt was so capable of taking charge of their work on the Coast as Geo. S. Long."

That was all well and good, of course. But Frederick Weyerhaeuser rested easier after he convinced Hugh Stewart to trade Eau Claire for Tacoma and serve as the Timber Company's accountant. Weyerhaeuser informed Long of this appointment in a matter-of-fact letter: "I engaged Mr. Hugh Stewart, a man who has been in our employ for the last 17 years, is still a young man, and you will find him competent, reliable and trustworthy in every respect." It was akin to double-bolting the door. Stewart had served Weyerhaeuser faithfully, with unquestioned honesty since he was sixteen, first at the headquarters of the Mississippi River Logging Company at Beef Slough and then at West Newton. Now a family man of thirty-two, he obediently prepared for the trip west with wife and four children, the youngest only four months old.

On the occasion of Stewart's death in 1924, Long would recall the circumstances of 1900: "Mr. Weyerhaeuser made but one request of the managing officer, that Hugh Stewart be made the chief accountant." While it is easy to imagine that such an insistence might have chafed, Stewart was as modest as he was efficient, and he would be as loyal to George Long as he had been to Frederick Weyerhaeuser.

With the arrival of Stewart, the office of three was complete and work began in earnest. Where to begin was no problem. The company had purchased 900,000 acres, but that was all that was known. The land's location and how much timber it contained were questions awaiting answers, the Northern Pacific having promised, as McKnight advised, "to make any mistake or error good to us."

Long many years later acknowledged that "neither by training nor experience was I equipped for things that were most important to be done." Part of the challenge concerned the very scope of the task. What became immediately evident was that his previous responsibilities, which he had assumed to be large, were insignificant by comparison. "I comprehended that this was an empire instead of a bailiwick," he wrote in 1925, and as such it required a very different sort of management.

The facts surrounding the Weyerhaeuser Timber Company were beginning to be understood, both in the company's office and in a great many offices throughout the region. Smaller operators had yet to be assured of a future timber supply. Making purchases from the Northern Pacific was one thing; dealing with a colossus of a competitor was quite another. As Long described it, the Weyerhaeuser empire "had an industry tied on to it, a struggling industry," one which "by force of circumstances or of habit or of necessity, was largely an industry where the mill owner had no timber." The fact was that these operators had been purchasing timber as they needed it from the Northern Pacific. Then they awoke one morning to learn that their source of supply was owned by a syndicate of lumbermen. Naturally they assumed the worst. But just as naturally, it was crucial for the Weyerhaeuser Timber Company to be accepted by them and their communities. There would be no future in trying to operate in a hostile environment; an absentee owner was regarded by all as fair game.

It is simply impossible to imagine an individual better equipped to deal with such circumstances than George S. Long. He personified integrity, able to move among the important and the less so with equal ease and respect. If this was the representative of a mighty syndicate, then perhaps the syndicate might not be so bad after all. As Long later observed, "The job that confronted me for the first ten years was not a lumberman's job, but a diplomat's." And it was a job made all the more difficult by the times—that of the muckrakers and "the growing disposition to what is generally known as a popular antagonism to big corporations."

The business climate was indeed undergoing rapid change during the years now known as the Progressive Era. Growing concern with monopolies and political corruption, child labor and sweat shops—and a host of other ills—would have far-reaching consequences. And, perhaps most important for Long and his colleagues, the public was increasingly anxious regarding wasteful use of natural resources. "Business as usual" may always have been something of a phantasm, but in the early 1900s business promised to be more unusual than ever before.

There was another fundamental consideration. The

Weyerhaeuser Timber Company investors understood that there was little likelihood of any early return. Whether Frederick Weyerhaeuser actually said, as has been widely quoted, "This is not for us, nor for our children, but for our grandchildren," most of the investors understood that they had invested in the future, in a wilderness, that they had sent an agent out to manage it, and that any profits were likely to be a long time in coming. Long's recollections years later suggest more confidence than he actually felt at the time. Nevertheless, it was true that "most fortunately of all, the founders of the company were . . . not impatient for immediate results and were willing to trust responsibilities to whomsoever they selected."

Long was immediately beset by overtures from a number of "small mill men" who were eager to buy "something that they of course claim is of no value to us, etc." At the same time, timberland owners were offering parcels for sale. In terms of these dealings, Long proved as adept as Frederick Weyerhaeuser, forever less anxious either to sell or buy than the one across the table. Of course, his 900,000-acre base gave him a tremendous advantage. He was always pleasant, but when a deal was struck, it almost seemed as if he were granting a favor. As for purchases, he assured Weyerhaeuser that he would make them only if "extraordinary cheap, without referring the matter to you, and such investments that I may make will be of a small character." And he requested "to have word from you soon as to the general policy to be pursued out here."

No very useful policy statement was forthcoming. Weyerhaeuser wanted to help, but he knew too little about the situation on the ground, and perhaps already realized that Long understood enough to avoid serious error. Years later Long would say as much himself: "I think many of the problems were new to most of the people who had been prominent in the lumber business in the Mississippi Valley."

In fact, time and distance demanded that Long take hold. His early answers to inquiries were prototypical, indicating a policy evolving. To an offer to sell, he responded, "Unless your land should contain more first class timber than your own estimate puts upon it, we would hardly feel as though we would care to purchase it at the price you name." He seldom closed the door, adding that if it was "well located and a good quality," a reasonable price would receive serious consideration, and he closed expressing thanks "for calling our attention to this matter." And to one interested buyer, he stated that it would be "the policy of this company to sell timbered lands to manufacturers in all cases where it would not seriously interfere with plans which we have of our own for manufacturing." In this instance, he was confident that a deal could be made, but he asked for time, explaining that they were not yet "in very good shape to put prices on our land." Those two responses suggest the relative strength of the Timber Company, making considerate and respectful exchanges all the more crucial. It was something of a tightrope, but Long seemed able to skip along it. Many resented the power of Weyerhaeuser; few resented its manager.

Negotiations with the Northern Pacific for additional purchases continued. Thomas Cooper was the western land agent for the railroad, and his representative, a Mr. Wade, began to go over the estimates in detail with Long. In the process, Long commented "in an off hand way," as he explained to Frederick Weyerhaeuser in a letter dated March 16, that as much as 30 percent of the lands purchased "were practically unfit for timber lands." Wade didn't agree completely but did allow that some 150,000 acres were without timber and another 100,000 acres were "worthless on account of having been logged off or burned" or having too little timber to be considered. In a subsequent letter to McCormick, Long expressed surprise at Wade's frankness. "He does not hesitate to give his opinion of any lands" and was generally proving to be "a man of remarkable information."

Clearly there was need for adjusting. Just as clearly, such adjusting would depend on information provided by cruisers. Long was impatient to get Timber Company woodsmen busy investigating and protecting. This was easier said than done. The work was physically demanding, with the ubiquitous blackberry bush a formidable obstacle. While fighting one's way through the underbrush of a fir forest, estimating had to seem, at times, trivial. It was upon those figures, however, that decisions involving thousands of dollars would be made, again and again.

And, of course, cruisers had to be trustworthy. One of the first hired was Myron Cole. Cole had grown up in the

Chippewa Valley but Long had not known of him there. Thus he inquired of a mutual friend what he thought of Cole "as to all around make up as a valuable man to watch for trespassers and to look after our interests in a general way in one or two counties. Is he discreet and observant, and faithful in a sense of being loyal to your interests?" Cole was highly recommended, although allowing that "his penmanship is poor (like that of many woodsmen) but legible." That was good enough; it was figures, not letters, which were important. Cole was hired.

Even the best of figures would be of little benefit, however, without a practical system of bookkeeping. Long decided to utilize what he called a "card system" whereby each forty-acre tract would have its own indexed card (later a folder), and all information pertaining to the tract would be recorded, everything filed together: cruising estimates, taxes, right-of-way agreements, logging contracts. In addition, Long oversaw the development of a mapping system that by means of a color code indicated, almost at a glance, the history and current status of each section. Thus, almost piece by piece, Long began to visualize the holdings, to understand just what the Weyerhaeuser Timber Company was.

There were limits to what Long could learn at a desk. On March 19, he set forth; Stewart was expected within two weeks, "and Miss Evans is due here tomorrow, so I will soon be fixed up so that I can chase around the country a little, getting acquainted with the assessors and everybody else whose acquaintance is well enough for us to make." In a relatively brief period, he would know a great many by name and reputation. And they him.

Years later, Minot Davis, the Timber Company's woods superintendent, enjoyed reciting a story that Long had told about those early days when he was visiting mill after mill in western Washington, watching, listening, and learning:

> He was fond of telling how he stopped one day at a small mill in Lewis County, and for some time watched the circular saw cut laboriously into the big fir logs. Used as he was to saving every bit of clear lumber, he was shocked to see the very best part of the log going into timbers. He sought out the proprietor, a gray bearded veteran in patched overalls, and inquired

the reason for such procedure. "Are you in the lumber business?" inquired the old man. Mr. Long admitted that he was. "Lived here long?" asked the mill man. "Just came here from Eau Claire, Wisconsin," replied Mr. Long. "Thought so," said the native, and after a moment continued, "Young man, whenever you see us doing something out here that's different from anything you ever seen done before, you can make up your mind there's a damn good reason for it."

While the story was likely apocryphal, its point was legitimate. Long learned the lesson and thereafter "set to work to find out the reasons why lumbering out here was different from the lumbering I knew." He was forever the student, appreciating that there was always more to be learned, and enjoying the process. Further, he readily altered old theories in the face of new facts.

Some matters, of course, were beyond theory, and these included directives from Frederick Weyerhaeuser regarding selling and buying timberland. Sales, he suggested, should "add about fifty percent to our estimates, and then sell at about $1.25 [per thousand feet of timber]," although he immediately added a most typical qualification: "I believe the less we sell and the more we buy, the more money we make," and he urged Long "to block out our land as fully as we can."

Since the Northern Pacific purchase involved only odd-numbered sections, the Timber Company holdings resembled a giant checkerboard. Weyerhaeuser's buying policy simply stated the obvious: they should try to buy as cheaply as possible all the even-numbered sections within the pattern. In his selling directive, however, he offered plans for the future: "Only sell at a good profit, as we expect, in the course of time, to manufacture the most of it." In fact, Weyerhaeuser hadn't shared this vision with all of his associates, some of whom saw a clear distinction between, for example, a timber company and a lumber company. But those few words would describe the policy that Long would follow in the years ahead—blocking out the districts with a view to eventual manufacture.

The opportunities for making money under Weyerhaeuser's suggested terms seem nearly inevitable, at least in retrospect. The original purchase price had been $6 an acre. The contract assumed an average of 16,666 feet of merchantable timber per acre, making for a purchase price of thirty-six cents a thousand. Were they to inflate the estimate of merchantable timber per acre by "about fifty percent," they would come up with 25,000 feet. And were this to be sold at $1.25 per thousand, they would realize $31.25 per acre, amounting to an appreciation in excess of five hundred percent. (In 1921 the Bureau of Corporations published a report estimating that the actual amount of merchantable timber was such that the $6-per-acre purchase price amounted to about ten cents per thousand feet.) The advantages of buying wholesale and selling retail were seldom more clearly demonstrated—that is, assuming there were buyers at such prices.

In fact, there weren't. Long subsequently reported to F. S. Bell that he had "tried it [i.e., making sales at $1.50 per thousand] on two different mill men and it had worked to a charm: That is to say, they did not buy." But it was a complex problem. Although the Timber Company might prefer to avoid sales, at the same time they couldn't deal cynically with potential buyers. Indeed, Long advertised "as policy," or nearly so, that the Timber Company was willing to sell timber simply because he thought it desirable that this be understood than to have it rumored otherwise. There were also tax ramifications: "If we get all the little mill men down on us . . . at this particular time they would have quite a little influence with the County Commissioners and Assessors." And it was also true that by putting a very high price on timber, they would risk attracting the attention of those same commissioners and assessors.

Finally, the purchase had not been speculative, at least not in the mind of Frederick Weyerhaeuser. It was his intention to remain in the lumber business, and he had acquired the timberland for that purpose. Long considered opportunities to buy and sell within that context. Would such transactions lend themselves to an eventual efficiently operating unit? His charge was to keep the long-range interests of the Weyerhaeuser Timber Company foremost in mind while recognizing the legitimate needs of the small independent operators.

The decision of Weyerhaeuser and his colleagues to invest in the Northwest naturally made others look in the same direction. As a result, Long received letter after letter from some he had known in Wisconsin and from many who had only heard of him and the new frontier. In answering them he always avoided undue encouragement. As for employment opportunities with the Timber Company, there were none. They were only looking after their timberlands, and until the commencement of logging and manufacturing they would have no occasion to increase their force. He often closed such letters urging the correspondent "to stay in Wisconsin as long as times are good as they undoubtedly are better there than they are out here." To one inquiry Long responded that western Washington would "be all right when I am an old man . . . but there is nothing for them to brag about at present." Certainly there were pleasant aspects to life on the coast, "especially if one can live on scenery, for it is grand," but he maintained that "Wisconsin is better than Washington, today and also for some years to come."

An immediate problem was taxes. Washington, unlike Wisconsin, had no town governments, so at least there were fewer officials with whom to contend. But there was little or no consistency among Washington counties in assessment policies: some were content to work with large blocks, while others preferred to deal with forty-acre descriptions. The original 900,000-acre Weyerhaeuser purchase included land in ten counties. Taxes were generally payable one-half due by the first of June and the balance by the first of November, with a 3 percent cash discount if paid before March 15. In short, sales and purchases might wait, but taxes would not.

Long well understood that the tax issue was sensitive terrain and that any expression of concern could backfire. He wrote as much to McCormick: "I have been thinking of some scheme whereby I could get in touch with the different county assessors, without exciting their suspicion that I was trying to get in with them." He simply planned to meet them individually, at which time he would request a list of owners of adjoining timberlands, a service for which the Timber Company would pay a small amount, "and in this way gradually get acquainted . . . and have them feel under some slight obligation to us." He was, as he admitted, tentatively finding his way.

This campaign of friendly persuasion with county boards and assessors would become the standard procedure. To one he submitted a copy of the legal descriptions of Timber Company lands in the county with the accompanying thought, "it might be of some convenience to you." Almost as an aside he mentioned looking forward to "a pleasant interview when we come to talk over valuation." To another he wrote, "We have every reason to expect at your hands, good, fair treatment, and that is all we want."

By early April, Long was beginning to feel somewhat more at home. He had three men in the field—Milford Jacobs and S. A. Graham cruising, testing the Northern Pacific estimates, and Myron Cole looking after other matters—and he assured Frederick Weyerhaeuser that reports would be forthcoming "as soon as I get sufficient [information] to make interesting comparisons."

Cruisers from the Lake States found themselves starting over in Washington. Not only did the western forest require very different knowledge, but the physical demands were much greater. The woods of the Douglas-fir region often involved very rough terrain, wet and slippery rocks, fallen logs, and some of the thickest undergrowth Long had ever seen. This was no parklike pine forest. He could only hope that the Wisconsin-trained cruisers would be up to the job. As he explained to J. T. Barber, in Eau Claire, "It seems that a man who can go into the woods out here, and do work at a rate that is satisfactory, has to be in tip-top physical trim; and it occurs to me, that inasmuch as we propose to bring some men from Wisconsin, we ought to get as young a class of men as possible, who have all the other desirable qualities that we need." And he told of a recent experience involving one who was ideal as to "ability, integrity, and intelligence, but a man who is simply too old to do the work."

The reports from those cruising the Northern Pacific purchase most interested Long, and no detail was too small to go unnoticed. A case in point was a disparity in two sets of figures for the same tract submitted by Cole. In his letter calling attention to this, Long noted that although the difference was not great, "it is a difference, and your attention is called to the matter so as to urge upon you the necessity of being quite particular and accurate in putting down figures, for you know, we are proposing to buy and sell on what you report." And so they were. Cole was apologetic in his reply (even if his spelling was somewhat unusual): "Sorry to learn of my making a misstake in coping of my report . . . for I relise the importens of being carful in such maters. Will try not [to] let it hapen again. A purson some times [gets] tierd and not feeling as well as he mint after traveling through the wet all day but I will try and not let it hapen again."

Back in Eau Claire, Carrie and the children were all packed, ready to begin their western adventure. Long headed east at the end of April, and would be gone less than two weeks. But when he returned, it was with a very different feeling than when he arrived in February. Everything was new and exciting to Carrie and the children, but to George Long there was already something comfortable and familiar about the scene and the ambiance of western Washington. He had found his home and his lifework, and he knew it.

FOUR

At Work in the Woods

THE MATTERS AWAITING ATTENTION FOLLOWING Long's "flying trip" back to Eau Claire were by now familiar: timber cruising, trespass, possible timberland sales and purchases, and taxes, always taxes. As for cruising, Ed Markham had agreed to terms of employment with the Timber Company. Long's plan was to "have two or three positions of a permanent nature . . . to give a certain territory to a man, and have him look after it, so that he could make his home at some central point, and depend upon not being very far away from his family."

Trespass was more often a frustrating and sensitive issue than a serious one. In the first place, the losses usually were not worth the trouble of proving the offense, at least not from a consideration of dollars and cents. Still the point had to be made, over and over again, that even petty intrusions would not be ignored. Long's approach to the guilty party was almost always respectful, couched in terms assuming that the trespasser was an honorable man who had simply made a mistake. His typical reaction was, in effect, "I know Mr. So-and-so and I know he would not intentionally do such a thing." Responding to one such "error," Long observed "that from the fact that your trespass has been unintentional, it leads us to take a more cheerful view of the situation," and the terms offered were reasonable, provided the trespasser made "an early settlement." Finding their honor at stake, those who might be otherwise tempted usually concluded that their honor was more valuable than a few trees across the line. Most also came to care a good deal about what Long thought of them.

As for timber sales, Long continued to be reluctantly

accommodating. The following exchange would be repeated many times over. A potential purchaser, declaring that "there was no timber of any consequence on this section," inquired as to its price. Long replied that a recent cruise indicated otherwise, "so we beg to ask if you were not mistaken. Look it over carefully, and make us your best offer." That would be the usual procedure—the offer, whether to buy or to sell, came to Long and not from him.

This is not to say that Long merely sat back and waited for things to happen. On the contrary. He continually sought information and did his best to keep abreast of purchase opportunities. In this connection, he subscribed to nearly every local Washington paper west of the Cascades, in part to keep current with land sales, but also simply to learn more about the communities themselves.

By June, Long had five cruisers busily occupied, two of them testing the Northern Pacific estimates, and the other three investigating lands that had been offered for sale. He thought that perhaps things were "beginning to loosen up a little now, and I am in hopes that I will have an active month in June, buying timber," or so he reported to McCormick. Another cruiser was soon added to look over lands for sale. Long began purchasing small tracts, and some not so small, including "an entire section in the midst of our holdings, twenty-eight million feet, for which we paid Six Thousand Dollars." He was, as he informed Frederick Weyerhaeuser, "coming in contact with parties that were in a hurry to sell."

These commitments of capital naturally raised some questions regarding the wishes of the company directors. An annual meeting was due to be held, but St. Paul sent no word concerning it. Thus Long requested that F. S. Bell let him know of any such plans, allowing that he was "in complete darkness as to what that date is, and necessarily will have to have a little more light before he can comfortably carry the burden of conducting an annual meeting of the Directors." Long continued, referring to himself in the third person: "He is just beginning to see the faint outlines of the great big shadow of the business that the Weyerhaeuser Timber Company will cast, when it gets out into the sunlight; there is a great lot of it in every conceivable direction."

He also inquired, in case there was no formal meeting, whether it would be feasible "to request a visit from a few of the active members, to talk over some of the features of the situation." Aside from buying an occasional crucial parcel, Long thought that the company would do well to delay any other purchases, noting that there were "a whole lot of people who want to get a little Washington timber, and when they get filled up, the situation will be less active, and we can operate to better advantage than we can today." But he closed acknowledging that the "avowed policy" was to buy, and that he was arranging immediate plans in line with that policy.

Long's practices became established policy. No clearer procedural statement would be made than the one included in a letter of April 28 to cruiser T. E. O'Neil, specifically responding to a request from a small mill operator to buy some much-needed timber, but in fact touching upon the whole:

> We want to have the mill man think that we are willing to pay attention to his request. . . . As a matter of fact, we really do plan to supply the mill men, whenever they have a reasonable want. If you find that there is much inclination to imagine that we want everything there [that] is for sale, you might arrange to have the boys work for a week or two, and then leave the country and let it cool off; of course, it may be well enough to fix any good trades you have, before getting away, but there is no occasion to show any anxiety.

In the first year Long made some mistakes, among them the occasional purchase of timberland more important to other operators. One involved an April bid at Shelton at the school land auction. This had been his first significant purchase and proved erroneous, being outside the Timber Company's sphere and within that of another, specifically the Port Blakely Mill Company. Long subsequently admitted as much to its manager, and expressed a willingness to "do the neighborly thing." The solution was simple enough: he turned the property over at cost and 6 percent interest, "including the expense of cruising and attending the sale, something like $25." Long also allowed that had he known Port Blakely was "figuring to get hold of this timber . . . we would not have appeared on the scene." At the same time, he hoped to learn more than the obvious from his mistake, proposing that the Timber Company cooperate in the logging, the logs going to Port Blakely, of course. Long explained his offer: "We have no experience in that direction, and a little experience would be of some benefit to us."

This was the first expression of concern about anything other than timberland. Long realized that the better he understood the process, the better he could assess the value of the resource. Port Blakely officials readily agreed to the conditions of purchase, including allowing Long's crew to participate in the logging, to see "how the timber cuts out, and more particularly as we would like to test our cruiser's work." Cruising was indeed an abstraction if unrelated to the end product, and Long wasn't much interested in abstractions.

Long's advice to cruisers was characteristic: "We again desire to caution you to make very conservative cruises, and not under any excitement drift towards finding too much timber; take plenty of time to have it right, and under no circumstances send in a cruising report that you are not absolutely sure will cut out . . . in fact we ought to get more."

Woodsman Cole seemed uncertain as to just what was expected. Wasn't it enough that he simply estimate the timber as he saw it? In response, Long listed other factors likely to influence future value: "Please understand that we are not criticizing your work. . . . We think our cruising should be on conservative lines, and should not include everything that can possibly be cut off from the land, even under the present condition of logging. We certainly hope the time will come some day when timber can be cut closer than it is now, but it is not safe for us to presume that it ever will be . . . and this country is so very far away from where they use common lumber, that we may never be able to get freight rates to ship it."

Even cruisers were expected to have a business sense. Not only were they to estimate the composition of the forest as to amount and kind and condition, but also how difficult it would be to log, the risk of fire, its accessibility, and how marketable it might prove to be. Long's immediate concern, however, was that his cruisers might get caught up in local enthusiasm: "Do not be influenced a particle by the excitement going on out in that country; if we change anything it will be the price;

that is to say, we do not want to get an exaggerated notion of what is on the land."

The effects of the land boom, that "excitement" that Long worried about, were clearly apparent by the fall of 1900. Long described one troublesome situation "this side of Mt. Tacoma, where we own all of the odd numbered sections," and where "we are having a lively little fight." An ex-Michigan lumberman was busy buying timber "in there amongst us," and he seemed willing to pay high prices. As Long reported to Bell, the Timber Company was already paying more than they should, "but it is simply a question of our getting it or the other fellow," and Long wasn't willing to let the other fellow establish a foothold, even if he had to pay fifty cents per thousand. "What we are trying to do," he explained, "is to get our work in on the best lands, i.e., where our lands are of the best general average, and to keep outsiders away from our most desirable localities."

With the passing months, Frederick Weyerhaeuser began to relax back in St. Paul. George Long was everything that had been promised and more. He was now the western expert among them, and all matters concerning the area were referred directly to him. Weyerhaeuser felt somewhat guilty about simply forwarding inquiries and problems from his desk to Long's, finally writing on that very subject. Specifically, he had recommended that a timber broker from Duluth, E. M. Runyan, contact Long regarding the sale of some western Washington land. After Long negotiated a brief working relationship with Runyan, Weyerhaeuser worried lest his recommendation had been given too much consideration. "I may have been making a mistake as regards parties coming here and offering timber for sale," he wrote, before adding, "The easiest way to get rid of them is to send them to you." He needn't have worried. Long knew that he was in charge and would not be fulfilling his responsibility were he to attach blind faith to suggestions made far from the scene. Thus even Weyerhaeuser's suggestion "not to buy any lands, excepting you buy them for less money than the lands we now hold from the Northern Pacific under option, taking locations and quality of timber into consideration," was not restrictive. Putting "locations and quality" into the formula allowed for plenty of flexibility.

To reassure everyone, himself included, Long responded

to Weyerhaeuser, outlining his understanding of the policy "to purchase lands adjacent to our present holdings." And, he continued, "When I depart from these instructions I will do so after first conferring with you about making investments remote from our present holdings. I take it that it will be all right to make investigations where there is a promise of first-class trading; and in all instances where I feel any such investment is desirable, I will lay the matter before you before taking any action."

In September, Long still had just five cruising teams in the field but hoped soon to double the number, the difficulty being the availability of competent men. Still, by October 1, a total of 28,191 acres had been purchased, an estimated 875,746,000 feet, at a cost of $275,450, for an average of slightly less than thirty-two cents per thousand. All of the purchases were accessible from previous holdings, and "all has been carefully cruised by our own cruisers." Weyerhaeuser expressed satisfaction with the way things were going. "Please keep on buying," he wrote in reply, "and if you have an opportunity to sell at good prices, to good parties, you may do so." The beginning augured well. George Long was pleased. Frederick Weyerhaeuser was pleased. His foresight once again seemed to have been consummate, from forests to managers.

FIVE

In the Market for a Sawmill

AT THE FIRST OFFICIAL MEETING OF THE TRUSTEES, ON February 9, 1900, George Long had been appointed "agent of the Weyerhaeuser Timber Company," and the specifics of his responsibility were listed. In addition, he was required to "execute and deliver" a $25,000 bond to the president, "conditioned for the faithful discharge of his duties . . . such bond to be paid for by this corporation." Finally, he was given the opportunity to "obtain title to one thousand shares of its capital stock at par," the original subscribers surrendering that amount "pro rata." The terms were as follows: "This corporation will loan to Mr. Long One Hundred Thousand Dollars or so much thereof as he may desire . . . with interest thereon at the rate of five per cent, payable annually." Five percent of $100,000 was exactly equal to his annual salary. The terms may have been impossible, but the intent was clear: the owners wanted to make certain their general manager would stay put.

Now, a year later, in January 1901, A. E. Macartney, longtime legal adviser to Weyerhaeuser and member of the St. Paul firm of Clapp & Macartney, sought to amend the earlier arrangement. That agreement should never have been made, Macartney argued, for several reasons, "particularly because it is liable to hang around so that the credit will be in the hands of fifty people" instead of those original few. He suggested that the arrangement apply to the new stock with interest payments deferred "until the $100,000 is all called." At an informal meeting of the trustees on February 23, W. H. Laird "moved as the sense of those present" that George Long could give his note for the $5,000, this instead of cash, "such note to be dated January

31, 1901." The resolution was adopted unanimously. And as George Long became ever more valuable with the passing years, the terms of his stock subscription were made ever easier.

Macartney would become increasingly important to Long, advising in matters legal and otherwise, a dependable connection in St. Paul. Also helping was F. S. Bell. Whatever assistance Long requested, Bell did his best to respond, and it was easier to approach him than to bother Weyerhaeuser. Frederick Weyerhaeuser was still in command, but the Timber Company, though foremost, was only one of many interests. Fortunately, Long had the happy capacity of being as independent as circumstances required. He did not grasp; he assumed. "I would prefer if you were here to help," he seemed to say, "but in the meantime . . ." On February 2, Long wrote to Bell indicating that he had been looking forward to a visit from McCormick, but Macartney informed that this was now unlikely. Then Long offered a fairly typical observation: "There are no particularly vexatious features about the situation here, although there are many things I would be pleased to discuss in detail." The fact was, of course, that only George Long knew the details.

Long's bookkeeping routines for the land records were now established. Some would call them quaint, but he seldom had problems locating essential information, and that was all that mattered. Similarly, he established a simple system for archiving office correspondence. From the beginning he kept copies, made with the use of a "letter press," of his outgoing correspondence—pages being bound when they reached 1,000, and these books accumulating on the shelves, volume after volume, until in the end they would number 136. (The copying process entailed placing a damp sheet of thin rice paper on top of the original letter in a press, and with the combination of moisture and some pressure, enough ink was absorbed by the rice paper to make a copy of the original.)

O. H. Ingram was a frequent correspondent. As a former competitor in the Chippewa Valley, Ingram was something less than a devoted follower of Frederick Weyerhaeuser, but Long didn't care. Indeed he and Ingram had been neighbors in Eau Claire; over the years they became friends, regularly discussing the latest news, gossip, and business developments. The practice continued after Long moved to Tacoma, where one subject of large mutual interest was the wisdom of undertaking manufacturing in western Washington. In the spring of 1901, Long informed Ingram that C. H. Jones "and various members of the Hewitt family," all connected with the St. Paul & Tacoma Lumber Company, had recently purchased a major interest in the North Western Lumber Company mill at Hoquiam. Long would have liked to have been a party to the arrangement, observing that "it is a very nice plant and favorably located," and then added, "It is an expensive one to manage, and I have never thought there was any money in the way it was handled, nor in the way it would be handled at the present time." But, he continued, "we have to commence out here sooner or later to manufacture, and it would be well enough to commence with one mill, and get acquainted with the details of the manufacturing end of the business." Long was simply suggesting what seemed a reasonable approach. One moved ahead cautiously, in the process learning what worked and what didn't, all the while adjusting in the direction of efficiency and economy.

A new development that would prove of long-term importance was introduced by a request from the city engineer of Seattle. It concerned the acquisition of rights-of-way through Weyerhaeuser holdings to protect a proposed water-supply system. The area in question was the Cedar River watershed. Long pledged cooperation, placing a value on the lands at $2.50 an acre, provided his company would not be damaged in terms of future logging operations in the area. He promised "to lay the matter before our Board of Directors promptly, and think we can assure you that they will be willing to carry out the spirit of this proposed agreement." The terms did prove agreeable, Engineer R. H. Thomson replying on June 5, "We are willing, in a friendly settlement, to give you $2.50 per acre, you reserving the timber and coal."

Plans were firmly in place that Robert Laird McCormick would join Long in the Tacoma office. McCormick, representative of the Laird, Norton families, was manager and a principal owner of the North Wisconsin Lumber Company at Hayward, Wisconsin. Although not one of the initial investors, Frederick Weyerhaeuser became involved in the 1880s at a time when others were unable to meet additional assessments. Since the Laird, Nortons were major shareholders in the Weyerhaeuser Timber Company, it seemed appropriate that one of their own would assume a responsibility in Tacoma, and R.L. was the obvious choice. He would come, eventually, and would be of great assistance to George Long, officially as Secretary of the Company but more important as a trusted friend and adviser. His departure from Hayward was delayed by the difficulty of finding a buyer for the North Wisconsin properties, and it was not until 1902 that the Edward Hines Lumber Company made an acceptable offer. In the meantime, Long and R.L. corresponded.

Early in July 1900, Long tried to bring McCormick up to date on recent purchases. These hadn't been "very satisfactory in the matter of volume," primarily because he had generally been unwilling to pay the asking prices. Few would disagree with that policy, although Long knew that by giving in on the price a bit, more acreage would come their way. "What would you think of a policy of our drifting into paying the higher range of values?" he inquired. "Accessibility is 90% of the factor of value," he reminded, and at present they were most interested in purchases that would not be immediately utilized, but would likely be wanted in six or seven years. This class of timberland was priced "somewhere from 40 to 75 [cents] per thousand."

Long also reported that the only Pacific Northwest millmen making money were those dealing in cedar, "especially the combination mill which manufactures lumber and shingles." The fir producers were not "sharing in the prosperity." Nevertheless, stumpage prices were on the rise, "because there are more buyers than usual and because people are buying for speculative purposes." There was no quick answer to the question of what to do. The easiest thing was to do nothing, but that wasn't Long's style.

McCormick's response was detailed and thoughtful. He reviewed the lumber situation generally, observing that from his Hayward, Wisconsin, vantage, given "the unparalleled activity, the wonderfully increased demand, the full list prices and prompt payments, I think our local judgment is liable to be decidedly rose-colored." And so it was. The white pine of Wisconsin and Minnesota was nearly gone, while the nation's "population and wealth and ability to use lumber is increasing." The lumber had to come from someplace. Southern pine would "push its radius of shipment a little farther," but, partly because of "the shiftlessness of the Old South," it could hardly meet future demands. Possibly Idaho might provide lumber to the northern tier, but that too seemed unlikely to satisfy the nation's needs. The much-discussed Nicaragua Canal would open up new opportunities for West Coast lumber, and McCormick thought that timberland purchases were certain to prove out eventually. He recommended, "When you can combine quality, location, and solidifying our timber, I would be disposed to raise prices to 50 or 60 without a moment's hesitation." Regarding other tracts, he expressed a willingness to pay more, provided the quality was "A 1," and he underlined this emphasis on quality, recalling that twenty years earlier in Wisconsin, "Norway [pine] was worth as much as White Pine [and] today White Pine has double the value of Norway."

Long agreed for the most part, but continued to be bothered by that "question of time." Thus he responded, "I am inclined to think we will have to let the Yellow Pine people have their 'fling' before we get right down to business the way it would be comfortable; and I am hardly in position to guess when that date will arrive."

For all of its extent and supposed strength, the Weyerhaeuser Timber Company at this time was solely a timber-holding company. From a lumber-industry perspective, it was not impressive. Though some of the owners were restless to become true lumbermen again, there were already many manufacturers on the West Coast, most of them small operators who seemed satisfied to run their mills and pile their boards, occasionally making some sales at slight margin. Why should the Weyerhaeuser Timber Company enter such an unpromising arena of too many producers producing too much lumber?

Curiously, even as turn-of-the-century journalists were fueling public fears of powerful lumber barons and trusts, lumbermen were discussing the advantages of reducing the number of producers, hoping to increase efficiency and profits. But in truth, there was little harm in their activities; the kind of monopoly that was possible in many industries wasn't possible in lumber. That industry was simply too fragmented, with too many independent producers who prided themselves on their independence.

In a letter of September 18, Long explained his outlook to F. E. Weyerhaeuser, Frederick's youngest son, who was assuming more of the day-to-day responsibility in the St. Paul office. Inevitably, efforts were made to organize "a combination that will control the situation and improve it," but Long wasn't much interested. Even if as much as three-fourths of the stumpage in western Washington was controlled by a single group of investors, assuming that this group might agree on a basis of cooperation, "the remaining one-fourth of the timber [would be] in the hands of people who are willing to realize on it at the present market price of logs." Furthermore, scattered through the state were "a whole lot of little mills that have timber enough to last from three to five years," acquired at low cost, and these operators would continue to make money "even if lumber sold at $1 less than it does now." By any reasonable accounting, given current prices for stumpage, these operators were losing money, not making it. But they tended to count real dollars, not theoretical ones, and many were satisfied.

Thus, Long remained pessimistic about any early improvement in the market, and, as to a combination, "I confess I do not believe the time is quite ripe to enter upon it with much hope of success." The better course, he advised, was for the larger investors "to continue the policy of adding to their holdings, and selling as little as is feasible, and thus bring about at an earlier date the time when they can control the market conditions."

Under prevailing market conditions, the only manufacturers with a chance for success were those who had purchased stumpage at bargain prices and who ran efficient plants. By the fall of 1901, Long appeared ready to enter the fray. He studied the Everett harbor, obtaining "reliable information concerning depths of channel, proposed government work, etc.," and made plans to inspect the Bell-Nelson sawmill there. Oscar Nelson had visited Long at his Tacoma office in late October, "indicating that Mr. [James E.] Bell would sell out and he would like to remain in and have something to do with the management." Long made no promises other than agreeing to give the proposition some thought. A month later he was still thinking.

The actual inspection of the property increased his interest, and he telegraphed St. Paul on December 18 as follows: "I recommend buying Bell–Nelson plant and shipyard, with all personal property except logs and lumber, at $115,000." The logs and lumber, including the Maple Valley Logging Company, would involve separate negotiations. Long subsequently explained his recommendation: "I look upon the mill site of the Bell–Nelson Company as very desirable, provided we can get the ship yard property, which would undoubtedly give us all the 'elbow-room' we need for an export mill. But so far as the saw mill which the Bell–Nelson Company now has, it is not the kind of a mill we would like to own, or run for any length of time." That assessment was accurate. It was the site rather than the sawmill that was of value. Long was advised to "close the deal at once."

On January 13, 1902, Long informed Weyerhaeuser that the Bell–Nelson purchase was about complete. He had engaged Bell on a temporary basis at a salary of $300 a month, and in addition hired E. M. Warren, "who you will remember as being connected with the grading movement in the Mississippi Valley Association." At present, Warren was busy "taking inventories and arranging details." Long assured lawyer Macartney that the legal questions had been satisfactorily handled. It had been decided that the three related corporations—the Bell–Nelson Mill Company, the Bell–Nelson Logging Company, and the Everett Shipbuilding Company—would be jointly purchased, although "we decided to have the Logging Company sell all their property to the Mill Company," and then had the shipbuilding company do the same, with the exception of "some few of the assets and retaining the corporate existence." The stock of the Bell–Nelson Mill Company was then signed over to Long, "all of it except one share, which was given to E. M. Warren." Bell and Nelson resigned their positions, "and I am now President of each Company and E. M. Warren Secretary."

Suddenly, the Weyerhaeuser Timber Company had a logging camp "in full blast," a sawmill, and a shipyard, which "is dormant, but we may possibly lease it." Although the Bell-Nelson purchase would be of major importance, Long tried to treat it matter-of-factly. As he wrote to Macartney, "This trade in my mind is just an ordinary one; nothing to be especially matter of congratulation, except we now at least have a mill and a mill site where the heaviest draft ocean vessels can touch the wharf; and in this respect it is a great improvement over the Barge Works . . . there being no water facilities at that point."

Long took a slightly different tack with the directors. "I am afraid none of you will like the saw mill," he acknowledged, after which he emphasized the point once more that the true value was in the site. New sawmills could be constructed, but there were a very limited number of strategic locations. The mill resumed production on January 15, and the two logging camps were both busy, "so we [are] in operation, and that is something which the public have been expecting us to have for the past year and a half."

Bell had been retained for a six-month period, primarily to supervise the logging. E. M. Warren was in the office, "in charge of the lumber and yard." The logs were brought by rail and dumped into the Sound at Seattle and towed from there to Everett. Long noted, "We will have to get up a little organization to handle the timber proposition through the Tacoma office, and just as soon as we can study out the best thing to do, will advise you of further particulars." Within days J. L. Bridge, Jr., had been placed in charge of Camps 1 and 2, and also given responsibility for the log rafting.

Weyerhaeuser was surely interested in such details, but at the moment he seemed preoccupied with the purchase of additional timberland under the Northern Pacific option. In that connection he suggested that Long should be in St. Paul on February 20, to "explain the Bell-Nelson deal and give us information in regard to the Washington and Oregon timber covered by the option." As negotiated, the price for Washington timber would be $6 an acre, and for Oregon timber, $5. Weyerhaeuser warned Long to expect some contention, noting that McKnight and Ingram, "and some others," were rumored to be opposed to any increase in their stock, and if so "we may not be in shape to buy the property." Weyerhaeuser put the matter curiously:

"We have spent so much money on it for examining and getting the necessary information that I wished our old stockholders to get the benefit of it."

Dragging "our old stockholders" along was nothing new for Weyerhaeuser, of course, and even he must have occasionally wondered how distant a return on the Weyerhaeuser Timber Company investment might be. Long conceded that "the general trend of timber is to advance in value," but he was careful to point out that it had to advance rapidly in order to justify payment of interest and taxes. From his perspective, "The investment on the part of the stockholders of this Company has been simply an investment where they are putting away surplus money where they never expect themselves to see it again; being an investment for their children and children's children. This in itself, is a recommendation to anyone holding Washington timber, who can, to wait for the good times to roll around which will ultimately bring Washington lumber into the market on a basis that will make stumpage valuable." (This statement of Long's, "being an investment for their children and children's children," is as close to the statement often attributed to Frederick Weyerhaeuser, "This is not for us but for our grandchildren," as I have found. Again, regardless of whether Weyerhaeuser ever actually spoke the words, that he believed them is clear beyond a doubt.)

Long arrived in St. Paul on schedule and the meetings began, dragging on for more than a week, in part delayed by the absence of Northern Pacific agent W. H. Phipps. Long finally headed homeward on the evening of March 3, reasonably satisfied with the progress made. Others were, too. McCormick subsequently praised Long's contribution, and expressed confidence that their "proposition is a good one for us . . . and sooner or later it will be accepted by the N.P. people." He added, "I think your time was well spent in coming East [and] in fact we could not have made an intelligent proposition without you."

Long was not quite so sanguine regarding the proposition, although he agreed that "if it goes through, it is a first-class trade in every particular." He also noted that Phipps had not objected to any feature of it in the course of "our interview with Mr. Weyerhaeuser," but Phipps had been agreeable in the presence of Weyerhaeuser on other occasions, only later to assume a different stance. Anyway, as Long responded to McCormick, "the only grounds" he had for fearing failure were that "it is such a good trade that I imagine somebody will try to upset it."

In mid-April, Frederick Weyerhaeuser reported that their offer had been rejected by the Northern Pacific board. In reply, Long repeated his earlier thoughts, that had their $6-per-acre figure been accepted, it would have been "a strictly first-class trade, in fact, cheap." He had no hesitation about paying $7 per acre "for the lands embraced in the list which was submitted with our proposition, if you cannot do better," and he would even be willing to go as high as $7.50 rather than lose out altogether. He reemphasized the importance of including only those lands "embraced in our proposition," which left out the four or five poorest townships as well as the scrip lands, "and all of the land which is now in contest between them and the government."

The Timber Company directors were scheduled to meet in Tacoma on June 19, an occasion which naturally made Long just a little nervous. For one thing, he wished that his list of accomplishments was lengthier, and he also knew that the visitors could not help noticing problems. Would they share his faith in the wisdom of the Everett purchase? What about the log supply for that mill? In that connection, he did inquire of Weyerhaeuser whether he should make some purchases of Snohomish River Valley timber at inflated prices. Weyerhaeuser advised him to wait until they could resolve the Northern Pacific deal. In addition, Weyerhaeuser had recently talked with neighbor James J. Hill who "told me the other day, [that] his Road [the Great Northern] has a large body of timber near Sultan and vicinity, which I know is very good." Hill further promised "to give us a fair price." In the meantime, Weyerhaeuser did not want it known that they would pay "anything like $30.00 or $32.00 an acre," and especially not when "you can buy at the prices you have been paying." As to a general policy, that was simple enough: "It seems to me the interest of our Company would be best served by keeping the prices down until such a time as we think best to stop buying."

Weyerhaeuser expressed some concern regarding the Tacoma meeting. First, he tried to prepare Long for the arrival of the large entourage. And second, he requested that Long do nothing to "advertise our coming any more than you possibly have to," explaining, "The less fellows I have calling on me who have axes to grind the more I will enjoy my visit."

The visiting dignitaries totaled twenty-six, and the meeting itself went well. The directors were pleased with the announcement that the Northern Pacific board had finally accepted their proposition. Many also enjoyed a tour of the Everett operations, both sawmill and woods. Long was a firm believer that the best way to entertain guests was to let them watch real work being done. Nonetheless, it was with considerable relief that he stood on the platform of Tacoma's Union Station and watched the train pull away. Now *he* could get back to some real work.

SIX

Fire

FIRE WAS AN ALL-TOO-FAMILIAR SCOURGE IN THE forested regions of the Lake States. For the most part it came seasonally, usually in early spring or late fall, the smoke hanging in the atmosphere as if part of the natural cycle. George Long well remembered the Wisconsin fires, and hoped that history need not repeat itself west of the Cascades. The fact was, however, that wherever the logger or the pioneer farmer went, the threat of fire increased. Logging closely simply didn't pay, and the defective logs, tops, and trimmings the operators left behind created a tinderbox. In 1900, western Washington loggers had only begun the forest invasion, and here nature was a more dependable ally. Summers were commonly dry, but long periods of high temperatures and low humidities were infrequent.

It was early evident that fire prevention would be a cornerstone of Long's efforts. Central to his policies was a reluctance to "open up" areas—that is, to permit entry for limited logging or any other purpose prior to the large-scale harvesting. But too often he was forced to compromise. Some of those compromises involved "squatters" on Weyerhaeuser lands. Rather than throw them off, making enemies in the process, Long usually allowed them to remain, provided they operated with care and consideration. Typical was the letter sent to Fred Shefferly, permitting him continued use of some five acres of cleared land "with the understanding . . . that you will use your best endeavor to see that no fire is started in the slashing of the timber . . . or if a fire is started, you will use your best efforts to put it out, so that it will not spread to our timber adjoining." The agreement was to run for five years, with no charge, but

51

would be voided if Shefferly failed "to take necessary precautions as outlined."

Although the danger from fire varied with rainfall, temperature, and humidity, the worry was constant. In the late winter and early spring of 1902, Long had his woodsmen post linen fire-prevention placards—warnings printed on linen cloth affixed to thin boards to be tacked on prominent trees—reminding of the extreme fire hazard and preaching the doctrine of individual responsibility. This was possibly the first use of such printed circulars in the region. In May he persuaded the editor of the *Pacific Lumber Trade Journal* to publish an article on forest fires, the article to be distributed to every country newspaper to "see if we cannot get them to reprint it . . . during the months of July, August and September." Long outlined his plan of action in a letter to F. S. Bell, observing that the "question of forest fires out here is a big one, and we ought to do all we can to head it off." In this effort, Long thought that the appearance of leadership coming from a trade journal would be more effective than from the Weyerhaeuser Timber Company. "What do you think of an expenditure of $100 or $200 in this direction?" he asked Bell, adding, "Personally, I think it would be a very good investment."

Now actively logging, Long had to practice what he preached. In early June he inquired of J. L. Bridge, Jr., whether it was too late "to do any firing of your cut-over lands," warning his foreman to use extreme caution if any such work was contemplated. And to a new cruiser working in the vicinity of Eatonville and Elbe, a popular area for tourists and sportsmen, Long allowed that visitors should be reminded about the fire hazard, but he believed the greater danger was posed by the settler who burned off his lands during the dry season and occasionally let these burn-offs get out of control. The settlers needed to be educated as to these dangers. The best approach, Long counseled, was "to get in their good graces," explaining the need to protect their own property "and thereby protect ours." He went on to say: "A great deal can be done by the representative of any Company using a little tact and diplomacy and kind treatment. . . . Then I would take particular pains to try and 'hook up' with just the right kind of people. . . . In all such frontier localities there are some people who have more influence than others, and it is this class we should try and cultivate, at the same time not ignore anyone." In conclusion, Long preached perseverance, especially as concerned the posting of the fire placards. No doubt some of these would be torn down, but this shouldn't diminish the effort: "Instead of feeling or expressing any indignation, ignore it," and put up two where there had been one.

The summer of 1902 was extremely dry, conditions made all the worse within the broad Columbia River gorge where eastern winds lowered humidities to worrisome levels. In early August, cruiser E. A. McDougall reported that fires "were burning very hard and getting into the green timber," and expressed fear that "if the wind blows hard today . . . these fires will do some damage." There were personal dangers as well: fires often moved unpredictably, entrapping the firefighter. McDougall also complained about a general absence of concern, that many refused to "get to work and put out these fires," no doubt thinking "that it would be easyer [*sic*] to buy the timber which they need if the fire had run through it."

Within two weeks, Long was reporting some fire damage to Weyerhaeuser Timber Company lands, losses estimated at between $10,000 and $15,000. Still, he seemed confident that the danger was receding. "We had a little rain last week," he informed McKnight, "which came at a very opportune time, as the country was getting exceedingly dry." He closed predicting no further trouble that year, "unless we should enter upon a dry spell that should last into September." And that, unfortunately, is exactly what happened.

On September 11, the forest could take no more. It burst into flames in a great many places simultaneously, and soon more than a hundred fires were burning over 700,000 acres. The next afternoon Long described the conditions in a letter to Weyerhaeuser: "For the past two weeks, we have had warm, dry, clear weather, and all day yesterday there was a stiff wind blowing from the North; a wind of unusual velocity for this country. At this hour, 2 p.m., the City of Tacoma is practically in a state of semi darkness, the sky having a pinkish overcast, and the general effect is exactly that of the reflection of a big fire at night. There seems to be quite a little smoke of a fog-like formation in the air, and taken all together it is decidedly a peculiar situation. We are getting all kinds of wild rumors about fires, some of which are authentic, and some which cannot be relied upon."

Although truth and rumors were difficult to sort out, it soon became evident that the greatest losses were in extreme southwestern Washington, specifically in Cowlitz and Clark counties. On September 16, Long advised the Thomas Irvine Lumber Company of probable damage to their Clark County timber. Earlier assumptions had been that the fires were largely confined to the logged-off lands, "and that generally speaking, the fire has not invaded the green timber." But while in Portland on Sunday, September 14, Long received differing and discouraging accounts, word coming from Vancouver [Washington] that "some settlers . . . had just come in, saying that that whole country had been burned over very badly." Long still hoped that the Timber Company losses were minimal, though it seemed likely that some of their Clark County lands had been burned. In that he was correct; he simply didn't yet know the extent of the destruction.

The rains finally came, and the fires were extinguished, although the wet and blackened forests would continue to smolder for many weeks. And it was weeks before Long knew the facts concerning Timber Company lands. The news was bad. Clearly the company had suffered greatly in Clark County. For Weyerhaeuser's benefit, Long described the course of events that would become known as the Yacolt burn: "The fire seems to have originated on the headwaters of the Wind River in Skamania County, a stream that empties into the Columbia River, and swept in a westerly direction, burning the timber on the head waters of the Washougal River . . . and swept on west to the South Fork of the Lewis River . . . and there in a northwesterly direction." When the figuring was completed, Long sadly acknowledged that the Weyerhaeuser losses included some 15,520 acres and aggregated "in round numbers, 500,000,000 feet of timber, 350,000,000 feet of which involved the Clarke [*sic*] County lands." (A word of explanation regarding the "e" appended to Clark may be appropriate. The two counties in southwestern Washington, Lewis and Clark, were, of course, named in honor of the famous explorers. In the first session of the Washington territorial legislature, however, Clark was erroneously entered as Clarke. And Clarke it remained until corrected in 1925 by an act of the Washington State Legislature. Thus when George Long created the Clarke County Timber Company, the obvious error in spelling was something the existing atlas demanded.)

It seemed a staggering amount. Still, they had to persevere. The effort now focused on limiting the loss, and the way to accomplish that was to log the burned district as quickly and efficiently as possible. Long had plans already in mind: "There is a railroad now constructed . . . which leads to Vancouver, Washington, across the river from Portland. It will be an easy matter to extend this road into the burnt district and take the timber out, and we presume this will be the policy that will have to be adopted. . . . The timber in these burned districts . . . is all held by strong people financially, so I apprehend that some general action will be taken where this situation will be handled to the best advantage for all."

In the midst of this predictable response to tragedy, Long recognized a singular opportunity to educate. Even before he knew the extent of damage, he was writing to editors of the regional trade journals and newspapers, requesting that they push for passage of "the kind of legislation that should be enacted to minimize the danger of a great fire catastrophe." To Victor Beckman, publisher of the *Pacific Lumber Trade Journal,* he reasoned, "I think if this subject can be made real acute now, while the forest fires are in the mind of everybody, and keep hammering away on it, you can get a great many practical ideas which could be placed before your association, and legislation asked which would be of a practical nature." In return for such efforts, Long promised to "pay you for your trouble," and reminded that progress was in the interest of everyone "directly or indirectly connected with the lumber business."

To George M. Cornwall, publisher of the *Columbia River and Oregon Timberman,* he was more specific, urging him to "start a line of inquiry addressed to all the prominent lumbermen and loggers that you know, asking them to suggest the line of proposed legislation that should be enacted; it would be well enough also to not ignore the poor settler who has to burn off his slashings. . . . My observation is that the opinion of 25 or 30 men on any subject will develop some new ideas that are practical that otherwise would be overlooked."

Nor did he stop there. Letters followed letters. "Are you

getting any responses?" he queried C. W. Jennings, editor of the *Pacific Lumber Trade Journal,* and "Have you succeeded in getting a copy of the Minnesota State Laws on fire?" One of the points that Long wished to emphasize was the destruction of green timber in the Yacolt burn. He feared that there was a general belief that fire affected only "burned over and logged off regions," when in fact it had consumed between "15,000 to 25,000 acres of green, standing timber" in Clark and Skamania counties alone, "and conservative estimate places the amount of timber actually destroyed . . . in the vicinity of one billion feet."

Meanwhile, the fire-prevention work went forward at full speed. In late October, Long met in Portland with "some of the prominent loggers and mill men," where it was decided to appoint a committee to draft a bill. He was impressed with many in attendance and urged Jennings to push their Washington colleagues in a similar direction, carefully avoiding "any publicity . . . until their work is in a completed shape." He further urged the editor to "keep 'harping' on the general need of forest fire protection, so the subject will not be allowed to 'cool off' in the minds of the public."

The more Long studied the question, the more convinced he became that Washington needed a Forestry Commission, with the state land commissioner serving as an ex officio member, the other members to be appointed by the governor. The commission would be responsible for the selection of a state fire warden, with other wardens, possibly one in each county, serving during the dry summer season. The state fire warden's salary would be a state expense, the salaries of the local wardens to be borne by the respective counties. Long allowed that this was just one possible solution, but he thought it had merit, and stated for the benefit of publisher Cornwall that the responsibility of the forestry commission should extend beyond "fire protection only, its scope and work being in fact to make any suggestions that would be for the general protection of the standing forest and for the growth of the young timber."

Although the Timber Company employed an attorney to "keep track of what is going on down at Olympia," Long continued to stay abreast of the details, as he explained to lawyer Macartney in St. Paul. "We have already done a whole lot of missionary work," and the Pacific Coast Lumbermen's Association had appointed a committee to draft bills for forest fire protection, one for Washington and another for Oregon. In January 1903 they were expecting "very good results."

Within the month, however, Oregon proved a major disappointment. The governor vetoed the Forest Bill. Long, writing to Cornwall, attributed the action to "a case of pique" on the governor's part because he was not allowed to appoint the commission. Whatever the reason, it was "a crying shame."

The Washington deliberations moved more slowly, but in the end to better purpose. In the process, Long became acquainted with a good many politicians. After the so-called Veness Forest Fire Bill passed the House, he sought the support of a number of senators in an effort to improve it. Among his recommendations was a requirement to equip all logging engines with smokestack screens during the dry season, and that timberland owners should have their "cruisers appointed deputy fire wardens to watch their property, at their own expense." The Senate eventually approved the bill, albeit "mutilated somewhat," and Long then turned his attention to its implementation.

The fact was that Long was feeling good, and not just about the political success. As he admitted in a May 1903 letter to his old employer McKnight, "In a general way the situation out here is quite encouraging, and the more I get acquainted with our investment the better I like it, aside from the one great element of fire hazard." Now that they had taken initial steps in the legislature to reduce that danger, he was utilizing every opportunity to publicize the act:

> I have been working through one of the trade journals, and have gotten him to devote several issues largely to the necessity of forest fire legislation, and he has printed for distribution 2000 or 3000 copies of the enclosed bill; and more recently I have gotten one of the machinery firms here in town to issue the enclosed circular, which goes to every logger in the state of Washington. I also had incorporated in the Bill the privilege for timberland owners to have their own cruisers appointed deputy forest patrolmen, and I have had each one of our cruisers deputized so that in going around among our timber we will at least be in position to regulate the actions of

careless people. We feel that we cannot let any avenue escape us that will have a tendency to head off forest fires.

Other actions were demanded, including those having to do with Weyerhaeuser logging. Long again cautioned about slash burning, urging his loggers "to make a vigorous effort to do this before the dry weather sets in." The new legislation allowed counties to declare a closed season on slash burnings, and Long did his best to encourage strict adherence to the spirit of the law. Thus he wrote to the Lewis County board, warning that a closed season which did not include July, August, and September "would defeat the object of the law." Admitting that he had his own company's interests in mind, Long again reminded, "When any of our timber burns, it means that much wealth taken from Lewis County." The county board finally agreed to a closed season from July 10 to September 10.

Thus the battle was joined. While there could never be total victory, perhaps future Yacolts could be avoided. If not, the industry had no future. It was as simple as that.

SEVEN

Frustrations Abound

IN THE FALL OF 1902, GEORGE LONG'S PRIMARY CON-
cern was organizing the log-salvage effort in the Yacolt burn.
Several owners were affected, and Long estimated that the area
included some 12,500 acres and close to 500 million feet of
timber. As yet he didn't know who owned the railroad line into
the territory: "Some people think it is a N.P.R.'s Co. venture,
and others say it belongs to J. J. Hill," he wrote Frederick
Weyerhaeuser on October 31. Regardless, he would make arrangements "at once to get this railroad to extend into the timber and to make as satisfactory rates as possible to put it into the
Columbia River at Vancouver." The other critical effort would
be to get the various owners together about the middle of November to formulate a joint program of operations. In such
circumstances, the degree of cooperation determined the degree
of success.

Concurrently, Long was trying to implement the details
of the recent Northern Pacific option agreement. There were
problems. Once again, agent Phipps was proving to be a hard
bargainer, as Long complained in a letter to McCormick. Mc-
Cormick advised him to involve Weyerhaeuser directly: "In this
particular matter he really did the trading, got the option, closed
it and it is his pride. I think he feels that he has avenues of approach to this whole matter of N.P. Lands that none of the rest of
us can reach, so don't feel at all delicate about writing him in
regard to any hitch in carrying out the proposition. . . . He does
not often neglect a timber trade. That is his specialty."

The Clark County salvage effort promised to be complicated. As Long reported to Weyerhaeuser, the burned-over dis-

trict included "a remarkable belt of timber, there probably being as much timber on this area per acre as on any similar tract in the state of Washington." The two townships most affected had been "exploited a great deal by different people as well as by people who had claims there, so when the owners were ready to sell, there were three or four very eager and strong buyers, including ourselves." With the prices already inflated, it was "agreed between ourselves, Mr. C. H. Davis and Mitchell & McClure to make a joint arrangement and buy up what was left." Mitchell & McClure subsequently dropped out, leaving the Timber Company sharing a joint interest in 3,560 acres with C. H. Davis. Michigan Congressman J. W. Fordney served as Davis's agent. The Weyerhaeuser ownership, including the shared interest, amounted to 8,760 acres, or about 203 million feet of burned timber. An additional 525 million feet of burned timber in the area was divided among other owners. Counting the green timber that would be harvested along with the burned, Long estimated at least 750 million feet "that will have to be logged, to make a clean sweep of the territory."

There were two options: assume responsibility for the logging, or sell the timber. Initially Long seemed uncertain, in part because he wasn't convinced that the Portland market would be able to absorb the additional timber. Also, it was no small undertaking. In order to be successful, he estimated that they would have to log at a rate of 150 million feet a year. For an organization with almost no experience in such work, that would be a major challenge.

Weyerhaeuser was again helpful in terms of rail service, thanks partly to his next-door neighbor. James J. Hill had originally owned the little railroad that ran near the burned-over district, but had recently sold it to the Northern Pacific. He promised to discuss the matter with Northern Pacific President Charles Mellen, requesting that he "give us a very fair lease on the property." Weyerhaeuser met with the NP "folks" on January 5, 1903, and found Mellen most cooperative, even offering to haul upward of one-third of the logs to Tacoma, "providing we can find a good market there."

Long did not await resolution of the ownership puzzle before proceeding with plans for a logging railroad into the burned area. He hired a well-known Seattle civil engineer, I. M.

Rice, to work with Lafe Heath, an experienced cruiser especially skilled at taking elevations. If they were to build a railroad, it had to be the best possible in design and construction. Mistakes would prove costly.

On February 9, Long headed back to St. Paul "to consult with our people as to the burnt district, and to lay before them all I know about the situation." At the moment he was feeling a bit harried, as he admitted in a letter to Congressman Fordney: "Personally, I think we are just in far enough now so we will have to go in a little farther in order to handle the matter right; and while all of my nature rebels against paying Holland & Rupp $65,000 for their burnt timber, I can see that to protect ourselves . . . and our joint interests with you and Mr. Davis, that that country ought to be logged at once, and that we ought to control the situation."

When Long returned to his office on the morning of February 19, he found a telegram from young Fred Weyerhaeuser on his desk: "Purchased Rupp jointly sixty thousand and taxes. Davis offers turn over Wolverton Rupp deals at cost, sell his timber eighty-five thousand dollars, about fifty-five million burned sixty cents, forty-eight million green one dollar ten cents. Will you consider?" Long considered, and recommended the purchase of the Holland & Rupp lands including payment of the 1902 taxes, but suggested that Davis's individual holdings be priced at fifty cents a thousand for the burned timber and $1.00 a thousand for the green. Charles Davis would respectfully decline.

Purchase of the Holland & Rupp lands left no doubt about a major Timber Company responsibility in the salvage operations. Long was eager to get started. As he predicted for the benefit of I. N. Gray of the Portland, Vancouver & Yakima Railway, within three months they would "commence logging on an active scale, inasmuch as we can get into the Holland & Rupp timber at once; and we will push our work there with the object in view of bringing out from seventy-five to a hundred million feet a year of burnt timber, until it is cleaned up."

An immediate requirement was finding someone to superintend the effort. Alex Polson, a crusty logger from Grays Harbor, recommended Peter Connacher, and Peter soon arrived to look over the situation and to be looked over in turn. Long found Connacher "fairly prepossessing," but wasn't totally convinced that he was "big enough for the job." Long would shortly be convinced.

Progress never seemed to be fast enough. Those in the St. Paul office felt some of Long's frustration as they read his report of March 12: "The great big word we ought to have before us in connection with this burned timber district is HASTE, haste all the way down the line. I think we are too slow; and while I know it is folly to be in too big a hurry, I think we should decide at once whether or not we are going to accept the N.P.R'y Co's charge on logs, and make immediate arrangements for a working basis with Mr. Davis, and get our railroad building underway, and get our machinery ordered; all of this ought to be done right now."

Everything couldn't be done immediately, but movement was evident. A contract was soon negotiated with the Northern Pacific officials in St. Paul, although F. E. Weyerhaeuser seemed almost apologetic when reporting its terms, admitting that "the only concession we have gotten consists in their doing all the railroad work . . . [relieving] us of the necessity of having any engines, machine shops, etc." In fact, Long found nothing very objectionable about the terms, "if we start out with the proposition that we are willing to pay $1.50 [per thousand] for the haul." But of course he hadn't wished to start out with such a proposition.

Congressman Fordney, representing Davis, had visited the scene. He and Long inspected the district and subsequently agreed on both the rail surveys and the details of cooperation. By mid-April, civil engineer Rice was authorized "to order a kit of supplies for grading purposes and quite a stock of grub and to put a lot men in there to work at once clearing out the right-of-way and doing preliminary grading for a track." Long informed McCormick, still stuck in Hayward, Wisconsin, that as of April 17 the supplies were in hand and men were being hired at Portland and Seattle, and some of them were on their way into the timber. As for the ownership-operation puzzle, he seemed confident that the costs could be divided in a mutually agreeable manner, and that before too long Davis would be ready to sell, although he had not "shown the 'white feather' yet."

If there were frustrations in Clark County, they could not

compare with those in the Everett plant renovation project. On the recommendation of Ingram, the company had hired J. D. Hills as mill architect. Long tried mightily to relocate him, but Hills seemed reluctant to leave his Menomonie, Wisconsin, home. Typical was a mid-November note in which Hills indicated that he had planned to start west that very day "but am delaid [*sic*] by continued cloudy rainy weather." One would have thought he was flying. And in what was unintentional irony, he closed expressing the hope "that every thing is mooving [*sic*] along to your satisfaction." If things were moving along, it certainly was no thanks to Hills, who further irritated Long with his billings, both in terms of amounts and method of submission. Finally an exasperated Long ordered Hills to submit his expenses each week, including specific notation of "time which you have put in for the Weyerhaeuser Timber Company." He wanted Hills to understand that this was standard procedure: "Everybody, including the writer, renders a weekly expense account." But Hills would continue to march to his own slow beat.

Given the circumstances, progress at Everett was disappointing to say the least. Responding to an inquiry from Ingram, Long allowed that they had found the old mill in such bad shape that "we practically are re-building it." As for architect Hills, Long minced few words: "While he may be a remarkably efficient mill man . . . it also looks as though he is remarkably expensive." He was, according to Long, "lavish in every direction."

The mill had been shut down for repairs in mid-December and, after assessing the situation, Long expressed doubt that they could resume production before mid-March at the earliest. He reported in detail to Frederick Weyerhaeuser: "We are putting in an automatic trimmer, an automatic slasher, have widened the mill and will put in a double cut band re-saw. Have been compelled to tear up the engine foundations and re-set them, and put up a new smokestack, and do divers other things, but it looks as though we will have a pretty good mill when we get done; it also looks as though we will spend more money than we originally contemplated."

Until its December closure, they had operated the sawmill for eleven months, sawing 26,312,159 feet of logs from which was manufactured 28,168,526 feet of lumber, for an overrun of about 7 percent. Profits totaled less than $40,000. A new mill foreman, Earl M. Rogers, formerly in the employ of the North Wisconsin Lumber Company at Hayward, Wisconsin, was now on the job. With new machines and a new manager, Long expected better returns in the future.

Meanwhile the St. Paul office was trying to improve upon the initial Clark County logging contract with the Northern Pacific. Little progress resulted, but it was, according to McCormick, "the best we could get pleasantly." No doubt negotiations with the NP officials would have advanced further had Frederick Weyerhaeuser participated. But he was unavailable, ill for one of the few times in his life, suffering from what was described as "nervous exhaustion."

As Fordney and Long had proposed, a separate logging company would be created, with the Weyerhaeuser Timber Company "working jointly with Mr. Davis." Long hired Connacher at a salary of $2,400 with a 15 percent share of the new company's stock. Initially, John Alexander, who had been serving as the Timber Company's agent/land looker in Oregon, also assisted in numerous ways. Back in Michigan, Davis was less than enthusiastic about the Connacher decision, but he agreed, in part because he was able to appoint one of his own employees to serve as office manager–bookkeeper in Clark County. He wrote Long a letter of introduction for H. C. Clair, "who has had charge of our [Saginaw] office and accounts for nearly ten years." Fortunately for all concerned, Clair's appointment worked out well from every consideration: he was capable, and his being on the scene tied Saginaw more closely to Tacoma, mitigating if not eliminating all sorts of suspicions and difficulties.

Clair moved into a beehive. By June 1 there were nearly a hundred men at work on the railroad, Long reporting to Davis that they had ordered eight logging engines, "donkey engines," and would be buying "in addition the gearing and cable that go with them, and erect logging camps, etc." A small portable sawmill was also planned, to provide railway ties and the timbers needed for bridge construction. Though it was important for Davis to understand that this was no small undertaking, Long tried to assure him that they were doing everything possible to be frugal. "We have been rather lavish in expenditures up to date," he wrote, "but are trying to watch all corners now, and when Mr. Connacher gets in there hope to have a rigid organization that will keep them down to minimum."

Connacher arrived on schedule, an impressive figure in size and stature, looking every inch the woodsman in charge. His first task, however, had less to do with leading his crew than with caring for them. A smallpox epidemic threatened, and from Tacoma Long could only urge that they take "good care of the victims and fix them up in good shape even if it takes a little money to do it."

Problems of a very different sort were affecting the Everett work force. There the teamsters had organized and were demanding that only union laborers be employed. So far it was a minor irritant, although Long counseled E. M. Warren that it wasn't to be taken lightly. They would not, of course, accede to the teamsters' demands: "Make no preference to union or non-union men . . . simply closing trades with the men you want to keep on account of their fitness for the work." He further suggested that Warren "treat them nicely in every particular, simply stating your position, and that is, that we have no grievance against the union, nor union men, but must reserve to ourselves the right and privilege of hiring whomsoever we please." The particulars naturally altered in the years ahead, but labor relations would occupy an increasing amount of attention and concern.

On June 9, Long went to Everett to check on matters himself. J. D. Hills had supposedly completed his work, and if things weren't running perfectly, the problems were presumed to be relatively insignificant. Adjustments and corrections were obviously needed, but Long nonetheless signed off on the work, happy to end the Hills association. Warren was then notified that "the mill, planing mill and everything else is now in your hands," Long advising that in making improvements Warren should "strike at the most vital points that need immediate correction, and let the little things wait until you can reach them naturally and more economically." But if Warren assumed that he was to have a free hand, he was mistaken. "One of the first things we will want to do," Long continued, "will be to decide as to the best way to fix up the trimmer, or rather the lumber carrier beyond the trimmer; and even before you change this, I would like to have you fully mature your plans with Mr. Rogers and submit them to me for joint consideration." George Long's thin hands rested heavily on many shoulders.

But Long also felt hands on his own shoulders. From his St. Paul office, Frederick Weyerhaeuser, feeling more like his old self, complained that it had been "some time since I have any correspondence." He was more than a little curious, having heard a rumor that the Everett construction had cost "at least $150,000, in which I hope they are mistaken." Unfortunately, they weren't.

On the happier side, and of more importance, Weyerhaeuser reported success in bringing to an end negotiations with President Mellen on the Northern Pacific purchase option. Mellen had seemed uncompromising for a time, and without Weyerhaeuser's personal involvement, matters had reached an impasse. Frederick subsequently learned that Mellen's intransigence had nothing to do with western Washington, but was due to a misunderstanding about "some iron lands in Minnesota sold through a mistake." Long was much relieved by Weyerhaeuser's assurance: "I am about closing our matters for good with Mr. Mellen . . . with the understanding that we have an option and the preference over any one else on what is left." On August 1, 1903, Long reported that under the so-called second Northern Pacific purchase more than 260,000 acres had been added in western Washington at a price of $6 an acre along with 120,000 acres in western Oregon at $5 an acre. As a result, Timber Company ownership totaled nearly 1.5 million acres.

By mid-August, a portable sawmill was set up in the Yacolt prairie, only a couple of weeks behind schedule, and soon the Twin Falls Logging Company was organized. Alexander and Connacher were instructed to conduct all future business under the Twin Falls designation, and to operate on "a cash basis, and no one need hesitate to sell [to] them as freely as they would sell [to] the W. T. Co." George Long was president of the new company, Connacher its general manager, and H. C. Clair the secretary-treasurer, "and the stockholders are practically Mr. C. H. Davis and the members of the W. T. Co."

While there was much to trouble the Weyerhaeuser Timber Company in the fall of 1903, prospects in the long term seemed promising. The annual stockholders meeting had been held in St. Paul on August 19, at which time the capital stock was increased by another $2.5 million, making a total of $12.5 million. F. S. Bell reported that the meeting was "all very cordial and pleasant, and everybody much interested in your report." Lawyer Macartney's account was similar: "Nobody made any sort of an objection either to the increase for this amount or to subscribing for their proportion." Long was naturally relieved: "It is quite gratifying to learn that the stockholders are in a pleasant frame of mind, and I really feel that they have no occasion to worry over the result of their investment."

In fact, however, the immediate future looked downright bleak. Portland log prices had dropped two dollars in recent months, to $6 per thousand. Long thought they might still realize a dollar per thousand at that price, but, as he wrote to Weyerhaeuser, "It seems too bad to sell logs for $6." The alternative was to "take the other 'horn' of the dilemma and manufacture them." Much discussion followed, mostly focusing on the relative advantages of a Yacolt or Portland sawmill site. The midwestern lumber markets were nearly as depressed, as McCormick observed: "The general feeling here is that the crest of the wave of prosperity has passed and people are anticipating cessation of work." Indeed he predicted that the Mississippi Valley Lumbermen's Association would soon authorize reduced prices, for the first time in five years. Even Frederick Weyerhaeuser, the eternal optimist, was talking of hard times and lower prices. Still, McCormick was confident that they would authorize the purchase of a Columbia River mill site favored by Long.

Bad times or no, the Twin Falls Logging Company was making haste, the portable sawmill "running very nicely," cutting more than 25,000 feet daily. Construction of bridges was under way, and Connacher was proving an able manager. "We like him first rate," Long wrote to Alex Polson. Connacher did need a bookkeeping assistant, however, and it so happened that McCormick had just the man in mind. "The gentleman's name is Peter P. Nelson, Swede, dark, about six feet two, dark mustache, good habits, quiet, a good penman, stenographer, accurate in figures, careful in everything." Nelson was soon on the job, another valuable addition to Twin Falls.

In the fall of 1903, Frederick Weyerhaeuser made one more trip west, desiring to see for himself the situation in Clark County and to consider, on the ground, whether the company should sell the logs or erect its own sawmill. They decided upon the latter course, and Weyerhaeuser agreed that Portland should be the site, but there was no definite commitment. By the time the logs began to trickle out, prices had increased somewhat. Long was encouraged, and turned his attention to organizational changes.

By mid-January of 1904, Long had negotiated an agreement with C. H. Davis and G. M. Stark for organization of the Clarke County Timber Company. This was to be a holding company comprising lands jointly owned by the Davis-Stark interests and those of the Weyerhaeuser Timber Company, with each contributing half of the total capitalization. There would still be problems regarding assessments, but the mutual advantages were so obvious that there seemed little purpose in delay over minor differences, protecting cents at the cost of dollars. Long suggested that initially they capitalize at $300,000, "and arrange for such funds as may be needed to buy other timber and to operate." The operating arm would continue to be the Twin Falls Logging Company, with the Lake River Boom Company assuming responsibility for the logs from the railhead to the Portland market.

With the organizational agreement, logging began in earnest. Lafe Heath arrived on the scene following the closure of the Maple Valley logging camps in December, and Long requested that he study the Clark County situation thoroughly. More than anything else, Long worried about log prices, warning Connacher that they should expect the log market "to sympathize further with the lumber market," possibly to such an extent that they couldn't even afford to operate. With that in mind, they should avoid undertaking any development that could be postponed to advantage. Long admitted that he envisioned no project to fit that description, but was simply restating the old rule of thumb: "Push your logging for all it is worth, cutting off all unnecessary expense in any and every direction."

Long was overly pessimistic. The log market would have to fall considerably below its $6 level to justify a shutdown, with Connacher demonstrating that they could log at an average cost of $2.50 per thousand feet. In fact, Long's real frustrations centered less in Clark County than in distant Saginaw, where he now had to keep satisfied another set of absentee owners.

EIGHT

Broader Horizons

GEORGE LONG HAD NEVER BEEN BUSIER THAN IN THE spring of 1904. The Everett sawmill was operating impressively; indeed, if one wished to see an up-to-date plant, that was the place to visit. But, given the price of logs and lumber, the satisfactions had more to do with machinery and men than with profits. And although the Twin Falls Logging Company was now beginning to salvage its burned timber, the Portland log market was discouraging, and the costs of logging were proving to be higher than expected. There were also the constant concerns with land management, timber sales, purchases and trades, fire prevention and taxation, all within the context of trying to please the absentee owners back in Michigan, Wisconsin, and Minnesota.

Ingram had recently written another polite letter, in which he inquired as to the health of their company and the industry. Long refused to sugarcoat, informing Ingram that the lumber market had taken a "violent slump" to such an extent that the mills still operating were "practically doing business for nothing." The railroads were no longer buying, and since so much of that business had involved timbers of special sizes, this was a great loss. California absorbed much of the western Washington product, "but everybody is 'dumping' it there, and common lumber is sold on the basis of from $5 to $7 on the wharf here for the California delivery." Finally, the foreign export market had declined, falling from a high of $14 a year earlier to roughly $9.50.

As for the log markets, the loggers on Puget Sound had organized and "acted sensibly and not put in many logs," which kept their prices artificially high. As a result, the average for Douglas-fir logs on the Sound remained at $6.75, whereas it had fallen to $6 on the Columbia, $5.50 at Grays Harbor, and $5 at Willapa Harbor. Long estimated that their Clark County logging was costing about $5 per thousand to deliver logs to the Columbia, "so we are getting $1 for our burned stumpage." He indicated that there would be no large push to increase operations for fear of further demoralizing that market. Any thought of a Portland manufacturing effort had been postponed, although he continued to believe "we ought to have a mill [there] and we are keeping in touch with the market and with desirable locations on that stream."

Even the mechanical operation at the Everett mill came apart, literally, for a time. On May 24, Warren reported "the most serious break that ever happened at this mill." One of the shafts on the band wheel "twisted like a cork screw and the wonder is everything under the mill did not go to pieces." Warren was most thankful that no one was injured. Although it took nearly a day to remove the twisted shaft from the wheel, the mill was back in operation within two days.

Narrowly averted disaster was the good news. The bad news concerned the order file. It was nearly empty in early June, and Warren awaited instructions "as to whether we had better continue or not." Long thought not, given the market conditions and prospects. True, one ran the danger of losing valuable employees to other operations, but the other operations were also closing down. Still it wasn't an easy decision, and some, such as McCormick, finally settled in Tacoma, opposed a closure. Long spoke of this disagreement in a letter to a San Francisco dealer. "We are making no money," he wrote, "and there is no particular reason why we should run . . . and [the greater benefit might be to stop rather] than to keep running and congest the market." In the end, he used McCormick's absence as an excuse for delaying the decision, McCormick having gone east to attend the Lumber Manufacturers Convention in St. Louis and also to visit the great Louisiana Purchase Exposition.

In the meantime, it was business as usual. Long proceeded with the purchase of a small shingle mill adjacent to the Everett plant, confident that the additional space would sooner or later be needed. He was less certain about how to deal with the Columbia River log market. To editor-publisher George Cornwall, Long allowed that "we would be willing to agree on prices that should be maintained, but our physical condition is such that we cannot well afford to shut down."

Working in burned timber reduced Long's options. They could store as many logs as possible, but that, of course, had practical limitations. "The lumber situation is really worse than the log situation," Long observed, "and I cannot reason out any solution of the present situation that is comfortable."

Long was also made uneasy by the dryness of the early 1904 summer. As had become custom, the Timber Company cruisers were appointed deputy forest fire wardens, or "special forest patrolmen." In mid-July there were reports of fires in Lewis County, and Long wasted no time contacting George Joy, stationed at Boistfort, urging that the fire be completely extinguished. He instructed Joy to seek the assistance of the Lewis County fire warden, "and if it is not promptly given, go ahead and hire men yourself." In other words, they would worry about expenses when the drought was over. Fortunately, rains fell across western Washington on July 14 and 15, reducing the fire danger considerably.

Long encouraged his crews at every opportunity, advising that expenses were to be of secondary concern. To Fred Conant in Eatonville he wrote, "Whenever you think men are needed, employ them," even if there was a chance "of having some man start a fire for the sake of getting paid to help put it out." Such chances must not be allowed to hinder the effort.

In mid-September, when the fire season was beginning to abate, Frederick Weyerhaeuser requested that Long come to St. Paul to help in the continuing Northern Pacific land negotiations. Long boarded the train with misgivings, not so much concerning what awaited in the East as with fear of fire in the West. While absent, Long received encouraging reports from McCormick, who wrote that everything was "running along very smoothly," even if they weren't doing any business. One reason for this was that they were all occupied "watching the fires." Then, on September 21, relief finally came with "a very nice rain, and it was quite general," and the weather since had been cool and moist, with the likelihood of more rain. In short, the cruisers were back cruising and the loggers were logging, and conditions were soon normal.

Carrie and the children had accompanied Long to St. Paul, and after the meetings there they all paid a visit to Eau Claire. McCormick urged Long to be in no hurry to return to Tacoma. "You may just as well have a good time and enjoy this trip," he wrote, adding, "Business is not crowding us, and I think we will be able to get along very nicely during your absence." Long followed his friend's advice. He and his family took their first extended vacation in many years. This included a visit to Indiana; George allowed that he had "thoroughly enjoyed the political excitement that was going on . . . as it seemed very much like old times." They continued to St. Louis to see brother Jim and the fair, afterward returning home by way of Denver, Salt Lake City, and Portland.

Long arrived back in Tacoma on October 31 and wasted no time in returning to the routine of work. The promise of better times signaled an end to any interest in shutting down the logging camps early. Thus Long advised Connacher to plan on continuing his Clark County operations through November, and Heath at Maple Valley was similarly instructed. The Everett sawmill was operating well; in round numbers, they had sawed 3.5 million feet of lumber in October and had shipped 3.3 million feet. Warren was more optimistic than most of his competitors as to future prices, refusing to sign a lumber association price list for export figures because he thought them too low.

A special item tweaked Long's curiosity at year's end. At the behest of the Timber Company and the U.S. Bureau of Forestry, Charles Chapman, a recent graduate of the Yale Forestry School, had conducted some forestry studies the previous summer. Long, having expressed interest in the results, received a thoughtful response from Bureau of Forestry timber-management chief Thomas Sherrard in January 1905. (The Bureau of Forestry became the U.S. Forest Service on July 1, 1905.) After noting that Chapman was still working on his final report, Sherrard did indicate that efforts had been concentrated on "determining the problems of management which were most important, and the most practical ways of solving them." From our late twentieth-century vantage, this evidence of happy cooperation between the Bureau of Forestry and a private lumber company may seem surprising. But at that time Long and Sherrard, with the encouragement of Chief Forester Gifford Pinchot, were committed to a consideration of forestry problems without the intrusion of politics. Sherrard listed the subjects studied: fire protection, reproduction of fir, financial problems in utilizing cutover for future crops of timber, and the deterioration of stands of mature timber. He offered some opinions: "We feel that high taxes and danger from fire stand most in the way of the practice of forestry on your lands. The fact that danger from fire increases each year, as more land is cut over, makes it most important that the fire question should receive very careful attention. Slash burning seems to be the most feasible way of lessening the danger from fire. The present fire law can undoubtedly be improved upon, and we are at work upon a simple and effective law for your approval."

As for raising a second crop of timber, Sherrard indicated that whether it would pay depended on various factors including "the rate of growth of the fir, its stumpage value, the rate of taxation, and value of the land." He thought it might be a reasonable investment, provided stumpage prices rose, but he urged more study "on the ground" before making any final recommendations. Long approved of this idea, whose application was yet very distant. Timber indeed could be a crop, but only if it turned a profit.

In short, growing trees concerned the future. There were worries enough at present, one of which was worker safety. In an activity employing sharp saws and axes, steam engines, large animals, difficult terrain, and immense trees and ponderous logs, even the greatest care and caution too often proved insufficient. One of the real hazards for loggers, and to a lesser extent the millmen, was their curious need to be manly, to take chances. But men were frail and fragile creatures compared with the powerful things about them, as was demonstrated again and again on farms, in forests, and in factories the country over. At the moment, Long was considering how best to provide for the health care of men in the Maple Valley logging camps. Doctors commonly contracted with organizations on the basis of a specified fee per individual, the men joining an "association" through a fifty-cent assessment, or some similar fee. Long was not convinced that this was the best arrangement.

For one thing, the skill of the physician was by no means assured. "I still cannot help but have a preference to making an arrangement with some first-class hospital," he wrote, instructing J. L. Bridge, Jr., to seek "the best terms you can make with one or more of the hospitals, and do nothing definite until it is brought to my consideration again."

Meanwhile, Long was pleased to report to major shareholder William Carson that the Weyerhaeuser Timber Company was "drifting along about as usual." One matter appeared near resolution: the purchase of a mill site, "about as good a location as can be found in Portland." Negotiations for the St. Johns property on the Willamette River were completed on March 17, 1905: "We will get about 30 acres of good bench land situated about 500 feet from the bank of the river; about the same amount of low land between the bench and the river, and then about 30 acres of hill land . . . [the cost] in the vicinity of $50,000." Some thirty acres could be sold off for town lots, "at a price which we think will realize anywhere from $500 to $700 per acre, but we will be slow in selling anything until we know just what we will need ourselves." In fact, the St. Johns property would never be used for a sawmill. It proved to be of continuing value, however, as a bargaining chip.

The 1905 spring weather was too good; there wasn't enough rain and the streams ran low. As a result, logs hung up in many places, especially in Clark County, where conditions were "magnificent" for logging, but "awful poor weather for us to get our logs off the bottom of Vancouver Lake." With logs stranded and Clark County tax assessments on the increase, there was much about which to complain. Long recommended that they not worry about paying taxes in time to benefit from the cash discount because he hoped to more than make up the difference "by getting a re-valuation by the Board of Equalization in August next." The increase, amounting to nearly 50 percent over the previous year, was "wholly uncalled-for, especially on the burned timber."

But the news was not all bad. The amended forest-fire bill had passed, and, as Long informed lawyer Macartney, "The nice thing about it is, that it went through just the way we drew it up here." F. S. Bell observed that the fire law seemed "a comprehensive and useful measure, almost a model for a timber state," and he also noted that the success in obtaining legislation "permitting corporations to hold stocks will be increasingly useful to us and to other extensive operators as the years pass." Macartney, however, worried that the lumbermen may have been too successful, suggesting that Long be quiet and "let other people take the credit" for its passage. In particular, rail officials Macartney had talked with were "very sore." Lumbermen, he warned, "and particularly the Weyerhaeuser Timber Company, do not want to incur the enmity of the railroads." Long wasn't concerned. While acknowledging that the Timber Company was "mixed up in the [Pacific Coast Lumbermen's] Association work," he said that the only legislation in which they had taken a personal interest "were the corporation holdings bill, the boom company bill, and the forest fire bill."

Frank Lamb, a Grays Harbor logger, was appointed fire commissioner under the new law. Long had recommended Lamb to Governor A. E. Mead, describing him as an educated man who had given "especial study to forestry and forest conditions, and is the author of the very commendable pamphlet herewith enclosed." Lamb wasted no time in proposing improvements, specifically the possibility of supporting the fire service by private contributions from timbermen. He recalled that Long had told him that Weyerhaeuser had spent $10,000 the year previous for fire protection. Lamb suggested that the private landowners unite with the state in one service for the entire work, thus avoiding confusion and conflict in the different sections of the country where the fire might originate. He further believed that if George Long led the way, they could secure a considerable appropriation from some fifty or more other timberland owners. Otherwise they would be limited to state funding of no more than $5,000 a year, barely enough to pay the state warden. Lamb cited Canadian examples where the expense was borne half by the government and half by the private owners. Lamb had also corresponded with the noted German forester, Dr. C. A. Schenck, who had suggested a plan of this sort. Long was delighted with Lamb's enthusiasm.

One subject that had come to Long's attention while in St. Paul the previous October would prove of far greater importance than initially imagined. Sam Hill, son-in-law to James J. Hill, had called upon Frederick Weyerhaeuser to offer a deal on some properties in the Klamath Falls section of Oregon, including "quite a bunch of timberland, containing sugar pine, yellow pine and fir." At the time, Long suggested that Hill forward his proposal to the Tacoma office and then advised McCormick as follows: "If you will look in the drawer in my desk where I am in the habit of keeping plats, you will find a plat made of the Klamath River district with the lands colored, showing different ownerships," adding his own assessment: "In my judgment I think there is in this district probably the best bunch of sugar and yellow pine left in the country, and I said as much to Mr. W. and he seemed quite interested in the idea of investigating it further, and if we could make a deal whereby we could pick it all up, to do so." In fact, Long was already looking forward to making a detailed inspection tour of the area in order to make plans "on the ground."

Early in 1905, Klamath opportunities once again surfaced, specifically concerning an option from the Pelton–Reid Sugar Pine Company for the purchase of $400,000 worth of real estate located in Siskiyou County, California, and Klamath and Jackson counties, Oregon, including some nine miles of logging railroad with equipment. The option ran until March 15, and Long was eager to study the proposition, again from the ground up.

Although the negotiations with Pelton–Reid proceeded pleasantly enough, Long soon found himself involved in more frustrating dealings. Hervey Lindley, president of the Klamath Lake Railroad Company with headquarters at Pokegama, advised Long to make haste in reaching an agreement on one strategic Klamath property. It was deemed essential to any future development and, in Lindley's opinion, "You had better get that [George] Mason matter under cover some way soon as his is cheaper even then than any and all the rest, and though a first-class crook and we will know it—yet he has what we want and a lapse of time will not help." The pine may have been beautiful, but the local politics was about as ugly and unpredictable as could be imagined. While Long eventually became accustomed to such reports from Klamath Falls, in February 1905 he refused to believe the worst.

Long would persevere. Nonetheless, Klamath was destined to be something of a stepchild, partly because of its relative isolation from the other operations, but also because it was predominantly a forest of pine rather than Douglas-fir. Long had never seen such a magnificent forest, and he would make sure that others, especially the doubters, saw it for themselves. Once they did, most were converted.

The first step was, of course, to get acquainted with the territory. So cruisers were dispatched to Klamath to make the necessary estimates of tracts under consideration. Long awaited their reports none too patiently. In a note to a cruiser operating out of Pokegama, he gently complained that he had received no news "since you went down where the air is pure and the sun always shines." More irritating was the lack of progress in the various negotiations. As Long described the situation to Myrick & Deering, San Francisco attorneys, "All three of the parties at interest with whom we have been figuring have their affairs in bad shape: namely, Mr. Lindley, whose property is the Klamath Lake Railway Company; the Masons, whose property is the Pokegama Sugar Pine Company; and the Cooks," heirs of John R. Cook, owner of the Klamath River Lumber & Improvement Company.

Still, Long remained optimistic. Earlier he had indicated to attorney Frank P. Deering that things were "in such a chaotic and uncertain condition that we did not care to dignify the mere mention of the trade to our Directors," but now he was "acting on the theory that if we appear to be ready to close it will keep the Mason crowd from flying the track as Mr. Lindley suggests." So McCormick left for St. Paul on the evening of April 15 to consult with the Weyerhaeuser directors about Klamath Falls. Although he would meet with some success—Weyerhaeuser and associates agreed that the Klamath pineland opportunity was significant—no final commitment was forthcoming. Thus they marked time, Long remaining confident that patience would eventually win out. In truth, Long wanted the Klamath properties more than anything else, but his manner of dealing was the same as always. "We are interested on our terms," he clearly said, if only to himself, "but only on our terms."

As spring turned to summer, it was time to prepare for visitors and the annual meeting. The formalities passed uneventfully. Frederick Weyerhaeuser had not fully recovered from the effects of a slight stroke, and "his family who are with him have virtually requested us to take up with him as little as possible matters of business," or so Long informed Lindley. Long wasn't misinforming, but in this instance the facts allowed him to buy a bit of negotiating time in Klamath Falls. In passing, he learned about some Klamath timberland owned by the Hopkins family, a tract that would become vitally important, but which in June 1905 seemed just another in a series of opportunities.

Although in retrospect the Weyerhaeuser Timber Company of 1905 seems to have been a small and simple organization, the problems confronting its manager were anything but small and simple. In part these resulted from the interest others took in Frederick Weyerhaeuser and the assumption that Weyerhaeuser had achieved success and power by questionable means. The times seemed to require a belief that businessmen operated above or outside the law, or at least outside the spirit of the law. Period journalists—the so-called muckrakers—recounted in delicious detail any rumors or indications of misdeeds by the supposed miscreant leaders of industry. Victor Beckman, whom Long had considered a friend, briefly jumped aboard the bandwagon and published an article in the *Pacific Lumber Trade Journal* representing Frederick Weyerhaeuser as a threat to his competitors; that article, to George Long's consternation, was reprinted in the June 30 issue of the *Mississippi Valley Lumberman and Manufacturer*.

Those who knew George Long and the nature of his efforts in the Pacific Northwest understood his hurt. F. S. Bell's reaction typified the feelings of the Timber Company's owner-directors: ignore it and hope it goes away, quickly. "It is natural to desire that Mr. Beckman be brought to a better sense of propriety," he advised Long, "and yet the only question that any of us raise with reference to your handling of the matter is whether the whole affair might not better be allowed to slip by without any further attention." In truth there was no effective way for Long to respond, other than to continue demonstrating responsible management in terms of the company, the industry, and the future.

He decided that it was time to reach a decision on the Klamath River properties. He informed Weyerhaeuser of the situation as of July 25, admitting that it was still puzzling: "There are a number of parties interested in there, two or three groups of which are badly mixed in their relations." But he had nonetheless agreed to purchase three-fourths of the stock of the Pelton-Reid Sugar Pine Company on a basis of a total value of $340,000. The balance of the stock was owned by some other interests and might later be acquired. The initial purchase amounted to 20,000 acres, 16,000 of them timbered. They had cruised the land, estimating some 224 million feet, of which 210 million feet was sugar and yellow pine, 21 million of fir, and 1.5 million of cedar. Long added that these figures had been compiled "by our most conservative cruisers." As important, "We will also get a mill site on the Klamath river, where the So. Pacific railway crosses, and the improvements and rights of the Klamath river for the Driving Company."

Since Long still rarely decided such questions without the counsel of those back east, he felt the need to explain to Frederick Weyerhaeuser the special circumstances surrounding these negotiations: "Now this one trade which we have agreed to make was made the other day when the time seemed ripe to act, and the situation is one where we look upon the investment as a good one regardless of whether or not we continue our investments in that locality, and it is for this reason that we made it without further conference with you."

This became the first of many purchases envisioned by Long in the Klamath Falls district. In fact, with hardly a pause he mentioned another opportunity to buy 10,000 acres, negotiating "with a view of buying the railroad which connects the timber with the S. P." And he expressed more than passing interest in the 40,000-acre Hopkins tract, which had been offered at $2.5 million. Although Long considered that price entirely too high, some distant owners must have wondered just where they were headed.

Among them was Frederick Weyerhaeuser. Was it wise to be making such large investments in an area so remote from their principal holdings, especially when the Idaho operations were requiring so much attention and money? And if Weyerhaeuser wondered, some of his colleagues had to worry. By 1905, however, Long knew an opportunity when he saw one, and regardless of distance and local complications, Klamath Falls was a first-rate opportunity. Most in St. Paul agreed.

Timber Company business, of course, did not stand still for the Klamath negotiations. Concurrent with those deliberations were the eternal summer concerns with forest fires. And as the season heated up, so did problems at Everett, exacerbated by a depressed market. In midsummer, Warren summarized the market conditions, reporting that he had only three cargo orders on hand, and these at less than satisfactory prices. And there was little hope for improvement. A Pope & Talbot representative predicted that it would be "very difficult for anyone to obtain $10.00 base price this year on foreign orders," and Warren concurred with that gloomy outlook. Rail shipments continued at about the same level, which was to say unimpressively, both in terms of volume and prices.

Lumbermen always had multiple worries when it came to marketing their product. The first concerned demand and price, the second delivery. Rail rates were a matter of fundamental importance, and lumbermen traditionally feared the power of railroad operators, particularly when one line enjoyed a regional monopoly. To place oneself at the mercy of a single railroad was regarded as stupid—if not suicidal. Frederick Weyerhaeuser was exceptional in his ability to manipulate the railroad interests. His unusual success, however, had as much to do with the extent of his dealings as with his skill as a negotiator.

At Everett, rail uncertainties constituted only part of the problem; even more troublesome were uncertainties in the cargo trade. It was largely for this reason that Long opened discussions with a San Francisco lumber wholesaler, Charles R. McCormick, who planned to enter the shipping business. In the summer of 1905, McCormick contracted to build the *Cascade,* financing the construction by selling shares in the vessel. In effect, the ship would be its own corporation. By August, McCormick had obtained all but one-sixteenth of the stock subscribed, and he offered that share to Long—the actual cost being about $6,200—promising in return to offer lower freight rates and to "always hold the steamer for your lumber, and . . . not put her in at any other mill unless you were unable to load her." With reluctance, Long agreed to the purchase, thereby involving the Weyerhaeuser Timber Company in steamship operations, the first step in what would prove to be a long history.

Just as reluctantly, Long was serving on the Freight Rates and Railroad Committee of the Pacific Coast Lumber Manfacturers' Association. At issue was a notice of the intent of the transcontinental railroads to increase rates to certain eastern terminals including St. Paul and Minneapolis, in exchange for reduced rates to the Missouri River territory. Meeting on August 24, the South Western Washington Lumber Manufacturers' Association, representing fir interests, resolved to oppose "by all honorable means any increase in the Fir rate into any territory." A meeting of the committee was called for August 31 to consider the same question, but Long sent his regrets, using R. L. McCormick's absence from the office as his excuse. In truth, because the interests of Weyerhaeuser were often larger than those of the other operators, what was good for them wasn't necessarily good for Weyerhaeuser. In this instance, as Long peered into the future, he thought he saw increasing opportunities in that Missouri River territory. Thus he advised the boys in Everett to "swing around and get in shape for a big Dakota trade."

As the end of summer 1905 neared, things began looking up in the Pacific Northwest. The weather was a bit drier than Long would have preferred, but they had not been troubled with fires, and the danger should soon be past. In addition, the lumber business was improving and orders were "quite plentiful, especially for car trade." Furthermore, the log markets on Puget Sound and Grays Harbor were "stiffening, while on the Columbia River it is not getting any worse." Connacher had put in 6 million feet during August, and the Lake River Boom Company was prepared to take care of any amount, Long predicting that they would go into winter with 20 million feet of logs or more, expressing confidence that they would get a good price for them before April 1. All in all, it was a good time to be a lumberman.

On September 16, Long announced the purchase of three-fourths interest in the Pelton-Reid Sugar Pine Company and the Klamath River Lumber & Improvement Company for $237,650. As he explained to Frederick Weyerhaeuser, on whom he had drawn to the amount of $200,000, "This purchase is the commencement of an investment which we hope to make in southern Oregon for the so-called California White Pine and Sugar Pine known as the Klamath River tract." It proved an accurate prediction, "the commencement of an investment." Some sales happened to coincide with the Klamath purchase and, although the two were unrelated, Long must have felt fortunate to be able to cite these while reporting the investment. One sale of 2,100 acres was for $170,000 cash and another of 1,240 acres for $84,000. Recalling the initial cost of $6 per acre, the profits earned were, by any standards, astronomical. Long described them simply as "good luck stories."

Luck may have played a part, but only a small part. Far more important was Long's attention to detail in combination with judicious foresight. In land dealings, Long might occasionally have missed an opportunity by delaying a decision until he had acquired all possible information. But he avoided the far more costly error of making ill-informed decisions. In large measure, this explains his steadfastness in his offers, because to do otherwise would suggest that he had initially misrepresented the facts. As he might put it, he merely wanted "to know the age of the horse."

Dealing from strength was a crucial advantage, of course, and though some lesser landlords occasionally showed resentment, Long seemed able to turn resentment into respect. In the process, he was establishing a clear identity. He was known far and wide as a man of his word, a man of integrity, whose reputation was only enhanced by his appearance and style. Doing business with George Long was a special experience: there was no need to question the details; Long would see to it that they were accurate, and you had his word on it. There is little doubt that Long took considerable pride in his reputation for honesty and fair dealing. There is even less doubt that he saw that reputation as essential to his responsibilities of representing the Weyerhaeuser Timber Company. While serving as the manager for the nonresident owners of its largest and most powerful organization, he was becoming a leader in a highly competitive industry.

On to Oregon

ALTHOUGH MANAGING THE MILLION AND A HALF ACRES of the Weyerhaeuser Timber Company in 1905 might have been a sufficient responsibility in itself, George Long came to understand that such a responsibility could not be readily limited. Like a magnet, his position attracted additional demands. There seemed no end to it.

Frederick Weyerhaeuser came west that fall in the company of his former adversary, Orrin H. Ingram. In part they were looking over their individual interests with a view toward possible consolidation; in part they were touring, considering new opportunities for investment. Of the two, Weyerhaeuser retained the more optimistic outlook, because he was simply made that way. Both headed homeward with increased confidence in Long's management.

Following their departure, Long set off on his own tour to California to consider a Humboldt County redwood property that had been recommended by Congressman Fordney, and to continue the Klamath Falls negotiations with George Mason in Los Angeles. On his way to California Long had stopped off in Portland, where he found Columbia River log prices on the rise, and he recommended that they advance their price to $8 on November 1. "We may not sell a great many at $8.00," he allowed, "but we have plenty of room now to store them and the price is sure to reach $8.00 shortly." Things were looking up.

The California tour was especially interesting in one respect: the redwoods. Long visited several mills in the vicinity of Eureka and, while admitting to McCormick that the redwoods impressed him, he decided that there was "something about the business that is not as good as it looks." Redwood sales had not kept pace with the rising market, and Long concluded that "either the eastern market is very indifferent about redwood, or else the people who are handling it do not know how to put it on the market."

The tour was also important in another respect: the Klamath negotiations. McCormick happily informed Charlie Weyerhaeuser of Long's success: "He has closed a trade for what is known as the Cook–Mason lands, about 12,000 acres, and has bought the Klamath Falls railroad." Thus they now owned all of the large tracts except Hopkins's, and they planned to negotiate for that.

When Long returned to Tacoma, one matter awaiting action concerned a new Washington state law permitting companies to own stock in outside corporations. Long thought the Timber Company should take advantage of the change and organize separate companies to handle the various activities. He had written to lawyer Macartney in midsummer, suggesting that it was an appropriate time to consider it.

The reorganization was effected in January 1906. Under the laws of Minnesota, five new corporations were created: the Weyerhaeuser Land Company, responsible for Oregon property management; the Weyerhaeuser Realty Company, for California properties; the Weyerhaeuser Lumber Company, the Everett facility; the Weyerhaeuser Logging Company, admitted for business in both Oregon and Washington; and the Weyerhaeuser Mill Company, in case there was any Oregon manufacturing. The five boards had the same directors, Frederick Weyerhaeuser, J. P. Weyerhaeuser, F. E. Weyerhaeuser, F. S. Bell, and F. H. Thatcher, and the stock of each was, of course, to be owned by the Weyerhaeuser Timber Company.

Among the other pressing matters were, as always, taxes. Long attended the State Convention of Assessors at Olympia as an observer, but ended up participating after all. The assessors, he wrote to McCormick on January 18, were "dominated by the Tax Commissioners, and their chairman said publicly on the floor that there was lots of timberland in Washington that was worth $200 an acre, and had been assessed for less than $10 per acre." That was too much for Long, and he "took the liberty of calling him down . . . as it was absolutely wrong." The chairman subsequently agreed "to get up before the Assessors and take back what he had said, and he did, in a very manly way." All things considered, the timber interests had fared well, "a very great deal better than the railroad people."

Despite the occasional frustrations, business was good and getting better. It was a happy time for most Pacific Northwest lumbermen. It certainly was for George Long. But even with good times masking problems, Long knew he ought to be paying more attention to the Everett operation. Too much had evolved there without benefit of study, both in the Maple Valley logging operation and at the sawmill. In particular, the accounting procedures connecting the two were in need of review.

One of the ongoing difficulties in an integrated organization such as the Weyerhaeuser Timber Company was determining internal costs and prices. For example, stumpage costs based on timberland purchase were artificially low, far below current log-market prices. Given the Weyerhaeuser situation, on an actual cost basis, even an inefficient mill would show a profit. But such operations ought not to be allowed to chip away at the profits from the initial investment.

Logs from the Maple Valley camp went by rail to Puget Sound, where they were dumped into the water and assembled into small rafts averaging 400,000 feet. Scaling took place after the rafts were formed, the whole raft being recorded on a single card with the figures then reported to Maple Valley and to Everett. The practice prior to 1906 had been to charge Everett fifty cents below the market price for Weyerhaeuser logs. In effect, this served only to confuse, making it seem that logging was inefficient and sawmilling efficient. George Long now informed the Everett mill that in the future they would receive logs "at the regular list price," explaining, "At the price we furnished you logs last year the camps have not been able to realize as much for stumpage as we have been selling the timber for, and we do not think this is doing justice to the camps."

Translating facts into figures was a frustration not only within the Timber Company operations; it was also an increasing concern in the Tacoma–St. Paul relationship. Frederick Weyerhaeuser had always been interested in details, but he was less committed to a statistical analysis than his son F.E. came to be, no doubt in part because F.E. had little in the way of practical experience to call upon. George Long recognized the need for a reliable system of accounting, but keeping track of the transactions of the Weyerhaeuser Timber Company was no easy task. In answer to inquiries from F.E., Long endeavored to explain his current bookkeeping arrangements.

Any system had to permit ease of accessibility and updating, and Long thought his card index and envelopes provided just that. "Ultimately," he asserted, "we hope to have our records here consist of the following:" a land book, a stumpage book, a tax book, and a transfer book. The land book contained the title record and land description "together with such information as may find lodgment against the land in the way of contracts, or leases, or sales." The purpose of the stumpage and tax books was obvious. The transfer book recorded every transaction in connection with the buying and selling of land, or the stumpage, and thus was "the foundation of the whole fabric." Long urged F. E. to come see for himself: "While it is rather full of detail, it is working very nicely and we think a good deal of it." Formally trained accountants might shake their heads, but Long would counter, "I am always able to find what I need." In the end, that was all that mattered.

In the meantime, Long continued work at consolidating the Klamath holdings. Agent John Alexander kept busy, becoming acquainted with the land, the owners and operators, and the communities. "As a general all-around proposition," Long advised, "our scheme will be to look over all the country 100 miles north, 50 miles east, and 50 miles southeast of Klamath Lake, embracing what is generally considered the timber belt." Cruisers would be dispatched to do the work, Long promising that Alexander would be relieved of the railroad responsibility just as soon as the new manager arrived.

That little railroad would be a large headache. E. T. Abbott, whom R. L. McCormick had hired to run the thing, seemed in no hurry to leave Minnesota, despite prods from Long. But Abbott's tardiness was not the only problem with the Toonerville Trolley. Though only twenty-four miles long and built primarily for "tapping the big timber belt," it was a common carrier between Thrall, California, and Pokegama, Oregon, a totally new business for Long and associates. They had purchased a used passenger coach, but delivery was delayed. On April 3, Long pleaded with the seller to speed things up: "We need it very badly, as we are hauling from 30 to 40 passengers a day in an ordinary box car, and their patience has about ebbed out." At the same time, one did not ride the Klamath Lake Railroad for its amenities.

Amenities also had little to do with the Timber Company's decision to relocate, moving its office to the lower floor of the building opposite the Tacoma Hotel, on the corner of Tenth and A streets. There were numerous reasons for the change, but a major one was the company's need to distance itself from the "influence" of the railroads. As long as Weyerhaeuser remained in the Northern Pacific headquarters building, there would be a tendency to confuse and combine the supposed interests of the two. Moving would not necessarily eliminate the assumptions, but it might lessen them.

On April 9, 1906, Long sat down at his old desk in its new place and prepared almost immediately for a trip to San Francisco with McCormick, primarily to consider additional purchases in the Klamath Falls area: the competition there had increased with entry of the McCloud River Lumber Company into the bidding. In San Francisco they stayed at the Palace Hotel, which, as Long would recall, "was gay and resplendent with the crowd attending the opera." When they headed homeward on Tuesday evening, April 17, the great earthquake was only hours away.

Attorney Frank Deering, with whom they had been consulting, provided early reports of the disaster. He assured Long that the city would be rebuilt, and he also stated the obvious: "If the Weyerhaeuser Timber Company ever intends to come to San Francisco . . . now is the time."

It is never pleasant to think of financial advantage accruing as a result of tragedy, but the fact was that the disaster would affect business in a great many ways. Acknowledging the "unanimous opinion" that the quake and fire would make "coast-wide business very active," Long prepared for the boom. He authorized Connacher to purchase two additional logging engines so as to increase the output, "taking advantage of the excellent conditions now prevailing for marketing logs." In the Columbia River market logs were selling at $9 per thousand,

and he instructed Clair to be patient in advancing prices: "wait until the time has surely arrived for it." He did later admit, "I am perfectly willing to be human and make advances when the situation warrants," but thought they should wait until the market settled down. In the meantime, "our policy should be to sell nothing in advance . . . and keep ourselves in shape to make advances which will be immediate when we do make them."

He gave a similar directive to Tim Guider, at Maple Valley. Guider had extended the logging railroad as planned, and Long now urged him to "make hay while the sun shines. I hope nothing will interfere with a good, steady stream of logs coming from your camp for the remainder of the year." At the same time, however, he cautioned against haste at the expense of efficiency. In that connection Long reminded Guider to be careful about the length of logs and any unnecessary waste. He advised that no more than four inches be added to the length of the sawlogs, and, "Also see if they are skinning the ground close, Tim, and cutting the trees low down, and going after everything in sight. If we can ever afford to cut timber close, it is now, and every foot you can gain on the stump is worth two feet at the top."

The annual meeting of the Weyerhaeuser Timber Company was well attended, with even Frederick Weyerhaeuser on hand. The most important tour was to the Klamath Falls district, and everyone left with a greater appreciation of that pine forest. As Long had promised, it was impressive. Less attention was paid to Everett, when in fact it deserved more.

The renovation of the Everett sawmill had been delayed by the failure of Allis-Chalmers of Milwaukee to deliver the machinery on schedule. Promised in March, it arrived in late May. The mill continued to operate throughout most of the ensuing confusion, closing only for a week or so in July. In early June, Warren reported that progress was "simply fine," emphasizing, "I never saw anything go faster or smoother than since this job was started." On July 11 he announced that the new band mill was "a great success, quality of sawing first class and she runs so steady that a common marble has stood on top of one of the columns since yesterday morning." Long, however, was less than totally pleased. Among other things, he noted that the cost of sawing for June had been a "holy terror," attributing this partly to running the mill while making the repairs. He advised Warren to "lie awake nights a little and see if you cannot reduce the cost of manufacturing pretty soon."

In the midst of these worries, word was received that fire had destroyed one of the Clark County logging camps. Apparently no one was to blame; indeed, Long praised Connacher for his "promptness and energy in getting right at the business of logging again and 'whopping her up.' " He also offered to help in any way possible, knowing full well that Connacher wouldn't call on him. Given the healthy condition of the log market, it was crucial that there be no interruption in work. There wasn't. In mid-December, Long was delighted to write Davis, "We are very glad that we are together on this subject, because the confusion of two operations in there I am sure would be very annoying and unprofitable to both interests." Connacher had recently been authorized to purchase four additional logging engines with the hope of increasing output during 1907 as much as possible, to perhaps 100 million feet. What with log prices firm at $12 and logging costs at $5.50, they were, as Long understated, "making very fair stumpage out of the burned timber."

Inevitably, as prices rose, there were accusations that some, including Weyerhaeuser, had taken undue advantage of the San Francisco tragedy. The *San Francisco Examiner* of July 28 featured a sensational story to that effect. Long was quick to assure agent Charles McCormick that such was not the case, at least with the Timber Company. "We really disapprove of any attempt to take advantage," he wrote, "and while I do not suppose we will get credit for this, and most certainly will not ask for credit, the fact remains that our attitude from the day of the disaster has been one where we felt and have acted on the belief and the theory that no advantage should be taken of the San Francisco calamity." As evidence, he noted that they had refused to advance log prices even when their competitors had done so. The Benson Logging Company began selling logs at $10 on May 1, and it was not until June 1 that Long approved an increase to $9.50. By the end of July, Benson was selling at $11, while Weyerhaeuser was committed to maintaining a $9.50 price through August. But the market prevailed, and prices naturally advanced with the demand. The letter was not to be shared. Long simply thought it proper that "you, who are our accredited agent in San Francisco, know just what the situation is here."

Long was also aware of their social responsibility closer to home. With considerable foresight and sound planning, Seattle was arranging for a public water supply that would be both plentiful and pure. In that connection, officials worried over the location of some of Maple Valley's horse barns, fearing the runoff would contaminate a stream tributary to the Cedar River. Guider wasn't enthusiastic about moving his barns, but Long ordered immediate cooperation. While the company might take the position that "we could do as we please about this property, as it belongs to us and nobody controls it," Long advised that "as a matter of decency and common interest in such an important matter as the purity of the water," they would move the barns. And move the barns they did.

One of the immediate results of the Klamath Falls tour during the annual meeting had been authorization for Long to proceed with timberland purchases in the region. He needed no encouragement. In late July he informed F.E. that they had bought 4,100 acres on the Klamath plateau, and had also enjoyed "a lively week in buying timber claims east of Klamath Falls where we found some excellent yellow pine timber." In addition, cruisers were working through "big blocks of the Booth-Kelly tract," Long estimating that they could probably put together between two and three billion feet of yellow pine stumpage east of Klamath Lake.

A major problem was simply locating the owners, scattered the country over as they were. Alexander continued to scour the Klamath district, and Long dispatched Julius Peetz to Klamath Falls to assist in recording all of the various transactions. Long also announced that two or three additional cruisers were on the way, suggesting that Alexander prepare "a campaign that will enable you to use O'Neil, Hubbard, Conant, Cole & McIntosh in cruising lands that we want to get information on outside of the Booth-Kelly tract." It was full speed ahead.

In mid-September, Long decided to check personally on the situation at Klamath Falls. In a letter to F. S. Bell, he told of visiting the McCloud River sawmill "to see how their logs and lumber looked," admitting that he had never seen any yellow pine (ponderosa) or sugar pine lumber other than a few lots in San Francisco. In the McCloud Company mill's log pond there were some eight million feet of "the most magnificent pine logs I ever saw," averaging two logs to a thousand feet, and the yellow pine lumber "had a closer appearance to high grade white pine than anyone could imagine from looking at the tree." Yellow pine, Long seemed to be saying, was the way to go, "an excellent drift for us to make."

After his tour, Long was able to report to Frederick Weyerhaeuser that they were making progress, predicting that by the end of October they would have purchased some 40,000 acres in the Klamath district, "at an all around average of between $1.25 and $1.30 per thousand." But a little "hitch" had developed in the largest single deal, this with the Booth-Kelly interests for some 137,000 acres, actually involving three separate tracts, of 18,000, 35,000, and 84,000 acres. The original discussion had presumed a price of $1.50 per thousand, and on the smaller two tracts, the cruises were reasonably close. But with the 84,000 acres, it seemed that "the Booth-Kelly people have been more liberal in their cruising on purpose, as it is the best and most important tract, with a view of forcing us into a cost price more than has been talked."

The differences were significant, Long's cruisers estimating about 1.2 billion feet, some 600 million less than the Booth-Kelly figures. An all-day negotiating session at Portland accomplished nothing, Long informing Bell that "there seemed to be no easy solution of the discrepancy." The passage of a month failed to improve the situation. As Long reported to Bell, "Mr. Kelly of their firm with one of his cruisers, and Mr. Alexander with one of our cruisers went in there and spent a week looking over the ground, but they did not have a very harmonious session." Long concluded that Booth-Kelly had raised its price by the only means available, increasing the timber estimates, since they had previously agreed upon the price per thousand. Long offered to raise the Weyerhaeuser estimate by 10 percent, but "they would not agree to it, and while we left them pleasantly and without any apparent friction, I am inclined to think they are going to make an effort to find another buyer at a higher price."

These were frustrating negotiations for George Long.

He wanted the timber, and he realized that it was probably worth the increased price. That, however, had not been the earlier agreement. Thus he was caught on the horns of a dilemma, forced to violate his own code. He explained as best he could to F. S. Bell: "Of course, if this high tide of prosperous conditions has come to stay with us this timber would be cheap enough at $2. per thousand, but I cannot believe that this will be true; and I think the only thing for us to do is to stay on terra firma in our ideas of values, instead of floating off into the unknown. However, I am quite a little disappointed in the turn affairs have taken."

Somewhat similar developments kept the Hopkins tract beyond reach. Long reviewed that case for the benefit of Sumner McKnight, noting that Hopkins was asking $2.5 million for the 40,000 acres, which came to some $3 per thousand feet, "but the timber lies very compact, and our 60,000 acres in the same district fits in with it very nicely; and while the price is high I am pretty well convinced that it would be a good purchase, as it is high grade timber with magnificent logging conditions, and we have our railroad right into it." He noted that they had been waiting "for old man Hopkins to either die or change his mind, but he doesn't die and every time he changes his price he marks it up."

If all was not going according to plan in Oregon, there were few complaints about conditions in general. For the benefit of former employer Sumner McKnight, Long observed that as 1906 waned, everything seemed "on the boom." He noted, for example, that stumpage prices in Washington had increased some 50 percent in the past year alone. And should the good times continue, he foresaw an additional 50 percent advance, making "the Weyerhaeuser Timber Company's holdings look about as good to me now as I had hoped to see them in ten years from now." While acknowledging that log and lumber markets were inflated and likely to fall, he still thought that stumpage prices might hold firm, "and from all indications it is going to be more than smooth sailing from now on."

With the encouragement of some of the directors, Long wrote to R. A. Booth in hopes of resolving the Klamath timberland impasse. Booth had suggested that Weyerhaeuser, if unwilling to buy the Booth-Kelly tracts, might consider selling its properties. Long, of course, wasn't interested, but replied that they made "it a practice to offer nothing beyond a section or two without first submitting it to our people for approval." Purchasing was quite another matter, and he expressed a willingness "to take up the thread of our negotiations with you at anytime," and ended with a question: "Do you know of any way we can get closer together?"

Although they would get together, getting closer came less easily. Before Christmas, Long and McCormick spent a couple of days with Booth and associates discussing two propositions: Weyerhaeuser would purchase the whole for $2,812,500; or Weyerhaeuser would pay $1,925,000, and deed 22,000 acres of Lane County fir to Booth-Kelly. McCormick and Long subsequently boarded the eastbound train on New Year's Eve, carrying the proposals to St. Paul for presentation to the Timber Company directors. In fact, they were still awaiting Booth's decision. Booth had returned to Buffalo, well aware of Long's request: "If at anytime prior to January 4th you should reach a conclusion as to which of the propositions you want us to submit to our Trustees," it would be forwarded to St. Paul. Long thought success was at hand, having informed Alexander that the agreement was "practically on the basis of the test [cruise] made by you and Mr. Markham," but he wasn't quite ready to celebrate. "There is 'many a slip twixt the cup and the lip,' " he reminded. And so there was.

The meetings in St. Paul in the first week of the new year, 1907, did not go smoothly. The problem had less to do with the desirability of purchasing the Booth-Kelly timber than with the details of payment. Although as a group the Weyerhaeuser shareholders had never been better off financially, many of them were hard-pressed for ready cash. Good times meant good investment opportunities, and these individuals seldom overlooked such opportunities. In fact, later that same month, F. E. Weyerhaeuser would warn Long to be especially careful since the Timber Company balance at the National German American Bank of St. Paul was down to $184,517.90, and they were unable to depend on Weyerhaeuser & Denkmann to cover their needs because "at the present time their deposit is quite low."

They could, of course, have spread their payments over

time, had the Booth-Kelly interests agreed. They wouldn't. Or the Timber Company could have borrowed. It didn't. Long was disappointed. In mid-February, he responded to another inquiry from Booth, explaining present problems without discouraging future negotiations. "Our crowd generally speaking have been compelled to 'dig up' a whole lot of money within the last year or two," he wrote, noting that they weren't willing to make "investments that pass the point where they have not the ready cash on hand." In other words, "they do not like to borrow." He closed: "This is rather strange talk for the W.T.Co. but I guess it represents the facts."

In fact, it wasn't at all unusual for the Weyerhaeuser Timber Company; borrowing was never easily accepted, even on terms where it was clear to most money managers that the borrowing could be done to advantage. Charlie Weyerhaeuser admitted as much. The stockholders of the Potlatch Lumber Company (Weyerhaeuser-affiliated operations in Idaho) had recently paid out $8 million in cash, and Charlie, Potlatch's president, observed, with more than a little pride, "We do not owe any one a cent that is due."

A month later, Long and McCormick met a Booth-Kelly representative in Portland, where F. E. Weyerhaeuser joined in the talks. Booth had been negotiating with his own associates in the Booth-Kelly Lumber Company and now proposed that they "convey to us the timber which had been under consideration for $1,925,000 and the 22,000 acres of fir timber land which we own in Lane County," with terms of $500,000 in cash, $500,000 at 5 percent interest due by July 1, 1907, "and the residue to be paid as fast as the lands were patented." In presenting the offer to Frederick Weyerhaeuser, Long made plain his own feelings: "We still think we ought to buy this pine," and if the stockholders could come up with the initial payments, "and we could get one, two and three years time on the balance, that in all probability the Tacoma office could raise the funds to meet these obligations as they might mature."

Frederick Weyerhaeuser must have smiled at that. It made it easier to say "no," knowing that it merely meant postponement. Regardless, George Long was not going to give up. The Booth-Kelly Lumber Company was in Oregon to stay— but so was the Weyerhaeuser Timber Company.

TEN

Good Times Are Not Forever

THE EVERETT SAWMILL CONTINUED TO SELL MOST OF what it manufactured throughout 1906, despite a troublesome shortage of rail cars. Although E. M. Warren had been trying to arrange things so as to be less dependent on the rail trade, there was no foolproof answer. Coastal shipments were momentarily encouraging, but McCormick & Company warned that San Francisco wharves were "all blocked up," Warren wrote to Long on December 20, 1906. And there was also news of a more ominous sort: the "yards were beginning to catch up on orders and in the future lumber would not be so scarce."

George Long had expected that the market would soon weaken—indeed, the boom had already outlasted his predictions—and he advised Warren to subtract 25 percent from year-end list prices on his inventory, "for I do not think you can count on the market sustaining full list values, even for the stock you have on hand."

The Everett inventory was of less concern than a long-considered management change. While Warren possessed many strengths, he had as many weaknesses. He was honest and hard-working, but his experience had been acquired in the upper Mississippi Valley, and much of what he knew of the West Coast trade was necessarily learned by trial and error. Long had big plans for Everett, now known as the Weyerhaeuser Lumber Company, and he realized that successful implementation demanded the leadership of a manager with specific West Coast expertise and demonstrated ability. With that in mind, he wrote to W. L. (Bill) Boner, manager of the Simpson Lumber Company's sawmill at South Bend, Washington.

95

In his letter, Long foresaw that the ultimate development of manufacturing at Everett would be on a very large scale: "I fancy the right man on the ground would be the natural one to have charge of the entire business." At present, he advised, there was only one mill, which they hoped to make into an export facility, "but we also have a large mill site there where we expect some day to have an up-to-date modern plant; one for cedar and one for fir and hemlock; and we will build it whenever the car trade . . . can be depended upon to take care of the lumber." In closing, he urged Boner to consider the opportunity and, when appropriate, they would "go into details a little more definitely."

Long apparently assumed that he already had Boner in his pocket. He surely didn't expect complications of the sort that developed. First came publication of the December 1906 issue of *Cosmopolitan Magazine,* featuring an article entitled "Weyerhaeuser—Richer than John D. Rockefeller." This had to be a painful ordeal for Frederick Weyerhaeuser, so private a person. The author, Charles P. Norcross, then head of the Hearst Washington, D.C., bureau and later to be famous for his assault on the "sugar trust," felt the need neither to interview nor to obtain the facts. He did just enough research to provide an outline, then happily filled in the holes with fancy.

The introduction clearly indicates the thesis: "Though practically unknown, Frederick Weyerhaeuser, Lumber King, Recluse, and Land-Grabber, is Lord over billions in vast forest tracts in the Great Northwest." The following passage is representative: "He is essentially a worker, but he works in the dark. He is a man of a thousand partners. His hand reaches the uttermost recesses of the wilds of the Northwest, and the highest mountain peaks are spots from which he could unfurl his banner to the air if he desired. Secrecy is his hobby. One partner has not an idea of what his relations with another partner are. His business is one of magnificent distances. His great wealth is in forests, and he seems to have acquired some of the impenetrability and brooding silence that are characteristic of them. There are many men rich through lumber traffic, but Weyerhaeuser is king of them all."

The veil was suddenly lifted, and the view beyond was less than flattering. Nothing would ever be quite the same again. Frederick Weyerhaeuser, his family, and his friends were terribly hurt. Life would go on, of course, but with an added weariness for a time, and a wariness forever.

Following the publication, letters poured into the Tacoma office, most offering sympathy. Long responded with care to many of them. For example, he thanked C. J. Winton of the Winton Lumber Company of Wausau, Wisconsin, for the "kind expressions regarding Mr. Weyerhaeuser . . . whose career in all these many years has never been tinctured with fraud or unfair dealings, but with all that is opposite to it, drawing to him a great many prominent men who have the utmost confidence in his ability, integrity and absolute fairness." As to specific charges, Long assured Winton that "the Weyerhaeuser Timber Company's record in the West is such that it is absolutely unassailable from any standpoint so far as acquiring any kind of property by illegitimate methods is concerned." But as always, making charges is easier than answering them.

Although Weyerhaeuser and Long might maintain that they were unaffected by it all, the truth is that they were each greatly affected, in both personal and business ways. Nearly twenty years later, Long would recall the experience, observing that while there was no substance to the charges, "at that time the public was very susceptible to that kind of muck-raking, so it made the job all the harder out here."

Long must have turned to other matters with relief. The winter of 1906–7 was unusually severe west of the Cascades, but even while the cruisers kept to the warmth of their cabins, plans for cruising went forward. Long hoped that at last they were in a position to examine lands in a logical and thorough fashion instead of jumping about, responding to opportunities for sale or purchase. Long's instruction to one cruiser was typical: take all the time he wanted on his next assignment, because the company wished "to get the very best cruise that has ever been made on a piece of timber." And then he offered what had become the standard command: "We want it looked over so carefully that we will never again have any occasion to want to go over it for the purpose of ascertaining how much timber there is or how the land lies."

There would, however, be some changes in cruising procedure. Long advised that they were beginning to pay more attention to hemlock, piling, and cedar poles, and urged the cruisers to take exhaustive notes regarding anything that might touch on such matters as the cost of logging and agricultural potential of the land. In the spring of 1907 he wrote a new set of "general cruising instructions," which detailed estimating, species by species, even indicating the manner by which grades were determined. In the section on land values, he noted, "We would like an expression from you as to the quality of the soil, the surface condition, and the value of the land by itself." Finally, Long directed: "It is quite important to report any possible sign of any kind of mineral or coal; and it is of even greater importance to report when the land you are examining has special value on account of its being a route for a railroad or a logging road which will have to be used for getting out the timber which lies behind it. Or if the land is situated on a creek or river affording a good site for a water power or a dam; these facts should all be noted."

To some it may have seemed that Long was obsessed with such details. But he knew that, despite the time he devoted to the Everett sawmill or to logging in King and Clark and Cowlitz counties, his primary responsibility was managing the wilderness. And the sooner he knew his wilderness, the better he could manage it. Little by little his maps began to take on varied tints, each color representing a different description. Soon, almost at a glance from his desk, Long could make those determinations so crucial in terms of buying, selling, and putting together future operating units.

In the meantime, the one manufacturing unit, Everett, demanded attention. Bill Boner had yet to make a commitment, and in fact was considering several other options up and down the coast, from British Columbia to California. Long had advised against California involvements and, following a visit there, Boner agreed. "As you stated," he wrote, "the money in the lumber business is going to be made in the North, and I have come back to try and get my small share." What Boner wanted was a share of the company, something that had not been included in Long's initial offer. Boner therefore rejected it, asserting, "[I] cannot bring myself to the point of accepting a salaried position," adding, "but I appreciate your offer so much it has been hard to say no."

The "no" was to be short-lived. In mid-July Warren

took ill, and an immediate change was required. This time around Boner accepted an offer in which considerations of salary and a share of ownership were combined. He went to work on August 1, Long indicating to O. H. Ingram that they were "quite in hopes he will prove just the man we want."

They would need a good man. Business had begun to slow in the late winter of 1907. Problems were first apparent in the Columbia River log market, in part because strikes had closed many of the sawmills. The loggers met on March 13 at the Portland Chamber of Commerce building. Attendance was good, Alexander estimating that nearly 90 percent of loggers were represented. An immediate result was creation of the Columbia River Loggers Association, and the only action considered was the only one feasible—reducing log output. Another meeting was scheduled a week hence. Given that Weyerhaeuser was logging burned timber, the thought of stopping or even slowing operations was distressing. Long explained his predicament to friend E. S. Collins: "We are just entering this year into a bunch of at least two hundred million feet of red fir [younger growth] that was badly burned, on which we have not attempted to log heretofore because of the price on red fir logs." Now, "we really cannot afford to let a day go by."

Collins, whose Ostrander Railway and Timber Company specialized in long timbers, responded quickly, maintaining that the loggers' efforts were strictly defensive, a reaction to conditions brought on by "the Portland Mill Combination." The combination, as he understood, demanded a two dollar reduction in log prices, with "all scaling to be done at Portland." In any case, Collins urged Long to attend the next loggers' meeting: "Mr. Long, you are needed at Portland Saturday night mighty bad—more than anybody else and more than you can possibly be needed elsewhere and it now looks as if our reaching the promised land depended on your doing the Moses stunt, so we hope you may be there with the goods."

George Long did attend the meeting, but he was no Moses. The Portland strike was settled, but unless the lumber market improved, no improvement could be expected in the log market, regardless of what the loggers were able to do cooperatively. "The proper caper," Long agreed, was "to restrict the output, but of course that is more easily said than accomplished." Still, they would try not to hurt the cause; in Long's words, "If there is any practical scheme to bring this about I think we ought to lend to the scheme all the help we can."

In this instance, bad weather contributed to the cause. As the Clark County logging had progressed, Connacher's crews were working at higher elevations, where it often snows rather than rains; and in the early months of 1907, it snowed. The company had hoped to cut 75 million feet for the year, but in April Long lowered his prediction by some 15 million feet.

Earlier he had advised Clair that "on a falling market the policy is to keep selling ahead of the fall," and in late May he again urged that Clair "push sales as much as possible at the prices you are now getting," $8 and $10. Clair tried to follow orders, but to no avail, and by early June Long sadly reported that the Cowlitz River was "full of logs, nearly from the canyon to its mouth." Unless the loggers halted operations for a couple of months, he thought prices might fall as low as "$6 logs in the Columbia— possibly not so bad as that, but much cheaper than now."

Despite such conditions, Long remained optimistic overall, believing, as he wrote to F. S. Bell on May 25, that "conditions in the country at large will not go off very much," and they would soon "settle down to a normal basis, which probably will be two or three dollars per thousand less on logs and three or four dollars less for lumber." Although the national economy seemed healthy enough, Long admitted that it was "no longer at fever heat." His instructions to Everett expressed confidence. "No need to get scared about market conditions," he wrote, adding that recent failures were probably caused by simply figuring too high. Now they should plan for prices two or three dollars less, and cut costs accordingly. "It would not reflect very creditably on our management if we are not able to get business enough to keep our mill going." That was warning enough. The Everett management, of course, was about to be changed. Boner would be an aggressive and capable leader, one who thought in the broad terms that Long so appreciated. But Boner couldn't produce miracles; he couldn't make good times out of bad.

And times were about to get worse. Connacher's crews were working as fast as they could to salvage the burned timber in Clark County. Time was critical because the dead timber was deteriorating, helped along by insect attack. But on June 25, the Columbia River loggers voted to suspend operations for one month, beginning July 1, and to meet again on July 27 "to ascertain the advisability of a further shut down." Included in the announcement was mention of a pending decision to raise boarding rates and reduce wages. Clair considered that a mistake, insofar "as already the labor leaders are talking about a loggers' trust." As for the shutdown itself, he and Long both had serious reservations, but they also understood that the costs of continued operation would be higher in terms of good will. And regarding a possible extension, Clair thought that he would "want to see more alarming features than at present to keep our camps closed."

The annual meeting took place on June 20 and, as had become customary, Long distributed a few copies of the annual report, which he retrieved at the meeting's conclusion. While the others followed, he read aloud those sections of the report he deemed worthy of special consideration. One such section prompted a vote to increase the capital stock by $2.5 million, "rather more to provide for the future than any immediate wants," Long observed. He reported that the total timberland holdings of Weyerhaeuser were 1,668,337 acres, much of the recent increase the result of closings made on the second Northern Pacific transaction.

By and large, Long's 1906 report was encouraging. To one who missed the meeting, he did acknowledge "a very noticeable falling-off, and we are beginning to realize that we have had rather too much of a boom." But, he hastened to add, "I apprehend that our stumpage will be all right anyhow." And that was the perceived strength: regardless of the ebb and flow of the market for logs and lumber, the stumpage stood firm as a financial fortress. That was the way it was, and that was the way it would continue to be, far into the future.

Long attended the loggers' meeting in Portland on July 27, but his position was in the minority, the loggers voting to keep the camps closed for an additional two weeks. This time Weyerhaeuser and a couple of other large operators refused to go along. Twin Falls Logging Company was back at work on August 1, Long predicting that "logs are going to be cheaper next year than they are now, so that I advocate putting in all the logs we can under the present prices." Those prices remained at $10 and $8, but barely.

Still, as the summer of 1907 came to an end, there seemed to be few major problems. Long reported to Mc-Cormick, who was in St. Paul, that log demand along the Columbia continued to be "fair," and Connacher was doing the best work ever, putting in "six million feet of logs during August, and thinks he will have his great big banner month in September." Long also noted that Boner had paid him a visit and "shows he is getting a good grasp on the Everett situation."

We have noted that Long kept tight hold on the few copies of the annual report, a practice which, in retrospect at least, suggests an obsession with secrecy. This obsession—if that's what it was—is best understood in relation to taxes. Clearly one group forever interested in the details of Weyerhaeuser's operations was the county commissioners and assessors. Just as clearly, Long saw no reason to volunteer information they might use against him.

At the same time, neither did Long wish to appear adversarial, and he welcomed opportunities to speak before the boards. At one Chehalis County meeting, he apparently impressed most in attendance. Among them was J. E. MacDougall, publisher of the *Chehalis County Vidette* of Montesano. MacDougall returned to his office, still applauding, and wrote immediately to Long: "Your talk on that tax matter was one of the best things I ever heard . . . good old-fashioned honesty is one of my hobbies, and during your talk I had a constant longing for the power of a Dickens to reproduce that meeting in print." MacDougall said he would never forget one phrase in particular: "It does not pay to lie—it is the evidence of a weak man." But if it did not pay to lie, Long maintained that it didn't pay to be completely open either. Apparently he had convinced MacDougall of the reasonableness of that position, for the publisher agreed that "the assessor would have just as much right to ask merchants for their annual inventory, as to ask timber men for their estimates."

But while Long enjoyed his share of successful exchanges with tax officials, he was hardly a match for the "epidemic of higher valuations." To an extent, he found himself hoisted by his own petard. He could not expect to make sales at high prices and then have the valuation process ignore this evidence. Still, he could argue with some merit that these sales reflected values of readily accessible choice parcels that had little or no relation to the vast majority of Weyerhaeuser lands. As Long sarcastically noted, "We are now blessed with a State Tax Commission in Washington . . . and they have been remarkably active in going after all kinds of corporate property."

The tax commission had, among other things, called attention to land sales in which the prices were excessive in relation to the valuation. As a result, the counties of Snohomish, King, Pierce, Mason, Chehalis, Lewis, and Clark notified Long of their intention to increase assessments. He thus found himself "pretty busy hustling around to meet the various Boards of Equalization" to get reductions from the proposed advances. The results were uneven. Pierce County had considered an increase of 125 percent, and reduced it to 50 percent; King County had proposed an increase of 130 percent, and agreed to 62 percent. And so it went. Long declined to apologize or agonize. As he reported to Bell, "We did the best we could, and while the result is not satisfactory . . . yet on the other hand it is comparatively satisfactory from the standpoint of what we were threatened with."

In other words, they could live with it: "We are beginning to feel the heavy hand of public sentiment against large corporate interests; and while I think the fact is generally recognized that we are treating the lumbermen and loggers very friendly and fairly, it is the public sentiment that our property is of vast value, and we are not paying our proportion of taxes." Whether the Weyerhaeuser Timber Company was paying its proportion after the 1907 reconsiderations may have been questionable, but clearly it was paying more than ever before.

McCormick, still in St. Paul, expressed the consensus there, that Long had managed as well as anyone could expect: "You are pretty well organized and simply have to meet a current of opinion that is very prevalent and very fashionable in recent days, and between County Boards and 'President Teddy' the people that have a dollar will have to crawl to cover or some fellow will get it away from him. There is one great satisfaction to me in our business and that is that we have no loop holes and no questionable proceedings that we are afraid of being unearthed."

Bell also consoled, writing, "Please accept our assurance that we feel confident that the matter has been handled with both tact and judgment, and as well as it could be handled under the difficult circumstances." Long knew that to be the case, but it was hard to lose any battle in an endless war.

The Puget Sound loggers resumed operations in early October, Long noting in a letter to McCormick that they had lowered their prices about $2 per thousand on fir, and that "cedar logs have dropped down from their 'giddy' heights," from $18 to nearly $12 per thousand. The Columbia River market continued to be good, "rather to my surprise," Long admitted. He also reported that Connacher's crews were cutting "about seven million feet a month, and are selling them just as fast as they are rafted at $10 per thousand, which is a very satisfactory condition." In addition, the Coweeman River logs were beginning to move, "but they are getting but $9 and $9.50."

Long encouraged the crews to push ahead as far and as fast as they could, warning that conditions were "apt to get worse instead of better during the next year." He shared his thoughts with Collins. After remarking as to the "magnificent shape" of the Columbia River log market, Long wondered how long it would stay that way. Then he advised that they should simply "keep cutting timber where you can make a good profit, and put the money back into remote timber at a lower price." And that was, to a large extent, the policy Long was following: cut where profitable and invest well in advance of profitable operation.

Bill Boner had gotten off to an impressive start at Everett. Long was pleased with the daily reports he received, all of which indicated that "the mill was doing very well, and that orders were tip-top for making a record." He suggested that Boner consider undertaking a serious study of the log and lumber scaling operations, admitting to their having "drifted along" in that phase of the business, and he was certain "the figures would not check out with the logs received by any means." It was just another instance of Long insisting they should understand everything as completely as possible.

Business slowed markedly in the fall of 1907. They were short of cash, especially the Clarke County Timber Company, and Long urged Clair to be cautious about extending credit. Be sure of the financial status of your customers, he advised, adding that he would "rather see the logs pile up than to have an unknown asset . . . accumulating on our hands." Clair had sold about ten million feet of logs in October alone, and some $100,000 was owing. Although there was no record of failure on the accounts, caution was clearly in order. In his report to Davis, Long indicated that he thought the Portland millmen were in "fairly good condition," but, he warned, "The lumber market is fast going to pieces [and] it rather looks toward a lower range of prices and restricted operations."

The ever-widening effects were quick to be felt. On November 8, Long suggested that Bridge visit the Maple Valley logging camps and discuss the situation with the men, advising them of the need to lower wages, "say 50 cents per day, possibly making an exception of your cook and one or two particular men who it would be difficult to replace." The facts were, he asserted, that they could buy logs on the market more cheaply than cut their own, and unless they reduced their costs, they would have to close down their camps. "It is all right to be liberal," he wrote, "but it is very foolish to be extravagant."

But things are seldom as simple as they seem. In this instance, the choice between reducing wages or shutting down the camps was not that easy. Boner's Everett mill needed the kind of logs that Weyerhaeuser's Maple Valley camps provided, and logging superintendent Tim Guider knew that the loss of good men could only hurt efficiency. He took his case directly to Long. In the first place, he argued, the Maple Valley wages were no higher than those of the other "good camps around the Sound." He expressed certainty that should wages be cut again, "we would lose our crew and it would take us a few days before we could run again, and the chances are we could not get a crew that would put out within 15 thousand [feet] of what we are now." Furthermore, and of obvious importance to Guider, such an action would "give the camp a bad name." The better approach was to finish the season under existing terms and then, if conditions failed to improve, begin the next season at the lower wage scale.

Long received much the same reaction from Connacher at Yacolt, and in the end he accepted the positions of his lieutenants. To Guider, he wrote, "We appreciate your honest effort," agreeing that they should "let matters run the way they are until the first of December." Then, should they decide on a reduction, they would give the men "ample notice." In return, he urged Guider "to get out of your crew their very best efforts, for we need all of the type of logs you are cutting now that we can get." To Connacher he gave similar assurances, conceding that it was "best to not agitate in a very pronounced way the question of wages." He did suggest the obvious: it was a good time to "drop out worthless men."

By mid-November conditions had worsened. The Cowlitz County Logging Company found itself without any money after the Merchants National Bank of Portland failed to open on November 12. Still, Long advised that they keep going as long as they could sell logs to "responsible people and get money to pay our debts, pay roll, etc." The situation the country over was every bit as discouraging. F. E. Weyerhaeuser made his own survey and determined that most of the "well-known lumbermen north and south expect to curtail their business next year 50%." Charlie Weyerhaeuser worried most about the Potlatch Lumber Company and its company town, predicting "quite a hardship" should they be forced to shut down the mill entirely.

In reply, Long could only offer the solace of an equally depressing situation in western Washington. The one exception to the general pattern was the Everett sawmill, which was "fixed pretty well with orders on hand for South America and Australia that will keep them comfortably busy for the next 60 days." Many mills had already shut down, and many more were running on a part-time basis. Most of the loggers on the Columbia River had closed their camps because they could no longer meet payrolls. Long assumed the Weyerhaeuser camps would keep working until December 15, "with the expectation of not starting up before the 1st of March, and not at all if things are too badly off."

The markets had begun to weaken about the first of August, according to Long, even before "the financial flurry came." Now there was "no new business to amount to anything, and the lumbermen are very much discouraged at the outlook." There seemed little prospect for improvement. "The lumber trade out here is not demoralized," he wrote a month later, "it is simply dead." Long was wrong about one thing: they couldn't continue to log until December 15. Largely because of the difficulty getting the cash to meet payrolls, Connacher shut down his Yacolt operation on November 22. (Long's reference to "the financial flurry" was to the so-called crisis of 1907. As always, there were many causes, beginning with insurance company losses incurred as a result of the San Francisco earthquake and fire. But it was the October 22 failure of the Knickerbocker Trust Company, New York City's third largest, which made clear the seriousness and extent of the financial crisis. And the Knickerbocker proved to be only the first of many dominoes to fall.)

The new year, 1908, saw a continuation of old problems and hard times. Long reflected on the nature of the lumber business, reminding himself and others that traditionally it moved by fits and starts. For the previous two years "prices of all kinds have soared skyward," but today, January 31, 1908, they were down and nearly out, with half the mills closed and the other half "not making a dollar." It was curious. Prosperous one day, and depressed the next. Investing in timber might be all right in the long term, "but it is not going as fast as many people think—there is too much of it."

Actually, how much timber there was just happened to be a subject of current debate. In a circular entitled "The Drain upon the Forests," Gifford Pinchot had predicted a timber famine, exhaustion of the supply to occur in nine to thirty-three years. Writing to a Portland representative of the Forest Service, Long expressed doubts as to the accuracy of Pinchot's prediction. While allowing that the Chief Forester certainly ought to have the facts, Long countered, "I have fancy." This would become an unending game, Forest Service chiefs warning of famines and George Long and his colleagues believing otherwise and wondering how best to respond.

At the moment, however, the forests were safe. There was very little logging going on, and among loggers there was no certainty just when they would resume operations.

ELEVEN

The Washington Forest Fire Association

IN THE SPRING OF 1908, THOUGH CONDITIONS WEREN'T improving very fast, spirits were less gloomy than six months earlier. Boner allowed that he was eager to resume operations, noting that his Everett crews were "getting restless and anxious to go to work." He outlined the sort of limited program he had in mind, buying some logs cheaply on the market "outside of the Association." They would need only some 50,000 feet per day from their Maple Valley camps, thus saving Long the worry about sacrificing company timber to little purpose. Long finally consented to a mid-April start-up of both the mill and one of Guider's camps. It was a beginning.

The whistle blew again at the Everett sawmill on the morning of April 20, Boner encouraged by receipt of a foreign cargo order at $11.50 per thousand. While this was indeed a hopeful sign, Long refused to be overly enthusiastic, warning Boner that "unless we were able to make ends meet we would shut the mill down again." In the meantime, the Clarke County Timber Company's logs were selling surprisingly well, the yellow fir at $9.

As logging gradually resumed, the old and familiar worries concerning fire protection were again in order. Long had thought about these matters a good deal, and shared his thinking with J. R. Welty, now in his third year as state fire warden: "It has been manifest for some time that to make the work of fire protection a practical one, there would have to be organized a system of patrol, and we hardly thought the State would contemplate entering upon so large a task. We hope in every way to work in harmony with the existing organization that the State has and to receive their aid in all ways that are practical, and to lend our aid to the State."

What Long envisioned was timber owners forming their own protective association, somewhat along the lines of what had evolved in Idaho. As he advised Major Hiram M. Chittenden of the U.S. Army Corps of Engineers, on assignment in Seattle, "We cannot hope to make a perfect system of fire patrol because the territory is so large, but we are going to do the best we can." This was a more aggressive approach than was Long's habit, but as he explained to Bell, "It became necessary to act promptly and quickly, so I have decided that we will do all these things without consulting anybody."

The Washington Forest Fire Association began to take shape, soon acquiring its own office in the Colman Building in Seattle. Selected to head the Association and organize patrols in the various districts was D. P. Simons, Jr. Described by Long as "a young man of excellent character," Simons had very limited experience and knew almost nothing about southwestern Washington. Accordingly, Long requested Ed Markham and George Joy to give "the benefit of your best judgment as to what should be done."

Writing to Ingram, Long noted that all of the larger interests had been "heartily in favor of this work" insofar as it simply seemed more reasonable to "have the work done under one head." By mid-June Long and the Washington Forest Fire Association had 2.3 million acres subscribed, with an initial levy of 0.5 cents per acre. Seventy-five patrolmen were in service. Long knew, however, that the proof of success was distant. At the moment, as he admitted to an Idaho colleague, "it is nothing but glittering generalities, except that we all realize that our Washington forests are apt to be burned up if we do not watch them carefully." He acknowledged that there were "some features in the by-laws which are not entirely practical," but was quick to add that the words weren't as important as "what we really do." At the very least, a start had been made.

Meanwhile, one who would be an important colleague in these matters was in the other Washington, attending a White House conference on forestry and irrigation. Frank H. Lamb of

Hoquiam was unusual in lumber circles—unusual much in the manner of his friend George Long. Both were thoughtful and reflective, and both were able to express themselves in an almost literary style. Lamb described the conference in great detail for Long's benefit, observing that there had been "altogether too much insane condemnation of those who are making a business of furnishing to mankind one of the great necessities of life and I was boiling over much of the time." And he was astounded at the general ignorance regarding the West. Indeed, when Governor Albert Mead of Washington arose to address the gathering, some around Lamb inquired what Washington he was from, "apparently not knowing there was a state of that name in the Union."

Lamb's general impression was that easterners, "after 300 years of waste and neglect of their own resources," were now eager to share their recently acquired wisdom with westerners, "and have us save all of our trees for water sheds and scenery." If there was a lesson in all of this, Lamb thought that it was that westerners needed to deal with their own problems. He suggested that they might have a state meeting, coincident with the next session of the legislature, aimed at developing "a policy of forest reproduction and the passage of laws that would enable private owners to hold land for that purpose." Lamb and Long would be willing partners working toward such a program.

As the summer of 1908 waned, the usual worries about fire increased, and the association began to urge actions that were well intentioned but not always well informed. Such was the case when Simons requested that Welty forbid all slash burning for the rest of the season. The results of this ban proved disappointing. Many ranchers were situated where the burning of slash presented no danger. In the face of the general ban, they tended to go ahead and burn "accidentally," in the process burning without the supervision that permits would have ensured. George Long instructed Simons to "write to Mr. Welty at once saying that inasmuch as we have had a shower, it might be safe to resume the granting of permits to burn slashings." He hastened to add that while he didn't wish to "butt in," he feared that "the slashings will be burned anyhow and the law violated." In this, as in most such matters, the public relations side of the cause was crucial. Unreasonable restrictions tended to detract seriously from those that were reasonable. Young Simons had much to learn.

On August 11 the trustees of the Washington Forest Fire Association notified their membership—138 owners of nearly 2.7 million acres of timberlands—of a second assessment of 0.5 cents per acre, "payable at once." As for what had been accomplished, nine thousand notices had been posted to advertise the danger of fire and the need to exercise care; and the crews had extinguished "some three hundred and fifty fires, large and small, about twenty-five of which were especially dangerous."

But nothing the association could do was as important as what nature did. In late August the rains came, allowing Long to advise Simons to "call the men in . . . and arrange for the proper disposition of their tools, badges and everything they have that belongs to the Association." At the annual meeting of the association on September 4, Long expressed his relief that the first season had "terminated with an excellent record every way, and while I think nature has helped us quite a good deal, I am also very much pleased with the work of Mr. Simons, and consider it specially fortunate that we have had such a favorable year, as it will give us a good send-off for future operations."

Not unrelated to the concern with fires was a nasty development in Timber Company lands in the Snohomish watershed. Without bothering to seek permission, the Thomas Irvine Lumber Company had constructed a logging railroad across a Weyerhaeuser tract. An angry Long wrote directly to Horace Irvine on October 20 to complain, politely but firmly. The Irvines' status as major shareholders in the Timber Company mattered little in this instance. Opening up an area, Long pointed out, was a complex decision. Although it was sometimes necessary to grant a right-of-way, there were always questions to be negotiated regarding the manner of construction with attention to the cutting of timber and disposal of slash. In addition, there must be an agreement to transport logs belonging to others dependent on that same access. Finally, Long always wanted the opportunity to consider other options, including the purchase of the timber in question for Weyerhaeuser.

Almost in passing, Long mentioned discussions with representatives of the Grandin-Coast Lumber Company, a large Mississippi Valley concern whose shareholders included many who were quite familiar to those in the Weyerhaeuser organization. At the moment, the Grandin Company's western investment arm, the so-called Everett Land Company, was studying its 15,000-acre holdings, which "are amongst our timber and yours." After observing that this was "a fine lot of fellows, and as deserving of good treatment as any group of people we are likely to do business with," Long made his point. He would insist that a right-of-way agreement with Grandin-Coast follow exactly the same lines as he had demanded of the Thomas Irvine Lumber Company management. What ought to be done, of course, was to construct a single railroad that would serve all three interests, "the ownership in the railroad to be in the same ratio as the ownership of the timber."

This may have seemed nearly beside the point in October 1908, but Long had a way of looking and thinking far into the future. And what he saw often came to be. In this instance, the mixed ownership would eventually evolve into the Snoqualmie Falls Lumber Company, an evolution that required years of difficult negotiations, particularly with the Thomas Irvine Lumber Company.

But such negotiations were distant. In the fall of 1908, George Long was busy with more immediate matters. In addition to overseeing the construction of his own home on Prospect Hill, he was supervising the initial planning for a new Weyerhaeuser Timber Company office building. The site selected was in downtown Tacoma, on the corner of Eleventh and A streets, diagonally across from the new post office.

This was an entirely novel responsibility, and for the most part Long welcomed advice. He forwarded two architectural sketches to St. Paul, seeking the ideas of others. He did include some of his own thoughts: "It is quite desirable for our private offices to be somewhat secluded, both from the work room and from the public, so on this plan we have located the private offices in the rear of the building, and they each have a magnificent outlook over the sound." He also recognized the need "for a special exit, as I frequently have visitors who I would like to pass on out without having them come back through the front way, as it is not an uncommon thing to have callers whom we do not care specially to have other waiting callers see just who they are."

The original intent was to construct a modest building. As Long explained to Ingram, "It is not contemplated that we will make it other than a nice two-story building to be used exclusively by the W. T. Co." Bell, however, thought that the "office of Weyerhaeuser Timber Company ought to be a pretty fine affair," though he admitted that opinions differ on what a fine affair is. He offered his own opinion: "I am anxious that the building be simple and dignified and express perfectly what it is." He couldn't recall "a single business building in Tacoma which can be said to have real distinction." A month later, Bell repeated himself, urging "simplicity in its design, so that as time passes no features may become to appear over decorated, fantastic, flamboyant, or out of proportion."

Mulling over construction plans for home and office was probably a welcome diversion as the year waned, what with log and lumber prices remaining stubbornly low. Still, there was a hint of improvement. In mid-October, Boner reported that they had enough work to keep them busy at Everett, "although the only money we are making is by declining orders." He wasn't much interested in doing business at present prices, largely content to "allow our competitors to fill up with all the business they want." The correct strategy was to prepare for the spring trade "with a nice line of stock and have it in fair shipping condition."

In early November, Long suggested that "it might be prudent for you to buy up a few logs" in order to have enough to keep Everett running. In fact, log prices were edging upward, and it was a good time to buy. It was also a good time to produce—and Long was in both sides of the business. To his own loggers he expressed more interest in making logs than selling them—for the moment at least. Long was pleased with the logging reports from Cowlitz County and congratulated Robert Barr, manager of the log rafting operations in southwestern Washington, for "working with your head as well as your body." Further, he sensed "a much better tone in business circles and consequently in the log market," predicting they would be able to get $10 "before the first of March for all we have. . . . This I know would make us all feel better and put a little heart into our business."

In mid-December Long received an important letter

from F. J. Davies, general manager of the Edward Rutledge Timber Company of Coeur d'Alene, Idaho, reporting on a December 8 meeting of the various timber protective associations of northern Idaho. At the meeting, the assembled had voted to call a more general conference, the purpose of which would be to propose legislation for "the protection of the timber resources of Montana, Idaho, Washington and Oregon." Davies invited Long to choose two colleagues who, with himself, would constitute a delegation of three from Washington. Long followed Davies's instructions—almost. He felt compelled to appoint Simons, as representative of the Washington Forest Fire Association, and he also asked Frank Lamb. That should have filled the delegation, but Long had yet an additional delegate in mind, Edward Tyson Allen, the district forester stationed at Portland.

Long knew E. T. Allen pretty well by now, and had come to respect him as someone quite special. Not only was Allen knowledgeable about forestry matters, he was generally impressive. He was personable; he was erudite; he wrote and spoke with exceptional skill. Allen also held Long in esteem, and the forester readily accepted the lumberman's invitation to attend the January 4 meeting in Spokane. Long wanted Allen there so that others might appreciate his abilities as a communicator. Long was eager to find someone who could work full-time discussing forestry matters on behalf of the industry. What he envisioned was a one-man educational campaign supported by the lumbermen of Washington, Oregon, and Idaho. He convinced those present that Allen would be the ideal spokesman for their cause.

Happily, almost proudly, Long recounted the results of the Spokane meeting to F. E. Weyerhaeuser. The timber interests of Idaho, Washington, and Oregon had, in his words, "finally wound up with an organization with the high-sounding name of The Pacific Slope Conservation and Forest Protection Association." Long listed its objectives: "To set the pace for all the different forest fire associations in each state and the numerous forestry, re-forestry and conservation associations that are springing up, many of them being organized by theorists who need a little handling."

Furthermore, he reported, he had sold most of his colleagues on the need for "a publicity man whose business it would be to circulate the right kind of literature upon the necessity for re-forestry and conservation, and for reduced taxation." Although they now had their man in their sights, Long admitted that he didn't know how much he would cost. With appropriate leadership they would begin "a first-class campaign; and the intent would be to get right down to the root of the matter and get a hold on public opinion," starting with the school children and working up, perhaps developing "something like a tolerant spirit, if such a thing is possible, toward the timber interests." Long was convinced of the "absolute necessity for doing something of this kind," and he only awaited the approval from St. Paul to get things under way.

In his comings and goings, Long missed the opportunity for a personal goodbye to Monemia Evans. She had been the first employee he hired, and had served faithfully for nearly a decade. In his farewell letter to her, Long noted that fact, the feeling that "we have lost one of the corps," and added, "We will all miss you very much." Personnel changes were still something new for Long and his organization. As the years passed, they would of course become commonplace. That, however, didn't make them any easier.

McCormick also missed bidding adieu to Monemia Evans. He was in Washington, D.C., lobbying in behalf of retaining the $2 tariff on lumber. He seemed hopeful that the Ways and Means Committee would eventually recommend retention of the tariff. F. E. Weyerhaeuser had earlier been on the scene and had expressed similar optimism. "Every day we have been in Washington has strengthened our position," he wrote. Four weeks earlier he had been certain that they would have "free lumber," but now thought they might "be able to get at least $1.50 and we are all standing 'pat' for $2.00." It had taken considerable effort, however. F.E. admitted that "we did a great deal of expensive entertaining, having in mind the thought that if we can present our case intelligently to the public we ought to win purely on its merits."

Long wasn't much worried yet about the tariff question, although he did allow that the policy of importing Canadian lumber and having "them cut off their forests while we keep our own, is absurd." At the moment, however, Long was frankly more concerned that F.E. respond to the question of hiring "a publicity man to educate the Western mind on the proper line of conservation re forestry and taxes." Bell had already agreed that "it was a very essential thing to do."

Long was embarking on a lifelong enthusiasm for public education as a means of building public support. Believing, like educators, that the place to start was at the beginning, he wrote to Henry B. Dewey, state superintendent of schools, trumpeting the need to arouse public sentiment to the cause of "preserving the forests from careless or malicious destruction by fires." Why not start such a campaign in the schools, specifically "in the country public schools, and create a sentiment that would assist in this work of protecting forests"? He committed the Washington Forest Fire Association to produce a pamphlet for distribution to the students, "written in plain language setting out the interests that were at stake and the danger from forest fires."

Long reported his contact with the state superintendent of schools to F.E., and told of similar approaches to Enoch Bryan, president of Washington State College at Pullman, and Frank Miller, dean of the forestry school at the University of Washington in Seattle. All of this contributed to Long's urgent request that F.E. permit the hiring of E. T. Allen, one who "could talk plainly and sensibly before county boards and to the school teachers; in fact, to everybody." While Long allowed that the decision could probably wait until June, he clearly didn't want to wait. "As the profound people say," he concluded, "it is the 'psychological' moment."

It may have been the "psychological" moment in Tacoma, but it hardly seemed so in St. Paul. In his unhurried reply—nearly a month had passed—F.E. agreed that Long was "undoubtedly moving in the right direction," and suggested that Allen be told that "our people are giving the matter careful attention, and some final action will be taken in June."

While all of this was going on, an interesting and, as it turned out, important development occurred regarding the plans for the new office building in Tacoma. The chairman of the building committee of the Tacoma Commercial Club, Robert G. Walker, called upon Long to ask if Weyerhaeuser might be willing to sell its lot to the club. A large office building would then be constructed, with the Timber Company assured of office space at low rental and long lease.

Long questioned the club's ability to manage the effort, but the consideration prompted new thoughts on the subject. Whatever happened, the club was clearly in need of offices. Thus, Long informed McCormick that he was beginning to think in terms of a larger building filled with rentals, which would make the Timber Company's own costs quite low. He asked McCormick to stop by St. Paul on his way home from Washington, D.C., to discuss the matter.

McCormick did as requested, and soon everyone was an architect, offering advice about this and that. F.E. seemed downright excited about the possibility of a larger building. He discussed it with his father and, according to F.E., it was Frederick who approved plans for a ten-story structure, providing they were "reasonably sure that the eight additional stories would pay a reasonable rate of interest on the whole investment." F.E. also surmised that the construction of a "handsome office building" might encourage "a somewhat more friendly feeling" from Tacomans.

While the deliberations on the Tacoma office building went forward, the Longs moved into their new Prospect Hill home at 45 Summit Road. It was the end of April, and the move was complicated by a siege of illness that put both wife Carrie and her sister in bed. As George would write to his sister Rhoda in Minneapolis, "You never saw things piled into a house quite so promiscuously, as we were compelled to drop everything and take care of the sick people." But they soon recovered, things got put away, and home life went on much as before. Away from home, however, there was a notable difference. George Long had been at his Pacific Northwest assignment for a decade and was now clearly the leading figure among its lumbermen.

TWELVE

The Western Forestry
and Conservation Association

LITTLE BY LITTLE, GEORGE LONG'S ROLE WAS ENLARG-
ing. As the years passed, he served with increasing effectiveness
both his company and his industry. Foremost were his efforts to
establish a new organization to serve as an educational and
public-relations arm for western lumber interests. Months earlier
he had handpicked E. T. Allen to be his lieutenant in this campaign, but he had yet to obtain approval of the St. Paul office.
When F. E. Weyerhaeuser chose to defer the question of Allen's
hiring, Long could only "keep Mr. Allen on the string in a way
that will not embarrass him nor commit us."

Finally, in the fall of 1909, the St. Paul office gave the
go-ahead. It was understood that this amounted to a long-term
commitment: if others failed to recognize the same needs, the
Weyerhaeuser Timber Company would nonetheless continue to
support the cause. Long immediately called Allen, who subsequently notified Gifford Pinchot of his intention to resign his
post as district forester. With Allen's acceptance, Long was convinced that he had his expert, "the best available."

When Allen left the Forest Service on November 15, the
details of his new responsibility were vague. Judge A. L.
Flewelling, a Spokane timber dealer, had been elected president
of the fledgling association, but Long was clearly the one pulling the strings. As a beginning he suggested that Allen get his
Portland office set up, promising to send a check to cover expenses. Then, Long continued, Allen should "map out at your
leisure your own ideas . . . of the ins and outs of this work

which you want to undertake." In short, Long looked to Allen to define the job.

At a general meeting of timberland owners on December 1, the new organization was properly named: henceforth it would be the Western Forestry and Conservation Association, "representing the allied local associations of Montana, Idaho, Washington, Oregon and California." The new organization was built on the existing fire associations, and presumably would maintain itself by assessing those same associations. In addition to Flewelling the officers were Frank H. Lamb, secretary, and T. J. Humbird of Sandpoint, Idaho, treasurer. Allen's title was "forester." There were five vice-presidents, one from each state, and also five trustees, with Long representing Washington. The avowed purpose of the new organization was, as he envisioned it, to deal "more directly with matters pertaining to reforestry than would admit in our fire association." But fire protection would remain a central concern.

In Allen's first paper as the association's forester, he offered what would become a litany: "If you own timber, two things are worrying you more and more—your fire risk, and a lack of public sympathy with your industry as compared with other industries no more useful to the community." He then proceeded to discuss the history of the fire-protection movement, recalling when each owner had acted independently, "with no advantage in either economy or efficiency through cooperation with his neighbor." After describing what had been accomplished in Washington, Idaho, and California, Allen urged his readers to cooperate and to study the problem. "The narrow, inexpert, or prejudiced man, be he lumberman, conservation enthusiast or politician, is the man you have to fear, and his name is legion," Allen warned, and then added, "Study the problem yourself and beat him to it." The surest way to start and "to show the public that you are doing it sincerely" was through the organization of a fire association, "and it will pay in dollars and cents besides."

The next document that Long received from Allen was a "declaration of principles and explanation of who we are." The mission was "to promote forest fire prevention, conservative forest management, reforesting of cut-over lands not more valuable for agriculture, improvement in taxation systems, preservation of stream flow, and all of the other things comprehended by the word conservation."

George Long had little to add, other than suggesting that Allen describe the western situation a bit more positively:

> In the states embraced in our association, there now stands at least fifty per cent of all the merchantable saw timber that is left in the United States, and the mere fact that this heavy stand of timber (the greatest ever known) was found on this coast, was because the climatic conditions were the best for the production of timber, so that here above all other places in the United States, we have soil and climate and conditions that are best for reproduction and best for reforestry.
>
> Then again, you can elaborate upon the idea that the Pacific Coast people can profit by the experience of the lumbermen and the people of the eastern country, and that is, to protect their timber from fire, and to handle it so wisely that we will have another new crop of timber ready to cut before the old one is gone, and that the forests out here can be made perpetual and the lumbering perpetual.
>
> I sincerely believe all this and I think it is one of the strongest cards to talk, and we should not fail to make this point pronounced in almost everything we say or do.

Meanwhile, there had been little to cheer about in the timber business. Inevitably when times were hard, disagreements over price and quality were exacerbated. That was true not only with distant customers but also internally, with Boner at Everett complaining about the logs he received from Maple Valley. If the loggers logged close, the sawmillers were bound to kick. Long understood and could sympathize with both.

Central to the disagreements was the old problem of how best to determine stumpage prices. In the fall of 1909 this seemed likely to become far more than an internal squabble. At the time there was much popular discussion about a corporation tax, a tax which would definitely require an assessment of property values including, of course, timber inventories. For Weyerhaeuser that was worrisome, and not simply because of the fear of a tax. The fact was, as F. E. was quick to admit, "We do not now know and never have known even approximately the amount of timber owned by each one of the companies." In addition, even if they could arrive at an overall stumpage figure, what was the value of that stumpage? Stumpage was commonly computed as a residual. The costs of manufacturing, shipping, and selling were known, as were the logging expenses. These totals, along with some allowance for profit, could be subtracted from sales, and the difference was assumed to be the stumpage cost. The actual determinations weren't quite so quaint and crude, but they certainly weren't sophisticated. And this would prove to be a question of large and enduring importance from both an internal and a tax standpoint, now and in the years to come.

The year 1910 opened on a promising note, although given recent history, many doubted that hard times were over. Despite Long's general optimism, he continued to preach caution, predicting that "logs are going to advance," but he wasn't ready to say just when. Long did recommend that they would do better to sell at present prices than try to force an advance. In his words, "I confess that I am perfectly willing to get ten dollars for logs, but I would rather sell a whole lot of them at nine dollars than not be able to sell them at ten."

One event, much publicized and of great importance to Long and his colleagues, was President Taft's dismissal of Chief Forester Gifford Pinchot. In a sense, the dismissal was no surprise, but it was nonetheless dramatic. In August 1909 an agent of the General Land Office, Louis Glavis, accused Secretary of the Interior Richard Achilles Ballinger of collaborating with groups seeking profits from plundering the public domain. Taft, vainly attempting to quell the embarrassing infighting, placed an order of silence on those in his administration. Investigations subsequently revealed that Forest Service officers had been directly involved in the preparation of the charges against Ballinger. Pinchot clearly had been insubordinate, and when Ballinger was exonerated, the Chief Forester was summarily dismissed. Some, including George Long, worried that Pinchot's successor might be less committed to the cause of forestry. In short order, however, it was announced that Professor Henry S. Graves of the Yale Forestry School was to be the choice. Although there was no indication that

Graves knew much about the West—perhaps he cared little about it—Allen felt that he was "straight in every way and thoroughly devoted to Pinchot and his policies." Lamb, Long, Allen, and many others heaved a sigh of relief.

In matters closer to home, Long was trying to ensure that the budgetary requirements of Western Forestry and Conservation were met. The Washington Forest Fire Association trustees agreed January 12 to advance $1,000 to the new association. In forwarding the check to Tom Humbird, treasurer of Western Forestry and Conservation, Long suggested that Humbird's "Idaho associations could make similar contribution on account" until they got together and worked out the details. Things never seemed to settle on any permanent basis, however, and George Long was forever advancing needed funds to Allen out of the Weyerhaeuser Timber Company till.

In the meantime, Allen was working away, composing new circulars, with Long serving as editor. Many of his comments are revealing. In one instance, Allen had referred to an "antiquated and suicidal system of taxation." While Long agreed, he wondered about the timing of that remark: "We can lead up to that a little later." Allen had also used the phrase "deforested"; Long preferred "cutover." "This suggestion, I know, will grate on your feelings because the word 'de-forested' expresses your meaning a great deal better, but do you know that there are a great many people who will get this circular who do not know what de-forested means, and they do know what cut-over means. While my suggestion would be in the direction of spoiling your English, it will give the reader an Anglo-Saxon term which he is very familiar with, and especially out here on the coast where the people are whom we want to convince."

Long may have wished that Allen would momentarily mark time, but he couldn't bring himself to insist on inactivity. Allen didn't enlist to be muzzled, and Long did nothing to discourage him. He did, however, advise that "in view of the all-round chaotic condition of forestry and conservation, resulting from the so-called Ballinger-Pinchot scrap, I think it would be well for us to be a little quiet for a little while," except for the fire-prevention campaign, especially as concerned timberland owners who were still uncommitted.

Meanwhile business seemed definitely on the mend.

"The lumbermen out here are feeling somewhat better and are doing better," Long informed Frederick Weyerhaeuser, wintering at his residence in Pasadena. Although lumber prices were still slow to improve, Boner thought they might as well begin "shipping out a little heavier than we are doing and not wait altogether for the full advance of the market." Long wasn't convinced. It was, of course, "not very desirable to pile stock up against a declining market," he admitted, "but if you consider lumber values firm and likely to increase, we would be very much in favor of keeping your assortment of dry lumber in excellent shape."

In truth, Everett was doing just fine, and even ended the bad year of 1909 in the black. Long shared the figures, giving full credit to Boner. "The Everett Mill made a very fair showing last year," he wrote to Ingram, "its cut being about 48 million feet and sales close to 50 million, with an actual profit of $100,000, it being the first time that we have really made much of a showing in the way of any profit at all." Given the conditions, Long urged Tim Guider to log for all he was worth: "Mr. Boner will be very glad to have you give him all the logs you possibly can for the next sixty days."

The business of buying and selling timberland went on as always. Long noted recent major expenditures in a report to F.E.: "Since the annual meeting, we have spent four hundred thousand dollars for the Pacific Empire Lumber Company tract, and we are going ahead now to agree to purchase twelve hundred thousand dollars worth of pine."

One Weyerhaeuser, John Philip, reacted negatively, almost morosely, to word of the pineland purchase: "I feel most of your stockholders want dividends, at least the younger ones do, and the older men are passing away, fast." He thought it "a good purchase," but added, "Some would say when you have so much why more."

J.P.'s sensitivity may well have been heightened by some recent publicity suggesting that the Weyerhaeuser Timber Company had benefited from fraudulent timber sales. Long reacted to an item in the *Tacoma Evening News* of April 9, 1910, purportedly disclosing wrongdoing in the sale of state lands. He felt obliged to write F. S. Bell regarding the article, since Bell's name appeared prominently. (Long had used Bell's name in some scattered purchases, as he had used others, simply because it attracted less attention than Weyerhaeuser.) The case involved a 1901 Timber Company purchase of 480 acres, in an area that "would be called inaccessible and remote at that time." In the same township at the same time they had bought nearly 5,000 acres from private owners at an average cost per acre of $11.50. Since the price they paid for state lands was $14.76 per acre, Long thought the public had fared reasonably well. Regarding the larger question, he was unequivocal: "While it is pretty generally supposed that a great deal of fraud and graft has crept into the sale of timber lands belonging to the State of Washington, I know it will be quite reassuring to you to know that the Weyerhaeuser Timber Company neither directly nor indirectly or in any shape or form, have acquired any so-called state lands or state timber where there was any element of graft or fraud or connivance either with state officials or state cruisers."

Bell needed no convincing on that score. But those who were skeptical would remain so—at least until they confronted Long face to face. Then, to accuse him of fraud and connivance became considerably more difficult. It wasn't so much a matter of trying to prove the point as it was imagining George Long to be so involved.

The annual meeting of the Weyerhaeuser Timber Company was called to order on June 29, 1910, in Tacoma. Attendance was good, and it included the elderly Frederick Weyerhaeuser. As had become customary, Long stood before the group, cigar in one hand and report in the other, reading aloud the most pertinent portions. He noted the "gradual improvement" in business conditions: "both the demand for lumber and for logs, as well as the market price, have been reasonably satisfactory as compared, at least, to conditions which prevailed during 1908 and the early part of 1909." He also cited some changes occurring within the industry, among them a tendency of large loggers to also become manufacturers of lumber, and of quite a few mills that had depended on the loggers for their supply of logs to purchase timber and engage in the logging business. Thus industry operations were becoming more integrated.

What probably interested the shareholders most were the compilations for the first decade of the company's history. For example, the Timber Company had purchased 1.8 million acres at a total cost of about $15 million, or an average price of $8.18 per acre. Meanwhile, it had sold 172,000 acres of timberland for nearly $11 million, or an average price of $64 an acre. In short, it had sold those acres for eight times the cost. It was, of course, difficult to make any exact conclusions—the variables were many—but the advantages of buying wholesale and selling retail were clear.

Expenditures amounted to $3.7 million for the decade, but computed as carrying costs they averaged only about 11.5 cents per thousand feet of stumpage. That seemed cheap enough. The problem was, as Long explained, that the "expense of carrying this property is increasing by leaps and bounds, owing almost entirely to the increase of the tax burden," and he cited supporting figures. In 1905, "the average tax paid on our holdings was $.1391 per acre." In four years, this had more than tripled. As he later observed, "About the only apparent menace to our property is the burden of taxation and the danger of forest fire." And as soon as the Weyerhaeusers and others had taken their leave of Tacoma, Long turned back to these subjects.

While the real problems of the summer and fall of 1910 involved those ever-present worries, taxes and fire, there were, of course, other matters of importance. For example, Long completed negotiations with the Draham Timber Company, resulting in the organization of the Mud Bay Logging Company, capitalized at $500,000. A short logging railroad would be built to move the logs directly to Puget Sound. This was to be a very advantageous arrangement for both parties, and it was presumed that between 25 and 50 million feet a year would be harvested, if market conditions were right. Long was pleased, as he informed those in St. Paul: "I am well satisfied that the connection is the thing for us to do."

Work was also proceeding on the new office building, the decision having been made to construct a ten-story edifice, the top two floors to be leased to the Tacoma Commercial Club. But Long's major concerns continued to be taxes and fire.

The summer of 1910 had been dry, but that was not unusual for western Washington. Many small fires burned, and that too was not unusual. On August 23, Long addressed a letter to the Timber Company trustees, reporting that there were "smoldering fires all over the state" and that they were facing "a very serious condition," for if the wind began to blow, those small fires could turn into enormous conflagrations.

The next day the winds began to blow, "the worst summer day for wind that I have yet seen on the coast except the day preceding the big fire in 1902." By August 26, though estimates of losses were still uncertain, it was apparent that much timber had been burned in the vicinity of the Tacoma-Eastern Railway, that fires had burned along the Cedar River valley, and again in Clark County. On closer inspection, it was the Clark County fires that were the most destructive. Nearly all of the bridges of the Twin Falls Logging Company were destroyed, bringing operations to a standstill. The danger ended in the only way possible. On September 10 a general rain commenced, and Long reported, with relief, "I think the trouble is over."

There was an unexpected development on the tax front. Long received a memorandum in early August from the Washington State Tax Commission requesting information as to "the specific purchases and sales you have made of timber and lands in the State of Washington, during the last six months." On August 3, Long replied, "I must respectfully decline to furnish the information called for," although he subsequently offered to give "very cheerfully . . . my opinion of values."

Though Long's refusal seemed adamant, he realized that his position was far from secure. The press soon learned of the situation, and early reports suggested that the tax commission intended to pursue the matter in the courts. Long informed F. E. that it was "likely" he would be served with papers, but he remained confident that they could not "be compelled to divulge such information, especially as the purposes for which the information is asked, are not directly or indirectly connected with the taxation or assessment of our own property."

Long was indeed "arrested," as he put it, on August 9; he explained to those in St. Paul, "They were kind enough, of course, not to request bail nor to put me to any inconvenience." It was nonetheless a little disconcerting, even though he was "still of the opinion that there is nothing else for us to do." The hearing was scheduled for September 9.

This was just the sort of fight that the Weyerhaeusers and their associates hated—it placed them on the public stage and in an unfavorable light—but they expressed complete confidence in the correctness of Long's position. As Bell remarked, "One instinctively remembers the old fable of the camel and the tent, and it is probably wise to keep the animal's head from getting in at all if that can be done." The largest fear, of course, was the inevitable comparison of past prices with present values, as well as the prices paid on large blocks (which included both choice and poor timberlands) compared with those paid for the best, small, accessible tracts. Bell assumed that the defense would be based on "your protection against unreasonable search and seizure."

Frederick Weyerhaeuser turned the matter over to lawyer Macartney, no doubt with the instruction to get it over with as quickly and quietly as possible. But there were no easy resolutions. Macartney believed that "the tax commission will get this information," explaining that "it may not get it in this particular proceeding, but most of the courts will all lean towards the tax commission." He recognized their interest in keeping the commission "from insisting they must know all about every party's business just as long as we can." But he subsequently wondered whether it wouldn't be preferable to give them exactly what they had requested—sales and purchases for the past six months—fearing that otherwise they might get permission to examine the Timber Company books.

Long was not convinced. "The desirable thing for us to do is to get along with everybody," he agreed, "and that is a stunt which I have studied with a great deal of care always, but of course, once in a while you have to turn around and scrap; then it is the part of wisdom to pick out the kind of scrap you want and win. I have not got just the kind I want on my hands now."

The tax commission hearing was postponed and then postponed again, but there were more than enough distractions to keep Long occupied. He had accepted the chairmanship of the Market Extension Committee of the Pacific Coast Lumber Manufacturers' Association, and a report was expected at the annual meeting of the association in January. The responsibility of the committee was to seek ways and means of advertising the advantages of buying Pacific Coast forest products. The cause was legitimate, but there was no consensus on how best to mount an advertising campaign. As always, Long's approach was practical. Initially he focused on the fir door, convinced that it was "going to be the staple door of the future." One method of advertising was to participate in trade fairs throughout the country, "taking particular pains to call the attention of all builders to the exhibit, and inviting inspection."

It was, however, the work of the Western Forestry and Conservation Association that interested Long most. As he wrote to Allison Laird in a December 7 letter: "I feel that we are moving in exactly the right direction by maintaining this association and by maintaining Mr. E. T. Allen as forester." He then reviewed Allen's accomplishments thus far, recalling that prior to Western Forestry and Conservation, the Washington Conservation Association's membership included "a whole lot of long-haired people who had a lot of crazy ideas and who were issuing all kinds of manifestos and whose every word and sentiment was freely printed and commented upon by the press." Now the situation was much improved, largely as a result of education and the increased participation by more knowledgeable types. The governor had even appointed a commission "to recommend certain laws pertaining to forest matters," and in Long's opinion, it was "the best kind of a commission." And who was most responsible for the improvement? Without a doubt it was E. T. Allen, at least according to George Long.

"Now the point is," Long continued, "that if this had not been done in the right way, it would have been done in the wrong way," and he noted that Allen was at that very moment hard at work in Oregon, trying to provide the needed leadership there. He acknowledged that while it was not going to be possible to prove the worth of Western Forestry and Conservation in "cold dollars and cents," he was confident that they could not spend their money "in a more judicious manner than maintaining this organization and maintaining Mr. Allen."

In the course of his efforts to keep his lumberman colleagues working in concert, Long received a curious letter from a Tacoma friend and neighbor, W. W. Seymour, then visiting New York City. Seymour had come across an oil painting that he knew Long would appreciate. Entitled "A Friend of the Forest," the painting featured a hunting party breaking camp. The white hunters stood impatiently by while their Indian guide, the friend of the forest, carefully extinguished the campfire. George

Long purchased the painting sight unseen. When he saw it, he immediately informed the artist, Frederick M. Spiegle, "It fully comes up to my anticipation." Thereafter it hung on his office wall, and Long soon had Allen distributing calendars featuring "A Friend of the Forest" on behalf of Western Forestry and Conservation. Long may have been unsophisticated when it came to art, but he knew that people loved trees—and he also knew what fires could do to forests.

THIRTEEN

The Changing Scene

ALTHOUGH GEORGE LONG'S TITLE HAD BEEN CHANGED from agent to general manager at the 1910 annual meeting, his responsibilities were as before—or perhaps as never before, for they seemed to increase with each passing year. As 1910 turned into 1911, he worried a good deal about business conditions, especially the domestic market for lumber. With the exception of California, which seemed in perpetual growth and development, the country's economy continued to mark time. Weyerhaeuser's Everett plant performed far better than most operations, being relatively efficient in terms of design, machinery, and management. But while this was satisfying, the fact remained, as Long would later observe, that Weyerhaeuser "had an industry tied on to it, a struggling industry," and the problems of any individual organization could be dealt with only on an industrywide basis. Cooperation was essential.

Nowhere was this more evident than in the fire-association efforts and in the fledgling endeavors that had come under the rubric of conservation. Long had fathered the Western Forestry and Conservation Association to ensure a practical approach to these concerns. He had handpicked E. T. Allen to head that association, and he continued to work tirelessly to convince his colleagues that even in hard-pressed circumstances they must contribute to it.

Long tried to avoid diluting his energies. For example, responding to a request from J. H. Bloedel that he assume a fund-raising responsibility in behalf of the Yale School of Forestry, Long noted previous contributions from the Weyerhaeuser Timber Company. Although he wished Bloedel and Yale well, in this

130

instance he declined to participate: "I am on one or two other begging outfits at the present time, and think I might overwork the dear public a little."

The Weyerhaeusers' devotion to Yale was too firmly established to question. Weyerhaeuser's general manager, however, appreciated the distance separating western forests from eastern classrooms, and thought that in the end western forestry required its own answers. In this connection, it is interesting to note Long's fascination with his natural surroundings. To him, the world was a laboratory, including the trees, plants, and shrubs on his own Prospect Hill property. For a time, he kept careful records on what was planted, where, and when. "The following was set out in the fall of 1909," the list including "daffodils, tulips, dahlias, hyacinth, Scotch broom, hollyhocks, cherry tree, pear tree, the Crimson Rambler, and all other roses on the west side of house." Later he noted that the fir, hemlock, and cedar trees were "native, being on the lot when bought, as was also the Dogwood." For several years George Long, Jr., was assigned the responsibility of keeping a careful record of growth rates of the Prospect Hill "forest," but somewhere along the way the son tired of this task.

On the subject of forest sites, in January 1911, Long urged cruiser Ed Markham to look for tracts where they might "do some educational work in demonstrating what can be done in the direction of growing a new crop of timber." Markham kept his eyes open, and Long did too, soon noting an area "back of Mud Bay on the county road leading to Little Rock, where it is apparent that there has grown quite a sturdy young forest in the past twenty-five years." Markham inquired about the possible benefits of thinning. Long didn't know the answer, but was eager to find out: "This is a mooted [sic] question among scientific forestry men, some of them thinking that Nature knows how to do that better than man, and others are very much impressed with the advisability of the thinning process." Forestry was as yet hardly a science. But answers would come only if questions were asked.

However interested he was in the woods operations and in forestry, George Long the salesman never forgot that lumbermen sooner or later had to make a profit on their product. He understood that some advertising was necessary, especially as concerned Douglas-fir in midwestern and eastern markets, and he never missed an opportunity to demonstrate the worth of this western species. He used it in his Prospect Hill home, and ordered its use in the "so-called Weyerhaeuser & Commercial Club ten-story office building," where the flooring was to be fir. Boner's Everett mill sawed to order, "three inch strips, showing 2 inch face, of the usual thickness . . . a good quality, i.e., a fine grain, etc." Although the construction had been "rather slow," the contractor promised that they would be settled in their new offices by the first of June.

The elderly Frederick Weyerhaeuser was less and less active, but he continued to be in relatively good health and spirits. Such, however, was not the case with R. L. McCormick. He had been ill for several years, but in late January, feeling stronger, he and Mrs. McCormick headed to Pasadena. They got as far as Sacramento, where R.L. was stricken and died. The funeral was held at the Congregational Church in Tacoma, and McCormick was buried in the local cemetery. The Tacoma office closed for the day, and the Everett mill shut down for half a day without any reduction in pay. In all it seemed a small tribute to one who had contributed much. No one could have been more supportive of Long and his policies than McCormick. While R.L. was anything but reticent, he never forgot that he was there to assist. For George Long, R.L. was a strong lieutenant in the best sense. Although McCormick's tenure with the Timber Company was relatively brief, his commitment was total; when he moved west, he moved to stay. The family ties would continue unbroken, with William L. McCormick replacing his father as assistant and eventually as the Timber Company secretary as well.

In Everett, Boner was far more focused on the future than the past, at the moment contemplating advantages afforded by the opening of the Panama Canal. Long needed no prodding on the subject, already assuming that they would eventually establish terminal yards on the Atlantic Coast. In early 1911, however, he counseled discretion: "I question a little the advisability of saying as much to any one until we look the ground over pretty well and see where it is that we want to go, and have the choice of first selections." If they did decide to make the leap, Long thought they ought "to get into the game in a big, broad way," and this would mean owning their own ships and ensuring that there was sufficient business to keep them constantly employed. What, for example, would they carry on their westbound voyages? The canal was still three years from completion, but this did not prevent Long from anticipating opportunities, as well as the concomitant problems.

Long included F. E. Weyerhaeuser in those early discussions about an Atlantic Coast distribution network. F.E. would become a strong supporter of the plan, but he too initially preached caution, anxious that the company not put itself into a position where it was competing in eastern markets with other companies in which the Weyerhaeusers were interested, particularly those located in Cloquet, Minnesota. At the same time he thought he could see the bigger picture: "If our affiliated companies were so organized that we could work together, it would be a very easy matter for us to have large distributing yards in various sections of the Country where we could furnish promptly anything that the consumer of lumber demands." For now, F.E. suggested that their J. E. Rhodes, recently employed by the St. Paul office to study sales problems, begin collecting information about East Coast opportunities. F.E. also urged Long to make a tour of the eastern seaboard, because "before making any such move we want to be very sure of our ground, as the expense involved will be very large." He was right about that.

The rains of winter had arrived on schedule, but the concerns with fire and its prevention continued. Although public attitudes were always important, they seemed even more so at the moment. The forest-fire bill that had been proposed by the governor's commission was running into opposition, especially from loggers. Long and Mark Reed of the Simpson Logging Company had become good friends, and they corresponded at length about the forest-fire legislation. Long was willing to compromise, but, as he stated to Reed, "There is hardly anything that is asked to be done in this bill but what ought to be done." He further observed that "after we have burned up a whole lot more timber," one of these days "the indignant public and a responsive legislature will make a good deal worse bill than this could possibly be conceived to be." That was the constant refrain: If we don't do it, someone else will, and it won't be done as well.

While Washington was having its problems with proposed legislation, Oregon was still trying to organize an effective forest-fire association. C. S. Chapman had accepted the secretary position, and in fact shared office space with Allen in Portland, an arrangement that bothered Long just a bit because it gave an impression of a lack of distinction between the Oregon Forest Fire Association and the Western Forestry and Conservation Association. But economy carried the day. Allen was trying to influence the selection of the state forester for Oregon. He noted "a certain element" who wanted "above all a practical woodsman and fire fighter" even if he had no public-relations skills. There were other problems too numerous to mention, and Allen observed that the association south of the Columbia was "hardly so automatic and harmonious in its operation as yours in Washington."

Most of Allen's time, however, was spent writing pamphlets, some for distribution to loggers and others to school children. His was primarily an educational responsibility. In that connection, Long thought it appropriate that Allen accept an invitation to address a session of the National Education Association meeting in San Francisco on the subject of forest conservation.

Allen had a more pressing concern—funding. On May 10 he informed Long: "Broke! Association balance 97 cents. My own personal balance $22.41." Long would once again see that Allen stayed afloat. Other matters were also resolved. Despite the initial misgivings, J. L. Bridge, Jr., was elected chief fire warden, replacing D. P. Simons, Jr. And, even though Long continued to have questions about his performance, J. R. Welty remained fire warden for the state.

But the real work of fire prevention had to be done in the woods, and George Long was always concerned that his own loggers and those who logged the Weyerhaeuser lands were doing their utmost to reduce the danger of fire. Typical was a letter he addressed to a Cosmopolis company granting permission to cut piling, with numerous restrictions. For example, they were required to pile up the brush and burn it before June 1, by which date all the logging was to have been completed. Further, Long demanded payment for the entire length of the tree, a provision that would encourage clean logging. Finally, he promised that Clyde Martin would be around from time to time to ensure that "the cutting is done in the general way as outlined in this letter."

George Long was busy, no question of that. George Cornwall knew as much when he inquired, respectfully, whether Long had finished his paper to be given at the loggers' meeting in Vancouver, Washington. Cornwall only wished to prod a bit, wondering if Long might be able to provide a copy in a week or so. The response was not unexpected. "You know very well," wrote Long, "that I will not do that, as I never know what I want to say, until I am forced to say it, which is usually about two or three days before the event is pulled off." He did request a program "so I can see what the other fellows are going to talk about, and I will try to break away from my usual sluggish plan of not getting ready in time."

The loggers' meeting was not Long's only speaking commitment. At a meeting of the Southwestern Washington Development Association at Chehalis in early June 1911, he addressed the assembled on "The Logged-Off Land Problem." He left no doubt that he was concerned with the plight of many settlers, observing that they deserved to "be rewarded for their efforts on equally as valuable a basis as labor is rewarded in other fields, or as a tiller of the soil reaps in other districts." He urged his colleagues to proceed "in a big, broad, liberal way," warning that "unless it is found to be practical to utilize these [cutover] lands, and unless it is found to be profitable to reclaim and cultivate them, your efforts will be misdirected, and more harm than good will be the result."

So, what should be done with these lands ill-adapted to agriculture? There was no question in George Long's mind.

> My answer is to re-forest them. No place else in America, and possibly no place on earth, is there such a stand of timber growth as exists and has existed in western Washington. . . . It is the nation's great wood lot, and it is the last stand where the nation's supply for timber and lumber is to be found. No where else on earth has there grown a crop of timber equaling it in quality or quantity. In no other known place can a new

crop of timber be grown under such favorable conditions, and some day, timber will sell for what it cost to grow it, as other crops do. Is there any reason why the business of lumbering cannot be made perpetual on the Pacific Coast, and especially in western Washington? I contend that fully fifty percent of the land where the forests now stand is better adapted to the growth of another forest than to any other purpose, and I question whether it would be possible to utilize these lands to any other purpose that would yield as much wealth to the state or to its citizens as it would to grow a new forest.

Although the message was hardly a new one, seldom had it been delivered with such conviction. The time for reforestation may not have been at hand, but in the vision of George Long it was within reach.

Meanwhile, the lumber prospects refused to improve. In late August, Boner complained that his affairs were "constantly growing worse." Thanks to "the Canada business for Common [lumber] and the California market for flooring," they were selling as much as they were manufacturing, but, Boner warned, "The price, of course, will not be good."

What was the answer? No one seemed to know. Boner wondered how much worse conditions might be, say, if they had three or four or a dozen mills to worry about. What then? The question wasn't just hypothetical, for he knew that Long envisioned an increase in manufacturing capacity. Boner continued, "It seems to me before you can develop saw mills, you must develop markets, or sell the line yard at their price, which under existing circumstances will never be profitable."

The only solution that came to Boner's mind was for Weyerhaeuser to enter the retail-yard business with a vengeance, "locating yards only in good junction points on large trade centers; make prices with no one; divide territory with no one; sell at a fair consistent price to everyone any place; and depend upon the large volume of business for your profit." There were, of course, important stockholders who were independently interested in retail yards, and they would initially question such a step, but Boner maintained that given the stumpage owned by Weyerhaeuser, their interests would best be served by the suggested plan. The timber, the mills, and the yards were simply parts of the whole. Boner must have realized that when discussing the subject of marketing with Long, he was talking to an expert. But the questions were such that even the expert was stumped, for the moment, anyway.

Long headed east in the late summer of 1911 with a number of stops planned, not the least of which was St. Paul. Included on the agenda were what he knew would be tough negotiations with the Thomas Irvine Lumber Company and the Cherry Valley Logging & Railway Company owners "regarding the proposed amalgamation of the logging proposition." Still, he hoped he could get away soon enough to join wife Carrie and daughter Margaret at Northampton, Massachusetts, where Margaret was about to enter Smith College. And since he was in the East, why not begin "looking up terminal yards along the Atlantic Coast." Thus he wrote to F.E., encouraging him to join in the tour, for as Long put it, "I think it would not be a bad idea to get a little bit acquainted with this subject."

Tentative agreement was reached with the Thomas Irvine Lumber Company, but experience had taught Long that closure would not come easily. The capital stock of the new company was to "be fixed by the total value of the timber, and whichever one of the two parties has the most timber would get the larger share of the capital stock." Long expected board feet to get translated into dollars in a big hurry.

There were other discussions in St. Paul of a more personal sort. Lawyer Macartney explained them to Bell in a rather homespun manner, suggesting that Long did "not seem to enjoy being so largely in debt to Weyerhaeuser, and desires to reduce the same if satisfactory." Long's debt involved his option on stock purchases, the stock being held in his name without payment of the principal, but with interest on that unpaid principal all the while accumulating. By the fall of 1911, because of his subscription to 1,000 shares, he was indebted to the amount of $100,000, with an accumulated interest of $63,298.67. It was now agreed that the interest would be forgiven, that the subscription to 1,000 shares canceled, that Long would be given 500 shares outright and would "sell, assign and transfer" the rest to the company. It was further agreed that his annual salary would be increased to $15,000 for the year ending December 31, 1911, to "be the same amount for each year thereafter, until otherwise ordered." In return, Long gave his promise to remain with the Timber Company for ten years. There was no question that all parties were pleased and more than a little relieved. As for Long, now that his personal affairs were settled, he could look forward to getting back to his company's business.

FOURTEEN

Politics and Mergers

FOR MANY AMERICANS, THE YEAR 1912 WAS ALL-consuming politically. Although four years earlier Teddy Roosevelt had handpicked William Howard Taft to be his successor, with the presidential conventions in the offing it was abundantly clear that the two were now hopelessly at odds. The reasons were many, not the least of which had been the Ballinger-Pinchot controversy and Roosevelt's conclusion that his beloved conservation movement had been betrayed. While Long followed the political scene with more than a little interest, as the summer approached it was fire that most worried him. And his worries were not limited to the forest, as he made clear in a letter to Boner. The Everett superintendent was urged to inspect his plant, making the rounds with nothing but fire in mind: "I think you will find something that you would like to clean up . . . particularly as to oil waste spots and places where dust collects around boilers and in machinery generally." Boner doubtless gave a bit of a shrug, but one suspects that he toured as directed.

Boner had earlier noted that business seemed on the verge of improving. Long agreed, amazed that any progress was possible "under such a political hubbub as we are now having." Long was nonetheless convinced that the market would get better by "leaps and bounds as soon as politics were adjusted so that people understood what was going to happen."

Through the poor business period, Everett had fared relatively well, although from the Tacoma desk of George Long the reasons for this seemed none too reassuring. The Everett mill's annual production was about 60 million feet of lumber, with an average daily cut of slightly more than 20,000 feet.

Over the past year or so, Boner had been buying his logs on the open market, not only because they were cheap and it seemed silly to use up their own resource at such prices, but also, as Long put it, "because we wanted to know how hard it was for the other fellow to buy logs and make a little money." And partly because of the cheap logs, they had made a little money. For the year 1911, after allowing for some $30,000 depreciation, Everett showed profit "of about $90,000.00 which is pretty good considering the times."

The relationship between Long and Boner was lively and interesting. Because they respected each other, their disagreements were honest and never taken personally. Their perspectives, of course, were different. In 1912 Long expressed momentary disappointment that installation of new planing mill machinery had exceeded the capacity of the power plant. "I guess it is an illustration that a man does not always guess big enough when he is building" was Long's version of the lesson to be learned. But later, when Boner proposed some extensive improvements, Long demurred, suggesting that perhaps they had done enough improving, "and I think you will have to jar me loose with a pretty big shock before I will be content to see you put in a new double cut band saw, or a new installation in your planing mill engine house." Still, he didn't shut the door, adding in a manner most typical, "I presume when a man begins to feel like he does not want to spend any more money on a saw mill for its improvement that it is a sure sign he is getting old and losing his grip." Regarding the business upturn, Long remained cautious, allowing that there were always dangers around the corner, and he only hoped that the "lumber industry will see that it is to their advantage not to ever again get as wild as we did in 1906."

If Boner was the enthusiast for the manufacturing end of things, Long was ever the enthusiast when it came to timberland purchases. At present, his interest again focused on the pine belt of the Klamath basin, particularly on a tract owned by the Western Pacific Land & Timber Company. Chief among the Western Pacific owners were John H. Queal and some of his associates in the McCloud River Lumber Company. They had begun investing in the district at about the same time as the Weyerhaeuser Timber Company, and the holdings of the two were intermingled, along with the so-called Hopkins tract and some lands owned by the Oshkosh Land & Timber Company. In late April, Long had sent a map of the area to the St. Paul office, with copies to Bell and Thatcher in Winona. In the covering letter he observed: "The all around proposition is, that these entire holdings of the different ownerships in this district, probably constitute as fine a selection of pine timber as there is on the Pacific Coast, everything being taken into consideration."

This required some vision, an appreciation for what could be rather than what was. Specifically, in order for that pine to be harvested profitably, a railroad connection was necessary. Their own Klamath Lake Railroad Company had simply come to them as an accidental accessory to a previous timberland purchase, and, as Long acknowledged, "there is no further use for it as the timber will never go out over it." But now the Southern Pacific had begun constructing a line from Weed, California, northeast to Klamath Falls, as part of the main line between San Francisco and Portland. In addition, the Hill interests were building south from the Columbia River up the Deschutes River, and their surveys extended on to the Klamath Lake. "In fact," Long noted, "one of their surveys terminated at a point about eight miles north of where our timber limits stop on Klamath Lake." As a rule of thumb, and as noted previously, lumbermen avoided situations that placed them at the mercy of a single carrier. Now, with both the Southern Pacific and the Great Northern lines promising connections, Long predicted that the "transportation facilities, probably, will be most excellent whenever the time is ripe for manufacturing."

Although negotiations for the purchase of additional properties were unpromising, Long remained optimistic: "I cannot get away from the fact that if we can ultimately get control of all the holdings in that district that it will be hard to find anywhere its equal for economical logging and for excellent facilities for manufacturing and transportation."

The proposed merger of timberlands owned jointly by the Thomas Irvine Lumber Company, the O'Neil Timber Company, and the Weyerhaeuser Timber Company proceeded apace. Previously, logging operations on the Irvine and O'Neil lands had been conducted under contract with the Cherry Valley Logging & Railway Company, owned by the O'Neil brothers and

James E. Gowen of Everett. Under the terms of merger, a new organization, the Cherry Valley Timber Company, would purchase the equipment and track of the Logging & Railway Company. That much was easily accomplished, but the rest took time and patience.

Long thought that the timber cruises had been done "on a very conservative basis," and an overrun of at least 25 percent seemed likely, on a total estimate of approximately 470 million feet. The parties agreed to differentiate on the basis of quality, that merchantable and flooring logs would be accounted at $3 per thousand, with cedar and spruce also at $3, "and that No. 2 fir, the hemlock and dead and down fir would be considered as worth $1.00 per thousand."

On that basis, the Weyerhaeuser share would amount to $528,860.70, the O'Neil Timber Company, $379,111.20, and the Thomas Irvine Lumber Company, $256,632.10, for a total of $1,164,604. In effect, these figures constituted the sale price for each of the companies to the Cherry Valley Timber Company. The new company was capitalized at $500,000, with Weyerhaeuser subscribing for $250,000, and the Irvine and O'Neil companies the balance. Long was relieved when the thing was finished: negotiating with outsiders was difficult enough, but not so difficult as dealing with wary friends.

Closer to home, Long was having some preliminary discussions with O. D. Fisher, who was serving as western agent for the Grandin-Coast Lumber Company. The talks focused on the intermingled timberland holdings of the Weyerhaeuser Timber Company and Grandin-Coast (with as yet unrepresented awareness of Thomas Irvine Lumber Company properties), largely northeast of the Snoqualmie River in King County. Some six years earlier, shortly after the Grandin-Coast people purchased these lands from the Everett Land & Timber Company, George Long had urged them to delay opening up their timber "until they had railway facilities for bringing it out, and then I thought the thing to do was to build a mill, as the type of timber was such that it could be handled by rail as well as any other way." Now, in mid-July of 1912, the subject had been reopened, initially by Captain J. B. White of Kansas City.

The immediate cause for reconsideration was that Grandin-Coast was closing down some of its southern and midwestern operations and was casting about for new manufacturing opportunities. None appeared more promising than Snoqualmie, and that opportunity could only be enhanced by cooperating with Weyerhaeuser, "to make a good sized operation, and to jointly build a mill, each concern contributing the equal amount of timber . . . and each contribute an equal amount of cash to construct the mill and start it off." Furthermore, as Long described the situation for the benefit of F.E., Grandin-Coast owned "a most excellent mill site" near Snoqualmie Falls, "comprising about 160 acres of level land, and right at the mill site they had a slough from 150 to 220 ft. wide and about a mile or a mile and one-half in length . . . with storage approximately for somewhere between eight and twelve million feet of logs."

In addition, the Milwaukee Road had built to within a half mile of this mill site, and the Northern Pacific was already operating on the opposite side of the river, not more than a mile away. Captain White was convinced that the Great Northern would agree to extend its tracks from its terminus at Tolt up to the mill site. As White would quietly suggest, "If it was thought best for the lumber to be marketed through the Missouri Lumber & Land Exchange Company, it would be all right." Missouri Lumber & Land just happened to be one of the major distributors in which White held a large interest, and access to the market by that means would be no small advantage.

Such a proposition was exactly what Long had envisioned from the beginning. Now, as he explained to F.E., he wasted no time in encouraging Captain White to think about making a joint effort: "I told him frankly that it was one which appealed to me sufficiently to take it up with my people in the East." And that is where the matter stood, even though the actual negotiations were in the earliest stages. The timber itself ran "more largely to old growth fir, more largely to cedar, and more largely to hemlock than is characteristic of timber generally in Washington." Long was convinced that all of it, hemlock, cedar, and especially the fir, "running as it does, largely to high grade lumber, could be more easily handled from an interior mill . . . and I am very well satisfied that the time is not far removed when we could safely enter upon development of this kind, unless we sit down and adhere to the policy of not becoming manufacturers in a large way."

From the beginning, everyone understood that investment and commitment to western manufacturing would be a major decision, but as F. E. read Long's letter, he could have few doubts that such a decision was forthcoming. Long was not one to go overboard about such opportunities. In this case, it wasn't feasible to suggest that those in St. Paul knew best; in fact, they had not known best for several years.

In two weeks, Long was able to respond positively to Captain White, having received word from "our St. Paul office which in a measure voices the tentative opinion of our people that the suggestion of making a consolidation of our mutual holdings back of Snoqualmie might be worked out in a way that would be satisfactory to all of us." In other words, it was full speed ahead. It is interesting that as they did proceed, the complications came not between Weyerhaeuser and Grandin-Coast, but rather from the Thomas Irvine Lumber Company, a development that could hardly have surprised George Long.

Although the Snoqualmie plans predominated, Long had not forgotten his pet project in the Klamath pine district. As he prepared for the late-July meeting of the Pacific Logging Congress in Tacoma, he solicited the attention of W. P. Hopkins, ostensibly because he thought Hopkins might enjoy attending the Congress, but, more important, because "we could have plenty of time to talk over things that we might be interested in elsewhere." Long made no mention of the Hopkins tract in the Klamath district. He didn't have to.

Hopkins accepted Long's invitation for a Tacoma visit, and together they toured a nearby logging camp. The two discussed many matters of shared interest in Oregon—fire-patrol efforts, the beetle pest problem, and the future of the Klamath basin timber. Neither seemed eager to be specific about plans or hopes, although Long did observe that if his friend's "faith and inclination ran toward the development of his property and going into the lumber business, that he certainly had a very inviting field ahead of him." And he further remarked, as if in passing, "It might be a good idea for some of us to get together and bunch our stuff and make a company." And that was as far as it went, for the time being, but Long was encouraged. As he closed his report to F. E., "If there is any way of getting hold of that property at a fair price, I would like awfully well to own it."

The Snoqualmie proposition was much further along, the boundaries of that operation becoming ever clearer, at least as they appeared on the map. Long realized the necessity of a thorough examination of the country. As yet he wasn't even certain where the mill would be located. F.E. had suggested, casually, that perhaps the numerous lakes scattered about might serve as sites for several small mills. Long nipped that idea in the bud. The proper approach was "to commence on the front and work back," rather than working here and there in the midst of the tract. The reasons for this systematic campaign were two-fold: it was less expensive to extend the logging railroad sequentially, as required; and the fire hazard would be greatly increased "by having the zone of cutting in the midst of a forest instead of on the fringe of it."

As for a central mill site, Long quickly became convinced that the one owned by Grandin-Coast at Snoqualmie would serve the purpose admirably, and in mid-September he and Fisher inspected it and found everything satisfactory. Long subsequently dispatched cruiser Fred Conant to study the site carefully, particularly the slough, to determine what improvements were needed to provide adequate log storage.

While his boss was busy thinking about Snoqualmie developments, Boner had been making his own plans for expansion at Everett, somewhat buoyed by recent firming of the market. Long was in no hurry, but he didn't wish to discourage. "I think we better get together pretty soon on the ground," he wrote, "and have another look and another think and continue our ideas of deciding what is the thing to do." Long seemed convinced that Boner's initial plans were too grand, that the mill he envisioned was too large. Long thought that a mill with a daily sawing capacity of 200,000 feet, with accompanying planing mill and dry kilns, would be the most economical plant. Then, if greater production was required, "you can run night and day at practically no more overhead charges [and] if there comes a slump in the market, you can still preserve your organization in good shape by running day times only."

In the midst of these reasonably pleasant deliberations,

occasional clouds darkened the horizon, including a growing militancy of some portions of the work force. Everett, like other cities of the period, was experiencing aggressive socialist political activity, led by the so-called Wobblies (Industrial Workers of the World). So far, the Wobblies held small sway in communities like Everett and Tacoma, but Boner and Long listened to filtered versions of the IWW doctrine, and they cringed. George Marshall, whom Long had employed to watch over tax affairs in Oregon, forwarded one such report from Eugene on October 20, 1912. "I just heard an IWW soap box orator out on the street advocating rebellious actions against heavy [timberland] owners," he wrote, adding, "He was surrounded by an apparent lot who thot [*sic*] so too." The Wobblies would not go away, but at the moment they were more a nuisance than a problem.

A much more enjoyable involvement was the cooperation between the company and the public sector, represented by the University of Washington's College of Forestry. Dean Hugo Winkenwerder appreciated Long's contributions as an occasional lecturer, but he valued him even more as a source of information, guidance, and assistance both practical and otherwise. In the fall of 1912 Winkenwerder sought support in developing a course in logging engineering and in "launching a movement to have the lumbermen endow a chair of lumbering." Although Long expressed a willingness to assist with the curriculum planning, he opposed the idea of endowing a chair of forestry. His reasons were clearly stated: he thought it crucial for the state to make the commitment in its own right. "The State of Washington really owns more timber than most people imagine," he noted, estimating its value at "not less than between twelve and fifteen million dollars." Furthermore, the lumber interests were already contributing a large amount to the support of the university through taxes. Why do for the state what the state ought to be doing for itself? It was a timeless question.

In late November, Long was summoned to St. Paul for a meeting of Weyerhaeuser Timber Company directors. The principal subject before the board was expansion of the Everett manufacturing facility. The decision was for delay. Boner, however, had acted prematurely, having reached a tentative agreement with Arthur B. Pracna of Seattle, a noted engineer, to make the plans and supervise construction. Now Boner was forced to renege as best he could. This cost a bit of money, but Long didn't complain. "It is certainly the only dignified and courteous treatment that we could afford him under the circumstances," he observed, adding, "You did just the right thing."

The plans for Everett had not been canceled, just deferred, and by the end of January 1913, Long was telling F.E. of recent discussions with Boner and a Minneapolis mill engineer. "Our ideas," Long indicated, "are to make two mills, one to be a fir mill and the other will be a combination of a cedar and a hemlock mill," manufacturing short lengths only, making "a pretty good smart mill . . . where the work would be just alike every day," which was "radically different from a fir mill." Long assured F.E. that they were going slowly: "We will have to feel our way pretty carefully and take plenty of time to make our plans and have everything shaped up right."

Not everything was shaping up just right. In Olympia, for example, some in the state legislature were expressing interest in reducing appropriations for fire prevention and control. Long urged Allen to contact such influential friends as Major E. G. Griggs, requesting that they lobby in behalf of an adequate appropriation. "While not opposed to economy as a public principle," Allen's letter began, "protection of Washington's greatest source of industry and tax revenue is not extravagance but the most businesslike kind of prosperity insurance, every cent of which comes back with interest."

Another topic of concern involved the general question of public-land management, or, as Long phrased it, "the imaginary controversy between the western idea of state control and the federal idea of federal control." He thought they should steer clear of this debate, "unless there is some absolute necessity for our butting in." His own preference was to support the federal position, but he doubted that this would be appreciated by "our own constituency of western legislators and western politicians, so my idea would be to let the matter alone, so far as the Western Forestry & Conservation Assn. is concerned." He also advised caution in supporting a suggested change in tax policies, one that would differentiate between classes of lands by use. He feared that once the principle was accepted, the bigger and more politically powerful classes, such as farmers and orchardmen, might be tempted to take advantage of the smaller classes, such as timberland owners. If the amendment became law, Long warned, it "would provide the necessary legal way to do [to] any one class what the other fellows thought was justifiable."

In that connection, he acknowledged that although the tax laws of Washington and Oregon were occasionally described as "antiquated, the great big underlying basic justice of them is that all classes of property are assessed alike and taxed alike, and it is not a very wise matter to make any improvement on this kind of a basis." Instead Long urged that they continue efforts to educate the public on the general subject in advance of proposing specific legislation. (Long would soon change his position on taxation, largely as a result of favoring a yield tax which deferred the payment of most taxes until the time of timber harvest. This in turn required that timberlands be treated differently from other properties.)

In February 1913 Long and his family enjoyed a rare vacation, soaking in the warmth and sunshine of southern California. All was not leisure, however, at least not for George. He met with Frederick Weyerhaeuser at the old gentleman's winter home in Pasadena and kept apprised of Tacoma matters via Hugh Stewart and Bill McCormick. F.E. planned a late-February arrival in Pasadena, at which time he and Long could continue their discussion, particularly as to the plans for Everett. In advance, F.E. indicated that he was now convinced they should go forward with the construction of the additional mills, observing that it was "advisable from every standpoint for us to gradually work into the milling business." The primary reason for doing so, as viewed from St. Paul, was to "escape the criticism of being purely timber speculators." No final decisions were made, but there was no doubt of their direction.

Naturally, many wondered what Weyerhaeuser's plans were for Everett. Responding to a query from editor Cornwall, Long acknowledged that indeed they did have a site available for expansion, "not very well adapted to the cargo business, but could be utilized to good advantage for rail shipping," and said simply that they hadn't made up their minds: "[We] have not gotten far enough along yet to indicate just what the type of the mill will be nor exactly when we will commence to build it";

those decisions depended largely on the condition of the market. He closed with a prediction of sorts: they would proceed in a leisurely fashion, and would probably have a new mill in operation some time during 1914.

The two major items on Long's desk in the spring of 1913 continued to be plans for Everett expansion and the timberland merger with the Grandin-Coast interests in the Snoqualmie Falls district. Concerning the merger, Long wrote to Fisher in early April, requesting that they meet to develop a set of instructions for the cruisers, in Long's words "to convey to the cruisers the varieties of timber to be taken under consideration, and how it should be classified. . . . Personally," Long continued, "I am somewhat inclined to think that it might confuse some of the cruisers to go into details as to what constitutes grades"; he would be satisfied if they simply described the species they found, and gave some attention to timber that was obviously defective. Fisher had no objection, on that or any other score. Indeed, he allowed that it would be entirely agreeable to Grandin-Coast if Long's old friend Lafe Heath were to assume responsibility for the cruising effort. Long subsequently asked Heath to do the job, with this noteworthy instruction: "We want the lands all cruised just as though they were one ownership instead of two, exactly alike in every respect and of course we want it done by most competent men." By the end of April, four cruisers were in the woods.

Although Frederick Weyerhaeuser was now on the periphery of business affairs, he remained aware and alert and full of opinions. In recent months, he had been following with interest the plans for Everett's plant expansion. And he had a suggestion. In an April letter he advised Long to go slowly, largely because of uncertainty regarding the new Wilson administration and its tariff policies. Long promised to do just that: "I certainly will take no decided steps toward going ahead with the work until first conferring with you."

Despite the promise to Weyerhaeuser, there is no indication that Long slowed the process. He spent nearly a week at Everett, consulting with a mill designer, "going over the ground quite carefully and analyzing his plans to see how well they fit the situation," he wrote to F. E. on April 28. No quick solutions came to mind, and Long admitted to being undecided as to the ideal arrangement. They were, however, making progress and he hoped "to be able to submit, at the meeting in June, concrete and definite plans to lay before the trustees."

Boner, of course, was more than an interested bystander, but he had distractions of his own, including what seemed to be a weakening of lumber prices. Long advised that for the next couple of months it might be wise "to play the game on the theory that prices are not going to be any higher and may go considerably lower." In other words, don't turn down any business.

Another thing that kept Boner awake nights was worrying about the IWW. Again Long tried to reassure, noting that in his experience it was the unexpected that happened. Moreover, he thought that the radicals had already lost the "strategic time to make trouble," since it was early May and there was a large inventory of logs. Now "if they should get ugly," he continued, "I think the loggers and lumbermen generally won't care so much, and as you say, it may bring relief to the lumber situation." The fact was, however, that the IWW leadership was less concerned with strategy than with philosophy. They were as committed as they were inept. In a sense, Long was proved correct in downplaying the problem, at least as it affected the Weyerhaeuser Timber Company. But Boner's worries were well founded. The Everett community would suffer terribly in the years ahead.

FIFTEEN

Klamath Falls Pine and Everett Production

IN LATE APRIL 1913, LONG PROVIDED BELL WITH A PLAT map that detailed the ownership of the Western Pacific Land & Timber Company bordering on the west side of the Weyerhaeuser Timber Company's Klamath County holdings. Long's intent was obvious. He wanted to purchase these adjoining properties, and though Bell hesitated, F.E. Weyerhaeuser seemed ready to push ahead. As evidence, F.E. indicated a willingness to pay as much as $1,250,000 for property that provided "so excellent a consolidation of holdings in a territory that should be developed so soon." He also expressed an interest in arriving a week or so before the annual meeting, to make a tour of the Klamath region. "Such a trip," he wrote, "would give us a good opportunity to discuss other matters free from interruptions that must come up in your office," as well as taking a careful look at the Klamath situation. Nothing could have pleased Long more.

Also making plans for the 1913 annual meeting was Frederick Weyerhaeuser. His personal secretary, J. W. Mahan, outlined these plans for Long, informing him that arrangements had been completed to travel west aboard the private car of the president of Great Northern. Although there were obvious pressures associated with the annual meeting, Long actually enjoyed performing—especially when business conditions were on the mend.

All who attended the meeting must have been impressed with what the Weyerhaeuser Timber Company had become and

151

with Long's command of the many pieces. Some of the original investors may have wondered occasionally whether they could even keep track of company developments, let alone control them. As a result, when Long stood up to read or summarize sections of the annual report, all listened attentively and respectfully. For it was he who was in charge, and none knew it better than Frederick Weyerhaeuser. This would be the last annual meeting for the old gentleman. Of course, he could not have known that. He did know, however, that the decision to hire Long had been about as wise as any he had ever made.

In addition to such expansionary projects as the Everett manufacturing plant, Klamath purchases, and the plans for the Snoqualmie merger, Long discussed a variety of other subjects, including the continuing concern with tax increases, logged-off lands, and the Tacoma building. Regarding the building, Long recounted its history and detailed present circumstances. The cost to Weyerhaeuser, including property, had been about $295,000, which under the terms of the partnership amounted to half the total cost. It was owned equally by the Weyerhaeuser Timber Company and the Tacoma Commercial Club, although, as Long noted, architecturally there was no "apparent line of demarcation between the two sides." It is interesting, however, that the ownership division was maintained in detail. The building, for example, had four elevators, "two of which stand on the side of each owner and can be operated independently of each other." There was but a single common entrance. The Timber Company occupied its own side of the first floor and the Commercial Club occupied both sides of the top two floors, paying Weyerhaeuser $500 per month for its half of those floors. The remainder of the building was nearly full with tenants. No one, not even George Long, could possibly have imagined the day when Weyerhaeuser would be the sole occupant and looking about for additional space.

Long much preferred to be dealing with timberlands rather than with tenants, and in the summer of 1913 there were plenty of timberland opportunities on the table. In late June, he closed a trade with the Buckeye Timber Company, purchasing land adjacent to the operation of the Clarke County Timber Company. The price was $600,000, of which Weyerhaeuser's initial share was $300,000, with lesser subscriptions to Willis T.

Knowlton, Charles H. Davis, and Gilbert M. Stark, the Saginaw lumbermen who had been partners in the Clarke County Timber Company. This company's notes were now payable on demand, and drawing 6 percent annual interest. The salvage logging effort that followed the 1902 Yacolt burn, born of desperation, had evolved into a very profitable venture.

At the top of Long's agenda were plans for a Snoqualmie merger with the Grandin-Coast Lumber Company. To that purpose he sought a meeting with Fisher as soon as possible after the Fourth of July, at which time they would "send in the checkers," to verify the cruising reports. They were proceeding in the only way possible, carefully and step by step, and nothing would interrupt that progress.

While cruisers and checkers roamed the woods, distant developments demanded attention of a different sort. None loomed larger on the horizon than the opening of the Panama Canal. Although the canal was still a year from completion, many were pondering its effects. One New York City lumber wholesaler had been in touch with the Weyerhaeuser office in St. Paul. Later he wrote directly to Tacoma, advising that "other strong people are looking around now for a location so as to have made preliminary arrangements and be ready for business and I should be pleased to have your interests on our plant." Long needed no encouragement. His appetite for East Coast opportunities was well established, but he wanted to know a good deal more before making any decisions, particularly if these were to involve a partnership with unknown outsiders.

Closer to home, Boner was requesting further consideration of plans for expansion at the Everett sawmill. Most in St. Paul were supportive. F.E. responded almost immediately, agreeing that a fir and cedar mill ought to be constructed on the barge work property. And he added, for emphasis, "Father seems to be favorably inclined toward this suggestion." Most others concurred, F.E. noting that Thatcher was probably the "most conservative along these lines," but even he was likely to be convinced of the need for some mill development on the part of the Weyerhaeuser Timber Company. The message overall, however, was clear: The directors would approve whatever Long recommended in terms of Everett expansion.

But Everett was in some respects the least of Long's

worries. The only question was when expansion would occur, not whether. Klamath Falls and Snoqualmie were less certain matters, although progress with the latter appeared to be on track. The check cruises were nearly complete. The next step was simply to consolidate the various reports. Long and Fisher appointed representatives who would review the figures, ensuring that no mistakes had been made in the calculations. Long selected his nephew, Fred Firmin, instructing that he "make a grand summary of the whole performance, each company's lands and timber being kept separate in this recapitulation . . . and all the miscellaneous ownerships being kept in one group distinct from that of either of the other companies." (Fred—or Alfred—Firmin was the son of Long's oldest sister, Florence, and John R. Firmin, who died when Fred was six.) Firmin was provided a desk in the "rear room," where Long normally held forth. He would have access to all necessary materials, spreading out all the pieces and disturbing no one. Long planned to be away, touring in Oregon.

There would, of course, be problems, starting with those other ownerships. Even if Weyerhaeuser and Grandin-Coast came to terms, that did not ensure agreement with the rest. A year earlier F.E. had expressed concern about a lack of cooperation from his old friends, those connected with the Thomas Irvine Lumber Company and the O'Neil Timber Company, owners of much of the timber located in the heart of the consolidation scheme. F.E. noted that the O'Neil brothers were "very anxious" to commence logging in the district. Thus he forewarned Long: "In looking over this territory it occurs to Mr. Irvine and myself that it will be a serious mistake to allow anyone to get into this township without at least giving the matter some careful thought." Long had already given the matter careful thought, and he knew that the serious mistake would probably be to allow someone like Horace Irvine, whose interests were narrow if not selfish, to frustrate plans for the completion of what could be a wonderfully efficient production unit.

As for additional worries, F.E. was sensitive, probably overly so, to public reaction—"the effect on the imagination of the public of the uniting of our interests with one of the large factors in the southern trade." There was also the question of who would manage the operation. In F.E.'s words, "If you are entirely satisfied that they [Grandin-Coast] can give us a management that will please us, I should be more inclined to make the consolidation than I would be otherwise." But Long realized that no manager, however able, could make a success of a poor property.

Now, in the summer of 1913, Horace Irvine wrote to Long on the subject of Snoqualmie, a letter answered exactly one month later, Long explaining that he had been out of town. That was partly true. It was also true that Long chose not to appear overly eager to negotiate. He knew from past experience the difficulties of dealing with Irvine, and for the moment he simply requested a delay: "If you will give me a little more time, I will soon be ready to make a start."

As summer turned to autumn, the question of Everett expansion was once again on center stage. Boner resurrected the subject in a September letter. With the preliminary plans drawn, it was now time to discuss specifics with a construction engineer. Boner had one in mind. A. B. Pracna had considerable experience in mill construction in the Pacific Northwest, and Boner thought F.E. might wish to interview Pracna before deciding. Long assured him that this would not be necessary, that F.E. would "leave the matter entirely to us." Accordingly, Long advised Boner to get Pracna in tow "and have him go with us and see some of his mills." This commitment was well received in St. Paul, F.E. reporting that father Frederick had just returned from the East, "feeling very well and as cheerful as one could hope," adding, "He is entirely agreeable to your going ahead with the mill development just as fast as you see fit."

However important plant expansion at Everett may have seemed, events in Congress would be far more so. Bill McCormick was then maintaining a Washington, D.C., office from which he played shepherd to all manner of interests, from aiding in the negotiations to purchasing additional Northern Pacific timberlands, to occasional lobbying, to merely keeping Long informed of significant legislative and legal developments. McCormick was preoccupied with the new federal income tax law, which included both a corporation and a personal income tax, the corporation taxes being retroactive to January 1, 1913, and the personal to March 1. Although the details were only beginning to be understood, McCormick was already convinced that Long should begin preparing for the organizational ramifications. For example, dividends from other corporations would no longer be exempt. Thus if the Weyerhaeuser Lumber Company of Everett paid a dividend to the Weyerhaeuser Timber Company, on which it had already been taxed, the Timber Company would in turn be taxed on that same amount. Furthermore, should those earnings be distributed to the stockholders, they too would be taxed, depending on their individual circumstances. Although this would be one of the simpler problems to be resolved, it wasn't a pleasant prospect.

Meanwhile, down in Long's favorite forest, the Klamath pinery, agent Jackson Kimball was worrying about insects. The district forester, George H. Cecil, had recently sent Long a confidential report on insect infestation in the Paulina Forest. Long, in turn, shared Kimball's concerns with Cecil: "I feel myself expressing the hope that he is unduly excited over the matter, and yet I have found him to be a very conservative and conscientious worker on all lines pertaining to the devastation of timber." He also used the occasion to communicate again with W. P. Hopkins, heir to that tract of adjacent timber he so wanted to purchase. Long advised Hopkins of the insect problem, that between fifteen and twenty thousand acres of "small lodgepole pine were absolutely destroyed," most of the damage occurring within the past two years. By way of reassurance, Long also reported that Kimball "at the present moment is undoubtedly in there with a force of 25 to 30 men, about half of whom are contributed by the Forestry Service and the others by the timber interests," the team working to burn the infested area over before the winter rains and snows commenced.

Putting the best possible face on the matter, Long saw a potential benefit, thinking the "dangerous situation" might serve to focus the attention of the timber interests in that country and lead to a more careful general campaign. Few owners had been willing to contribute toward the control efforts, and Long thought the time was ripe for progress. Kimball was less sanguine. As he saw the situation, the major obstacle would be to get "the government to clean their lands, because relatively our portion is very small." The Forest Service had "only a few men who know anything about this big business, and because they are planning on disorganizing the Paulina National Forest their interest in the matter seems rather apathetic." It was a problem that would never be completely resolved.

In late October 1913, out of the blue, Long received a note from F. S. Bell that was more important that it seemed initially. F.S. was dutifully sharing a letter from his son, Laird, who wanted to do a favor for one of his University of Chicago Law School friends, William H. Peabody. Thus was introduced an individual who would play a large role in the future of the Timber Company. Peabody had graduated from Williams College, where he achieved the greatest success on the athletic field. He was admitted to Chicago's law school and stayed there long enough to become friends with Laird Bell. Years later, Peabody summarized this stage in his career: "Two years of law at Chicago for amusement and then he headed for the woods." In truth, the transition wasn't quite that easy.

Peabody was married, and, according to F.S., "started in to a business of manufacturing collars, which was the family business of his wife's people," hardly the sort of career agreeable to one whom Bell described as "a fellow of character and great force, physically and mentally very vigorous, a pusher and a mixer." That was a fair description. In the fall of 1913, however, he had one major deficiency: "He knows nothing about lumber or timber." These were always awkward moments for George Long. "I hardly know what to say," he wrote in reply, "except in a general way that I presume we ought to be picking up young fellows and educating them and preparing them for our future wants, and then again, I sometimes dread the ordeal of taking them on too thick, because it keeps you about as busy keeping track of them and making men out of them as attending to your other business." It was never easy to assume a responsibility for a friend of a friend. For now, Long simply suggested that Bell have Peabody write to him directly.

In late October, Long received a query from the Commissioner of the Bureau of Statistics and Immigration for the state of Washington, requesting information on logged-off lands that might be adaptable for dairy farming. Long offered no encouragement. The fact was, as he had recently indicated to the Southwest Washington Settlers' Agency, that very few tracts held promise for farming or even grazing. The Settlers' Agency had questioned his discouraging descriptions, to which Long replied, "As to your disappointment in that our lists did not give you much available agricultural lands, we beg to say, that we wish they were good agricultural lands, but the Lord made them and not the Weyerhaeuser Timber Co."

Long was preparing for another meeting of the Southwest Washington Settlers' Agency, his topic "more or less regarding the use of this logged off land for forest growth." Given the tax policies then in effect, reforestation was hardly a practical alternative, but that didn't prevent Long from recommending reforestation as an optional use for the cutover lands. He outlined a plan in a letter to T. D. Rockwell, formerly a tax board member and still active in Olympia's legislative halls. "What I have to say now," Long continued, "will pertain simply to the necessity of some kind of legislation that will allow logged off land to be reforested and to allow small second-growth timber to be exempt from taxation until it has reached a more mature stage. Assuming that the State may not enter into the business of growing timber by acquiring logged off land, or even if the State should do this, I still think provisions ought to be made where the owner of logged off lands and the owners of young growth immature timber should be relieved from the burden of anything but a nominal taxation, until a new crop of timber has grown, or until the young timber has developed into a maturity that warrants its cutting."

What Long was proposing was a yield tax, to be paid when the timber was harvested, at which time the state would receive "25 per cent of the value of the timber stand." Long wasn't above lecturing: "It is absolutely the duty of the [Federal] Government and State to try and enact such legislation as will foster and protect the young standing timber and assist in the growth of a new forest on the logged off land."

Rockwell wholeheartedly agreed, as did many others. One who joined the chorus of support was E. W. Ferris, just completing his first season as Washington's state forester and fire warden. He asked for permission to use portions of Long's address, "The Logged-Off Land Problem," in his own campaign for new policies. "I shall take this matter up in my annual report," he promised, "and in every other way that I can in order to get the people to thinking about and getting back of the proposition so that good results may be obtained from the next session of the state legislature." The state forester advocated that they commence educating the public and keep at it "until our great state is assured of the perpetuation of our present greatest asset—timber."

George Cornwall, publisher of *The Timberman* and eternal lobbyist, was another supporter of reforestation. He reported to Long that he had encouraged introduction of bills in both Oregon and Washington legislatures on reforestation, acknowledging that the idea expressed "in relation to cut-over lands, was the result of your own." From Cornwall, that was a considerable admission.

While Long was thinking about logged-off lands and taxation policies, Boner worried about markets. There was a new movement afoot aimed at a very old objective—curtailment, limiting the manufacture of lumber, while endeavoring to expand marketing opportunities. Among those involved were "Messrs. Bloedel, Talbot, Ames," and a group of attorneys. This time Boner thought he saw some evidence of progress. "Heretofore," he noted, "these meetings always went along very nicely until we came to a point of percentage of allotment for each mill, and then the bristles came out so strongly on us that we could not get together." They had decided to leave the question of allotment "in the hands of a directory consisting of nineteen members, representing each district interested in export."

Long doubtless heaved a large sigh upon reading Boner's letter. Such cooperative efforts invariably placed Long on the horns of a dilemma. Although he seldom thought the results were worth the effort, the benefits of good will were important too, even at a cost. It was often a balancing act, and occasionally one went along simply for appearance's sake. Long tried to explain to Boner why he hesitated, "why people have never taken very kindly to organizations of this character." There were two primary reasons. The first involved the legal question, "that in this day and age of the world, hardly any kind of an agreement of this nature is looked upon by the Government as being legal." The second was that "ordinarily our crowd feel that in the go and come of ordinary business, they can handle their own affairs just about as well as someone else can handle it for them."

Still, he didn't order Boner to pull out. Instead he suggested terms for cooperation, focusing on ways of increasing the volume of lumber consumption rather than raising prices. "In other words," Long advised, "quite a good deal of energy should be given to seeking wider fields and considerable energy given to the subject of getting a fair price instead of a very high price." And so the Douglas Fir Sales & Exploitation Company came into being. (Exploitation carried a somewhat different meaning in this context, having primarily to do with advertising and promotion.)

Boner was already thinking about ways to increase sales, such as enlarging the size of their own sales organization "to take care of our new plant." Long replied immediately, observing that there was no need to be in a hurry to add to the "selling force with the idea of taking care of the further product of your new mill, for it will be some time before you have much product." For the moment, Long thought they had little to complain about, at least concerning their volume of trade. "The main trouble is not the volume, but the price," he said, adding that it was "a little hard to keep out of the trade when customers need lumber." He assumed that Boner would continue to accept orders, but not promise shipment before February. On the general subject of marketing, Long had some words of warning regarding the new Douglas Fir Exploitation Company, recalling the "olden days" when Chicago was the dominant distribution center, and "they absolutely controlled, by their joint action, the selling price of the manufacturers who sold to the wholesaler." He also remembered that it was not until the mills began to do their own distributing that they started to make money: "As a matter of fact, the man who owns the raw material, and who manufactures it, should in this day and age of the world, cut loose as much as possible from the middle man, when he is simply a broker between the manufacturer and the wholesaler."

Long did agree with Boner that they should proceed with plans for the new plant. First, they had to reach a final decision on the engineer who would design the mill, and Long knew that F.E. had some thoughts about that. Accordingly, he suggested that Boner might wish to accompany him on the eastern trip and then they could all sit down in St. Paul and arrive at a consensus.

Long had hoped to go east in time to attend the annual

Conservation Congress in Washington, D.C., but minor surgery on his face prevented this. His interests in the forestry sessions, however, would be well represented by Allen—of that he was confident. The Congress would suffer as a result of political machinations, most notably those involving Pinchot's opposition to the report of the Resolutions Committee on Water Power. Many lost some of their enthusiasm for this general approach to the subject of conservation. For example, J. E. Rhodes, now secretary for the National Lumber Manufacturers Association, observed that the directors of the American Forestry Association and others would favor holding future meetings "entirely divorced from the conservation of other natural resources." Rhodes shared another observation: "Mr. Allen's splendid work in preparing the forestry program together with his complete mastery of the subject is generally recognized by the lumbermen and foresters, and he can have anything in this work at their command." Once again, George Long had reason to be proud of his lieutenant.

On December 3, Long met with representatives of the Grandin-Coast Lumber Company, including Captain White, the senior Mr. Grandin, and O. D. Fisher. The cruising on the joint Snoqualmie properties was complete, and now they discussed the details of the final merger. As Long had foreseen, a new corporation would be organized with a stock distribution based on the value of each property, and with an additional stock subscription to furnish the necessary cash to construct a mill, railroad, and so forth. Under such terms, Weyerhaeuser would have approximately two-thirds and Grandin-Coast one-third. There had been an earlier indication that the Grandin-Coast members might wish to increase their share of ownership by stock purchase, but Long wasn't interested in that proposition, preferring that each group be content with their proportion based on the value of properties merged. Although the subject would be discussed at length in St. Paul, Long's provisions were accepted.

Long, his wife, and Boner left on the morning train on December 9. This would be a busy trip, what with the meetings in St. Paul, a visit with daughter Helen, a student at the National Cathedral School in Washington, D.C., and a tour of possible Atlantic Coast yard locations. The Longs spent the holidays in the East, not returning to Tacoma until the second week of the new year.

One subject given lengthy consideration in St. Paul was plans for the expansion at Everett. As it developed, the planning went better than the execution. It had been agreed that W. A. Wilkinson & Son of St. Paul should assume an overall architectural responsibility, with A. B. Pracna of Seattle serving as the consulting engineer. Pracna, however, objected. He gave Long his reasons: "I am positive that this arrangement would result in endless friction between the Wilkinson crowd and myself and a split-up in the midst of the job." Pracna was still interested, but only on his own terms. As a further complication, Boner's expert on mill design and construction turned into "quite a 'booze fighter.' So much so that yesterday I let him go." In short, although the decision had been made, as yet there was no fit to the pieces.

Pracna soon received a letter from the head draftman of the Wilkinson firm, offering his services, "as well as the services of Mr. Onstad." The two were, in effect, changing over to Pracna's side. Actually, it would be Al Onstad, the electrical engineer, who would be the most important piece in the puzzle, and Everett would be his initial involvement in what would become a career with Weyerhaeuser. In January 1914, however, Pracna urged confidentiality writing to Long "merely for the fact that it bears me out in my assertion of a few days ago"— that Wilkinson & Son was in financial trouble. Long kept F. E. abreast of these developments, also informing him that they remained "hopeful that we will work out of this just what we want, which means to take the best there is out of the Wilkinson office and cut loose from the undesirable." In the meantime, he assured that while it must seem to others like slow progress, in fact Boner was preparing the way, clearing titles and arranging matters whereby within a few weeks "we can safely count on getting what we want and proceed with our work."

Long was much encouraged by movement in other sectors, most notably at Klamath Falls. He knew that his enthusiasm for those pinelands was not shared by everyone in the company. For one thing, there had been many political problems in Klamath County, and those in St. Paul were always leery of messy involvements, even if it was Long and agent Jack

Kimball who did the actual suffering. But politics in Klamath County were changeable. Opportunities for timberland purchase might never occur again. Specifically, the Hopkins tract was being offered for $3 million. At the recent meetings in St. Paul, a figure of $2.5 million had been cited as the upper limit. Long wasn't about to surrender for a mere $500,000, and he reviewed the proposition for F.E. and others.

The tract totaled some 40,000 acres, Long estimated approximately one billion feet of timber, "which would run about 25% sugar pine, 50% yellow pine, sometimes called, as you know, California white pine and 25% of the various varieties of other woods, largely red fir of an ordinary grade." He then noted that seven years earlier they had offered $2 million for the property, and Hopkins had turned it down. Now Long indicated that he would not hesitate to pay $2.5 million. His greatest fear was that someone else would close the deal, thereby negating his plans for the region. "If we ever have a revival of the boom conditions of 1906," he warned, "it would not be difficult to get a bunch of fellows who would look favorably on taking over these two properties at a much higher price than we have ever thought it advisable to pay." He closed emphatically: "We want the Hopkins tract." And F.E. and others knew by now that what George wanted, he usually got.

SIXTEEN

War and Lesser Worries

GEORGE LONG HAD GROWN UP IN A PART OF THE WORLD where the premier softwood was eastern white pine. He understood his colleagues' abiding preference for that species, but recognized this preference as dangerous for Pacific Northwest lumbermen. "The facts are," he wrote friend Charlie Briggs, "that fir lumber is more adaptable to all kinds of purposes to which lumber is put, than any other wood now left in our country, and possibly it is adapted to even more uses than the famous white pine. . . . I look upon the exploitation of fir as being one of the best investments that can be made."

A week later, addressing a gathering of lumbermen in Portland, Long predicted, "Just as white pine has been supplanted by yellow pine, so will Douglas-fir supplant yellow pine." Even more important, Long foresaw that Pacific Northwest reforestation would "be carried on more successfully and in a more scientific manner than in any other place in the United States, first because it is nature's most highly favored woodlot, and second because the lumbermen of the Pacific Coast today have a fuller realization of the problems of true conservation, and of the necessity for reproduction, than in any other part of the United States."

Long's efforts to promote Douglas-fir were clearly motivated by the sales prospects in markets made accessible by the imminent opening of the Panama Canal. Eastern customers were largely uneducated in the advantages of Douglas-fir lumber. "However good a wood may be," Long asserted, "it can never become a great commercial success, unless there is enough of it to establish the fashion so to speak." Certainly there was enough

Douglas-fir, but at the moment it was "an imperative business necessity now to educate all that part of our country, and of foreign countries as well, as to the merit of the wood."

While the promotion of timberlands and lumber was important, Long was more immediately worried about income tax questions. F. E. told of plans for their attorneys to meet in St. Paul on January 22, 1914, at which time they would decide how to treat distributions received from lumber companies, and he suggested that "no returns be made until we can advise you as to their decision." The answer finally arrived on Saturday, January 31. The figures indicated that in 1913 there were no net profits, and that stockholder distributions for the year were "from capital assets and not from profits."

If taxes were a depressing subject, Snoqualmie Falls was anything but. There they were on the verge of moving beyond mere propositions. Long had employed Walter J. Ryan, a civil engineer, to begin topographical mapping of the sawmill-site vicinity. Minot Davis, now and then accompanied by Long, came to check the progress, but it was soon apparent that Ryan needed little supervision. It was just as well. Long had plenty to do in Tacoma, and Davis was soon on his way to Oregon to check the cruises on the Hopkins tract.

Long continued to be hopeful about adding to their Klamath County holdings. Plans to establish an Atlantic Coast yard, however, were effectively tabled. Long admitted as much to one of his New York City contacts, observing that trade conditions discouraged any early thoughts about broadening the market. As for the future, that depended on the length of the view. Long predicted that the movement of lumber into the Atlantic seaboard would be much slower than many thought, "but of course, it will come sooner or later." And when it came, Long wanted to be ready.

Thinking ahead came naturally to Long, whether it had to do with Atlantic Coast yards or with Everett's mill expansion. He shared his long-range optimism with Boner: "We have a great chance here to dream and scheme and plan and I confess that I am delighted with the job of doing it and I think that we are going to get something that will be very satisfactory to us all. . . . My notion is that there are a great many different things that we can do with our lumber that have not yet been attempted by any lumberman on the Coast and things which no man can attempt unless he has abundant facilities, and that is what is delighting me in our Barge Works Plant, i.e., that we have room to put in facilities that will meet almost any situation, and that is what I am perfectly willing to do in our construction."

Expansion of the Everett plant was only one subject Long discussed with Boner. Bill Peabody had arrived in February and began following company cruisers around, in Long's words, "getting acquainted with timber and woods work." Now Long thought that it was time for Peabody to gain some sawmill experience. He therefore inquired whether Boner could "use him some place around the plant so that he can get a little idea of the manufacture of lumber." Specifically, Long suggested that Peabody might "keep time and look after the receipt of the material there at our new mill, when we commence to build it." That is what Peabody did, an unusual opportunity for one new to the work, to see a mill from its inception. As Long subsequently assured Peabody, "I am inclined to think you can make your plans, if you see fit, to consider that you will be located at Everett for quite a little while." Peabody saw fit.

In the midst of Long's many concerns, disturbing news of a personal sort arrived, first in the press and then by means of telegrams and letters. Frederick Weyerhaeuser was seriously ill. Although he was nearly eighty years old, it was difficult for his friends and associates to imagine a world without him. To Charles Davis, his Saginaw, Michigan, friend, Long wrote, "It seems very useless to use words to express what we all hope regarding Mr. Weyerhaeuser, for of him we fully realize . . . there is none other like him, either for his bigness or goodness."

Frederick Weyerhaeuser died on April 4, and Long immediately headed to Rock Island, Illinois, to attend the funeral. Frederick had relinquished the day-to-day details of business to son F.E., but no one would ever replace his spirit.

In Washington state, the business Weyerhaeuser founded lived on, though generally at a slow pace. The situation was heating up, however, in the political arena. Allen told of a recent meeting of the so-called Conservation Congress in Chicago where, he claimed, his presence "was all that prevented divorcing forestry from it [the Conservation Congress] to be turned over to the American Forestry Association." He later provided details of "a determined effort among lumbermen, foresters and others in the East to get some live active organization to deal with forestry, tax and lumbering problems wherever this will have public influence." The American Forestry Association was looked upon by many as the best choice, but Allen wasn't convinced. "My contention with them," he wrote, "was that so long as the Congress can be kept alive it affords a clean unattackable medium for doing certain things that we cannot do so well acting as a closed corporation, although I certainly did not oppose all possible strengthening of the American Forestry Association."

Despite Allen's opposition at the Congress, the Forestry Association offered him $1,000 a year to serve as its western agent. He would accept, he advised Long, only if it would not interfere in any way with his present work and "provided they are willing to leave me wholly unrestricted as to what I should do." Immediately, Allen worried about the Oregon political situation, noting that it was very unsettled, and there seemed to be "a stampede from all political directions toward State economy and reduction of taxation which is pretty certain to bring about some attack on the State Forestry system and probably considerable disturbance in tax legislation next winter."

Long had also been watching the Oregon political scene, sharing some of his concerns with Hopkins. Long began with mention of the Hopkins timber, assuring that there was still strong interest to purchase even though he could not yet make a specific offer. In part, or so he claimed, "the times are such that hardly anybody feels justified in going ahead and making investments, or making very definite plans for future operation." Concerning their mutual interests in Klamath County, Long reported that Kimball was doing everything possible to get other timberland owners committed to protective measures, "whether from fire or beetles, and likewise has been earnest in his efforts to try and include the Klamath plateau in some scheme of grazing." At the moment, the sheepmen were not enthusiastic participants.

Such was not the case with a crowd that gathered outside the Timber Company office in mid-May to view an immense piece of bark that Lafe Heath had peeled off a Douglas-fir. Heath estimated the age of the tree to be 2,200 years. Long was impressed, admitting that "it certainly is the biggest thing I ever saw of the kind." They displayed it in the front window, where, in Long's words, "You would think the crowd standing around looking at it were reading the baseball news or else about some battle down in Mexico." While he was not sure just how Heath had arrived at the figure of 2,200 years, when others made similar inquiry, Long would simply say that Heath knows "all about it and it makes it so."

Meanwhile, Long kept current with Everett's expansion program. Progress was slow, but that was partly the result of continuing to manufacture in the midst of the construction. There couldn't have been much interference from the old sawmill; Boner reported that on May 21 they had sawed 301,890 feet of logs into 332,079 feet of lumber. May 21 was indeed a "banner day," but Long worried that their log inventory was insufficient: "I wish you had ten million feet instead of two million," he wrote, urging Boner "to buy all logs you can at $5, $8, and $11," and sell as little lumber as practical at present prices. As for the construction work, Long offered the usual advice: think big. He recommended, for example, that they build the "boiler house big enough to take care of any probable increase."

One reason Long was impatient to finish the construction was because he was eager to demonstrate to shareholders that reasonable profits could be obtained from manufacturing. He discussed the Weyerhaeuser situation openly with a fellow lumberman: "Our people broach, very reluctantly, the proposition of drifting into manufacturing. We bought an old mill at Everett ten or twelve years ago, and after spending five or six profitless years with it, and likewise spending a whole lot of money for rebuilding the plant, we finally wound up with a good deal better mill of course than we started in with and with a record where we have made some money for the last three or four years, but still we haven't a good mill, and the consensus of their advice to me has always been, that if we ever do start into manufacture, don't get an old plant, but build a new one that suits you." Further, the reasons for the increase in the Everett investment were because they had a first-rate manager in Boner and "a very excellent piece of ground."

Annual meeting time was again at hand. There would be one notable difference: Frederick Weyerhaeuser would not be present. Weyerhaeuser was eulogized but the meeting focused mainly on the future, not the past. Long began by observing, "Construction of a new mill at Everett and plans for another new mill at Snoqualmie Falls, indicate on the part of the management, a belief that the time is approaching when this company not only would be justified in engaging more largely in the manufacture of lumber, but that it may become a necessity for us to resort to this method of disposing of our property to the best advantage." To those who were listening closely, Long was announcing the inauguration of a new policy. Some had assumed that the Timber Company would remain forever that, a dealer, buyer, and seller of timberlands. Now, clearly, they had acquired more than could be managed along those lines.

Long described what had been the practice in western Washington and Oregon, there being two opportunities for profit between the stumpage owner and the retail lumberman, one accruing to the logger and the other to the manufacturer. He noted that "a great many mills draw their supply of logs by purchasing them from the logger, who is supposed to make a profit over what his stumpage costs him. If we can save some of the profit which the logger makes, and demonstrate that we know enough about manufacturing and marketing to make a manufacturer's profit, we certainly can make our stumpage yield us a great deal more by manufacturing and logging ourselves, than we can to sell the raw material to the millmen." It all seemed logical enough, on the surface anyway. "I do not hesitate to recommend," Long continued, "drifting into the business of manufacturing to the extent at least that we can make a profit over and above the average selling value of stumpage, that will be sufficient to take care of all of our fixed charges, in the way of taxes and miscellaneous expenses." He ended with a cautious but confident recommendation to proceed slowly "yet just as fast as we can perfect organizations that will certainly handle the business right."

Some who heard Long's remarks were less than enthusiastic. They knew that his proposal meant increasing the investment and deferring profits. But to the four Weyerhaeuser boys, all in attendance, such a vision seemed in keeping with their father's legacy.

Midsummer of 1914 was a busy time, even by Long's standards. His primary concern was the plant construction going on at Everett and Snoqualmie Falls. Although the appointment of W. W. Warren as Snoqualmie's manager had been politically astute because he represented the Grandin-Coast interests, Long worried that he was ill-equipped for the responsibility, knowing too little of western logging and lumber manufacture. Fortunately, Warren was the first to appreciate his own deficiencies. He thus visited operations and mills in the region, following an itinerary suggested by Long. And he received other advice from Tacoma: "One very important thing that every man has to consider, is the particular kind of timber that the millman is manufacturing and the method of marketing it. One man is frequently justified in doing what he does, because of his peculiar timber or particular market, and his scheme may not fit another. Then again, one man will reach conclusions by reasons of his own, when he buys logs, usually buying those which are best adapted to his market, and yet all his arguments and conclusions may not fit another situation where the man manufactures all the timber that stands on the ground."

As for the logging end of the game, Long maintained that the successful operator must possess two prime qualities. He must know the methods of logging and he must be able to select and place his men to the best advantage: "Then, if in addition to these qualities he has got a good deal of steam in him, you will get effective logging and economic logging."

Warren had commenced his study tour of western sawmills. As Long explained to J. P. Weyerhaeuser, successor to his father as president of the Weyerhaeuser Timber Company, he hoped that Warren would "become thoroughly acquainted with all types of mills and machinery" and also "thoroughly acquainted with logging conditions on the Coast before he makes any move at all." Warren was under no pressure to hurry, Long estimating in midsummer that the training would consume the balance of the year.

Organization of the Snoqualmie Falls Lumber Company was now completed; the deeds from the Weyerhaeuser Timber Company and the Grandin-Coast Lumber Company covering all of the timberlands merged into the Snoqualmie Falls Lumber Company were turned over to that company in exchange for certificates of the paid-up capital stock to the amount of $2 million. The initial division, allowing one share to each officer, was 8,269 shares to Grandin-Coast and 13,726 to Weyerhaeuser Timber. As Bill McCormick noted in his covering letter to Fisher, "All the objections to the title of both companies have been satisfied." Few transactions of such significance have ever been finalized so easily and pleasantly.

Although Snoqualmie was still a plant only on paper, Everett was not. A few of the shareholders naturally worried about the expenses of its expansion in the face of market and political uncertainties. Long tried to be reassuring, but the fact was that they had already passed the point of no return. He wrote in response to one inquiry: "If any of us had known what conditions are now, I think we would have hesitated to commence this structure, but everything is under way now to an extent where we cannot economically drop it."

The incredible days of summer 1914 passed, and as they did, a stunned Europe found itself at war. The only certainty was that this war would someday end, albeit far later and more sadly than any could foresee. Like so many others, Long tried to look beyond present difficulties. In a letter of October 6, he told Bell about progress at Everett, noting that the new fir mill would have a daily capacity of about 300,000 feet, and the planned-for combination cedar and hemlock mill would produce some 125,000 feet per day. The old mill, Mill A, cut an average of 250,000 feet; so, in all, the daily requirement for logs would be close to 700,000 feet. In recent years, the Everett mill had been purchasing most of its logs on the open market, but Long had about decided "that we should abandon this practice now and be cutting our own timber, at least to a very large extent."

One of the obvious sources for logs would be in the Snoqualmie River valley, where the Cherry Valley Timber Company was now operating. Although the Weyerhaeusers were heavily interested in Cherry Valley, the Thomas Irvine Lumber Company was also involved, and as Long explained the relationship, "There are times when the Weyerhaeuser Lumber Co. [Everett] buys freely of them and other times when they do not. We treat each company somewhat as a free lance and let each secure its best source for buying and selling."

Long thought that Weyerhaeuser needed to acquire all outside interests—the Cherry Valley Company and the Thomas Irvine Lumber Company and the O'Neil Timber Company—and thereby "have the Weyerhaeuser Timber Company own and operate all of the property, by reason of its advantageous location for delivery to the Weyerhaeuser mills at Everett."

Long envisioned a logging operation sufficient to meet the needs of Everett and efficient in terms of water transport from forest to mill. He put the matter simply: the Everett branch ought to "own the boom and likewise log their own timber, very largely at least." But what appeared logical from one perspective might seem less so from another. In this instance, Long foresaw problems from a familiar source. After informing Bell that he intended to try to acquire the Thomas Irvine interests, he admitted, "I have never tackled a harder game than that will prove to be, because they have been getting a large price for their stumpage by reason of this large operation"—namely the Weyerhaeuser business at Everett. The Irvines were sitting pretty.

Bell was less than enthusiastic about purchasing the Cherry Valley timber. First, he wasn't convinced that the timing was right. The Cherry Valley owners, he wrote to Long, "do not need the money and have a business framed up which is running satisfactorily and as profitably as any other; so having no reason to sell we do not see how they can be induced to sell except by a price which represents all there is in the timber." He then reminded Long, unnecessarily, "There is no reason to think that our friends and partners would be any more pliable than average men." In short, if the Weyerhaeuser Timber Company had money in hand, there had to be other investment opportunities "with twice the chance of a good margin"—opportunities that involved those who "feel pressure to sell under present and prospective conditions." Bell also expressed belated misgivings regarding the Everett expansion, noting that if they had known that all Europe was going to war, they might have put off that building. But he was quick to add that they must "have faith to believe that things will right themselves after a while and that the time to build a mill is before you need it." That, of course, wasn't intended to refer to Snoqualmie, Bell explaining, "I am quite sure we don't need that mill within the immediate future as conditions have now developed."

The basic question was this: Was the Weyerhaeuser Timber Company to be a dealer in timberlands or a manufacturer of forest products? Generally speaking, the Laird-Norton interests, which Bell represented, opposed deferring dividends for the sake of expansion, particularly in new undertakings. The Weyerhaeusers, however, were always willing to defer and usually willing to expand if such activity would enhance the original investment. It was the Weyerhaeuser path upon which Long strode so enthusiastically. He certainly agreed with Bell that it wouldn't be easy to purchase the Cherry Valley timber at a reasonable price, but he felt that the final results of the difficult negotiations would be worth it. "The proposed amalgamation offers another magnificent body of timber," Long explained, then added, "It was to round out a magnificently shaped proposition, as well as to get the economic conditions of operation down to a fine point, that tempts me strongly to entertain the idea of owning that property."

Long refused to be pessimistic when it came to arranging properties to the best possible advantage, ensuring an integrated basis for future units of production. But to the likes of Bell, he merely indicated that he would "gradually feel out Horace [Irvine] on the subject of making a sale, just to get his viewpoint, but of course making no overtures that would in any way commit us." Long then summarized what was obvious to everyone: "We are putting up a magnificent mill, and I suppose we will be criticized for spending too much money and making everything too good, but the more we reflected on it and the more we thought it over, we felt that that which we are doing was right." Weyerhaeuser managers for generations to come would recognize that philosophy and spirit.

But as the 1914 summer waned, turned to fall, even George Long was hard-pressed to think in future terms. There was plenty of reason to complain about the present. Conditions were so bad that Long called a halt to activities at Snoqualmie. Warren must have been disappointed, but he protested hardly at all. After noting that it would likely be a year between the start of construction and commencement of production, "and we would have the advantage of building during a period when material of all kinds as well as labor would be much cheaper," he admitted that it would be unfortunate to have a new plant that would remain idle for a year or so. George Long just didn't like to leap before he looked, and at the moment he could not see very far.

SEVENTEEN

Marking Time

THOUGH MINOR COMPARED WITH THE WARTIME SUF-fering of millions the world over, the economic hard times in America were painful. Work at Snoqualmie Falls was limited to clearing the sawmill site. Walter Ryan continued his survey, mapping out the system of logging railways that would eventually penetrate the timber stands. But there would be no construction for some time to come. Many blamed the continuing depression on the politicians, mainly the Democrats. One Montesano lumberman predicted that it would be "a thousand years before we have another Democratic president." But there were other factors. In Long's words, "The world at large is all upset and maybe it isn't altogether the fault of the Democrats that we are having dull times and a low price for lumber and logs."

The stagnation in the lumber industry was by no means limited to the Pacific Northwest. Indeed, F. E. Weyerhaeuser had recently observed from St. Paul that "the lumber market in this part of the Country is more demoralized than it has been for a great many years." That assessment merely prefaced F.E.'s purpose in writing. One George Thompson owned and operated a number of line yards west of the Twin Cities, and F.E. wondered whether they might consider a proposition for cooperation with Thompson. He wrote to Long in that connection, inviting him to bring Boner to St. Paul without delay: "It will be necessary for us to decide the question of line yard operations very soon, and an expression of opinion from you would be helpful." On Monday evening, February 1, Long and Boner boarded the eastbound Great Northern.

Before that departure, Long wrote at length on the subject in letters to Bell and Thatcher. He began typically, "What I don't know about the retail yard business would fill a book," but then proceeded to discuss the matter in great detail: "I am quite confident that our company will have to engage in the manufacture of lumber in a large way and this of course means the distribution of it. . . . Taking it as a fact that we are going to make a lot of lumber in the future, that we ought to prepare ourselves for its distribution in the broadest and most economical ways in that part of the territory which is best adapted for our lumber and to the use of large volumes of it." If that was the marketing future, sooner or later they would have to enter the marketing business. The pertinent question was whether an association with Thompson was the best opportunity available.

A majority viewed an involvement with Welles-Thompson Lumber Company of Minneapolis as most desirable. Welles-Thompson yards numbered about sixty, nearly all of them located in eastern North Dakota. The proposal assumed purchase of Welles-Thompson's real estate and stock by the Weyerhaeuser Timber Company, but there was as yet uncertainty as to the role George Thompson would play in the new organization.

Long instructed John Kendall, manager of Potlatch's retail operation, to determine as best he could the worth of every Welles-Thompson property. That need was obvious. But he also requested additional information: the desirability of the location; "what the community at large thinks of the company and their method of doing business"; and an assessment of the "character of the resident agents and managers." Long closed in the familiar manner: "It seems rather ridiculous to me to make these suggestions to you . . . but to repeat in a nutshell what my idea is, I would say, that aside from the question of passing on the value of the plants, I would like your impression of every other feature pertaining to the business, as you see it, that would be beneficial for us to have in view of our possibly becoming buyers of these yards." Kendall was advised to work quietly, avoiding any unnecessary publicity. And upon completing his survey, he was to "come direct to Tacoma and make your full report to me personally."

Long subsequently discussed the matter with F.E., noting that the negotiations had shifted from a plan to purchase stock in Thompson's company to one of purchasing the property itself. Certainly something had to be done to facilitate sales, and no other opportunities came readily to mind. The purchase would use up any cash surpluses that the Timber Company had accumulated. Long may have preferred to begin construction at Snoqualmie Falls, but there was already more lumber than they could market under the present arrangements, and the future looked bleak. Although Long could occasionally manage a hint of optimism, it was always qualified: "If they quit fighting in Europe, we will have another rift in the clouds and things may not be as bad as they look now." But, of course, they would not quit fighting in Europe.

And while they fought, Long took the time to conjecture about what lay ahead for the Weyerhaeuser Timber Company. He wondered, for F.E.'s benefit, where they would be in three, four, five, or six years, "or any time in the future when we engage even more extensively in the manufacture of lumber, as I fear we will have to." He puzzled over how they would market what they manufactured. "Should we not be in position to control quite a little territory by our own yards?" he inquired. That was the big question, not for this year or the next, but for all the years to come. And if that future demanded the operation of their own retail yards, then the question awaiting answer was, "Is Thompson a good man to look after the bunch of them for us?" Long had already decided: "I am somewhat of the belief that if we can see our way clear, without any financial embarrassment, to take on these yards and to continue our plans for building mills, that it will be best to try out the experiment of the retail yards, and if we do not like it, we probably can dispose of them without any considerable loss, and if we do like it and find it necessary, we have got started in the game just that much earlier." In short, they could profit from the experience of others at no great risk to themselves. That seemed to make sense at the time.

In April, Kendall finished his survey of the Welles-Thompson properties and reported his findings to Long. As expected, it was agreed that the Timber Company would purchase the controlling interest in what was to be called the Thompson Yards, Incorporated. All that remained was to complete the inventory. Long expressed the hope to Thompson, now president of Thompson Yards, that everything would "go along smoothly and to the mutual satisfaction of all parties and that the year 1915 may prove to be reasonably satisfactory in this venture which our people have taken on."

The year 1915 saw a change in the Weyerhaeuser annual meeting and report. Until then the Timber Company had operated on a fiscal year ending May 31; the new arrangements provided for the Weyerhaeuser year to be the calendar year, thereby allowing an earlier date for the annual meeting. Bell announced that he and some of the other shareholders planned to arrive in Seattle on April 27, and they were looking forward to seeing for themselves what Everett had done with all that money. Long forewarned Boner to "expect quite a delegation to be with you Wednesday morning, probably arriving about ten o'clock." Nothing further needed to be said. Hosting dignitaries was no new assignment for Boner.

The Everett visit and the annual meeting went smoothly. There was, of course, some disappointment about the current state of business, but such disappointment went far beyond Weyerhaeuser. Indeed, in relative terms, the Timber Company had done well. Much discussion centered on the Thompson Yards involvement, and it was early apparent that the approval would be hesitant at best. As Long subsequently informed Thompson, "Some of the stockholders of our company are not at all enthusiastic about this venture, and therefore, we who took an active part in bringing it about, feel no doubt a good deal as you do, that it is up to us to make good, so you can count on us to do anything that will help float the ship along in placid and prosperous waters."

Lots of things were happening in the spring of 1915. The new mill at Everett was nearing completion and Long certainly wanted to be on hand when the whistle blew and the saws began to whine. But he was also eager to make an Atlantic Coast tour in the company of Harry Hornby, mill manager of the Cloquet Lumber Company in Minnesota. He wrote to Hornby, inquiring whether they could delay their departure until June 1. Long could then "steal off three or four days" and attend his daughter Helen's college graduation in Washington, D.C., after which he could meet Hornby, "say at Tonawanda,

Carrie Robinson poses in her wedding gown for an Eau Claire, Wisconsin, photographer. Carrie was nearly thirty-five at the time of her June 6, 1889, marriage to George Long, but she appears much younger here. *Courtesy Mrs. Carol Feige*

George, Jr., looks a bit startled here in this October 1896 pose with his handsome father. *Courtesy Mrs. Carol Feige*

In Eau Claire in 1898, George, Jr., looking unusually demure. Although the picture is obviously posed, books, especially the classics, were truly an important factor in life at home. *Courtesy Mrs. Carol Feige*

An 1898 outing in Eau Claire with Carrie and young George, at the time when George, Sr., was wondering about his future in the lumber industry. A move seemed likely, but he could never have imagined the details. *Courtesy Mrs. Carol Feige*

George Long in a formal pose. This was taken in St. Paul, Minnesota, probably about the time he accepted the Weyerhaeuser position in 1900.

Father and son, snapped on what must have been a Tacoma summer afternoon in 1901. *Courtesy Mrs. Carol Feige*

George, Sr., with Junior and Rover, 1904. *Courtesy Mrs. Carol Feige*

The Bell-Nelson Lumber Company Mill in Everett, purchased by the Weyerhaeuser Timber Company in 1902, the company's initial manufacturing facility. As a mill it was unimpressive, but the site was magnificent, especially its deep-water dock.

The same Everett mill site, known as Mill A, after renovation, remodeling, and expansion. Circa 1925.

A 1906 view of the Maple Valley logging camp #4, one of the Weyerhaeuser Timber Company's early operations and suppliers to the Bell-Nelson mill in Everett.

Frederick Weyerhaeuser and friends on a 1906 tour of the Klamath Falls, Oregon, forest. Immediately to Frederick's right, again nearly in profile, stands George Long. Ed Hines is next, with F. S. Bell looking over his shoulder. The handsome chap to Bell's right is George Lindsay.

At first glance, this might be a study in early-century hat styles. But for our purposes it is lumbermen gathered at the 1909 Alaska–Yukon–Pacific Exposition. In this scale individuals are difficult to identify, but F. E. Weyerhaeuser is standing tall front and center. Up one row and a little to the right, George Long poses, cigar in his left hand. He seems almost to be staring at R. A. Long, the slight patrician figure left front in the black bowler. These two would be serious negotiators a decade or so hence. Others of interest include Frederick Weyerhaeuser near the top, a little left of center, one of few with a full white beard. To his left is Captain J. B. White of the Missouri Land & Lumber Exchange and O. W. Fisher of the Louisiana Long Leaf Lumber Company. The other three Weyerhaeuser sons are also present, along with son-in-law Samuel Sharpe Davis.

James J. Hill holding forth at the Alaska-Yukon-Pacific Exposition in 1909. Even without seeing the text, we can safely assume that he spoke of progress and development in which railroads played a leading role. Some in the audience had apparently heard the message before. *Special Collections Division, University of Washington Libraries, negative no. UW 3011*

This 1903 snapshot by friend and cruiser Lafe Heath catches George Long looking at still unfamiliar woods from a skid road, the cross logs notched so as to encourage a straight ride for the logs being skidded. The location is in the Cedar River Valley near Cedar Falls, King County, Washington.

A portion of the Long clan sitting on the porch of the new Prospect Hill home, 1910.
Left to right, George, Carrie, Aunt Nettie (Carrie's sister), nephew Fred Firmin,
Fred's sister Kate, and the Long daughters Helen and Margaret.

The beautiful Prospect Hill home in Tacoma, one of George Long's few indulgences. Circa 1910.

Lafe Heath, looking every inch the woodsman, has his picture taken by George Long. 1903.

Although this picture postdates George Long's era by a few years, the scene artfully depicts the initial task that befell him and his woodsmen—determining just what lands the Timber Company owned.

An early photo of the Northern Pacific Building, home to the Weyerhaeuser Timber Company office from 1900 to 1906. This Tacoma building still stands on Pacific Avenue and has been recently renovated, but the wing to the right that held the Weyerhaeuser office is no more. In the background across the waterway is the St. Paul & Tacoma Lumber Company's mill.

The Friend of the Forest by Frederick M. Spiegle. In 1910, George Long purchased this painting sight unseen, and thereafter it hung in his Tacoma office. Long was no connoisseur; in his mind the moral more than made up for any artistic deficiencies. *Courtesy George S. Long, III*

One of the Washington Forest Fire Association's pump engines in action. George Long, father to the organization, understood that such demonstrations were of little importance. What was crucial was preventing fires. As the photo suggests, accessibility was largely limited to logging railroads. It wasn't until the advent of truck roads that prompt and effective response was possible.

This 1917 photo pictures a giant western redcedar being falled in the traditional manner. The undercut, which controlled the direction of the fall, was made by axes alternately swung by the fallers. Then the task was completed with a crosscut saw, the men standing on spring boards so as to work above the swell—the lower flared portion of the tree, often defective and pitch-filled. Observing the action is Cutler Lewis, the first logging superintendent for the Snoqualmie Falls Lumber Company.

Weyerhaeuser directors and executives posing at Everett's Mill B, shortly after it opened in 1915. From the left: George S. Long, F. S. Bell, C. A. Weyerhaeuser, W. W. Warren, T. B. Davis, W. H. Boner, H. H. Irvine, F. H. Thatcher, P. M. Musser, W. H. Peabody, R. M. Weyerhaeuser, E. B. Wight, O. W. Fisher, C. R. Musser, and Sumner T. McKnight.

Looking north on A Street in Tacoma at the time of completion of the Tacoma Building. Note work continuing on the 11th Street bridge (bottom right). At the other end of the block—separated from the old turreted Tacoma Hotel by the totem pole—is the building that housed the Weyerhaeuser office from 1906 to 1912. The Tacoma Building would remain home to the Weyerhaeuser Company until 1971, when the new Corporate Headquarters facility opened in nearby Federal Way, although the Tacoma Building continues to house company offices. The building cost a total of $600,000, including property. *Courtesy Washington State Historical Society*

In 1918, looking up the hillside from the Snoqualmie Falls Lumber Company mill site at the community buildings. The boarding house/hotel occupies the center; the company store is at the lower right; the hospital is outlined on the ridge; and the YMCA/Community Center is to the left. Note the "Safety First" reminder at the top of the stairs just this side of the train depot.

Life in company towns didn't always have to be dreary. George S. Long did his best to combat the tedium. At Snoqualmie Falls, the community center provided social opportunities quite like those available elsewhere, such as this 1920s dance. For a time at least, these couples didn't feel apart from society's mainstream.

And if you didn't dance, perhaps you played baseball. This team represented the Snoqualmie Falls Planing Mill in 1924. We don't know their record, but we can see that they had their fans.

The logging camps were usually at or near the scene of operations. This one belonged to the Snoqualmie Falls Lumber Company. The whole affair was portable, moving via the logging railroads to wherever, somewhat primitive but also neat and efficient. Circa 1920.

George Long visiting a Snoqualmie Falls logging camp with W. W. Warren, the large man with the flat hat, circa 1920.

Six of the seven Weyerhaeuser second-generation men on a visit to Snoqualmie Falls, circa 1920. From left: Professor Richard Jewett, husband of Margaret; Charles A. Weyerhaeuser; Reverend William Bancroft Hill, husband of Elise; Frederick E. Weyerhaeuser; W. W. Warren, Snoqualmie Falls manager; John P. Weyerhaeuser; and Rudolph M. Weyerhaeuser. Only Samuel Sharpe Davis, husband of Apollonia, is missing.

In July 1921, even this experienced group must have been impressed with the size of the log. George Long, on the left in the second row, is leaning against the giant, and George, Jr., is at far right, bottom row; next to him is Rod Titcomb, soon to be manager at Snoqualmie Falls and subsequently George Long's successor in Tacoma.

A dramatic photograph of E. T. Allen, George Long's hand-picked literary spokesman, a forester who could communicate and did so effectively for the Western Forestry and Conservation Association for many years. Circa 1925.

The S.S. *Hanley* off-loading timbers at the Baltimore port, circa 1925.

The Long-Bell Lumber Company and Weyerhaeuser negotiated frequently through the years. This Longview get-together took place on May 27, 1924, at which time Weyerhaeuser's Long was considering possible mill sites. Standing left to right: H. H. Rock (a Longview banker), Sumner T. McKnight, Wesley Vandercook (Long-Bell), Bill Ryder (Long-Bell), and Rudolph M. Weyerhaeuser. Seated on a pile of Long-Bell lumber are F. E. Weyerhaeuser, F. S. Bell, Dr. E. P. Clapp, Horace H. Irvine, Charles A. Weyerhaeuser, George S. Long, William Carson, C. R. Musser, and S. M. Morris (Long-Bell), with J. P. Weyerhaeuser in front.

To the casual observer, piles of lumber are only interesting by reason of size. This big one, featuring the loading elevator, is at the Baltimore Yard, the first of Weyerhaeuser's Atlantic terminals.

Colonel James Long receives an adjustment from Bill Peabody aboard the S.S. *Hanley*, one of two ships purchased by the Weyerhaeuser Steamship Company in 1923. Although brothers, the colonel and George were, cigars aside, about as different as could be.

Although his expression might suggest otherwise, Long enjoyed nothing more than showing visitors the Timber Company at work. Here he and his friends take a break. Charlie Weyerhaeuser is peering over George's back, and the Reverend Bancroft Hill is in the center. Hill was married to Elise, one of Frederick Weyerhaeuser's daughters. He preached and taught at Vassar College, but enjoyed western junkets.

Watching over the Longview construction phase in 1928 are (pointing) Harry Morgan, Sr., and Al Raught, Jr. The picture is posed, but their authority and ability were unquestioned.

More of Longview construction in 1928. The power house is in the foreground and Mills 1 and 2 behind. The distinctive natural feature in the background is Coffin Rock, eventually to be leveled.

Interior construction of the planing mill at Longview in 1928. Note the size of the timbers.

Klamath Falls was different from the fir mills in more respects than species. Especially in the winter, it had a forlorn and isolated look. But its forest impressed George Long like no other.

Fred Firmin in profile along with Jackson Kimball, George Long's tough lieutenant in Klamath Falls for so many years. Circa 1925.

George Long looking straight at the camera for a change. This was snapped at Everett, circa 1926.

Some key personnel outside the Everett mill office. Flanking the dapper mill engineer/ designer Al Onstad are Bill Peabody, Everett manager, and Harry Payzant, mad designer of machines.

Inside Mill B, Everett manager Bill Peabody is supposedly pointing out inch guidelines on one of the 4-Square machines to Colonel W. B. Greeley, then secretary of the West Coast Lumbermen's Association. Greeley is trying his best to look interested. Circa 1929.

F. E. Weyerhaeuser inspecting a load of Douglas-fir common upon its arrival in Minneapolis–St. Paul. Although he and George Long agreed on nearly every important issue, when it came to the marking of lumber their opinions often differed, and that was particularly true concerning the 4-Square advertising program. Circa 1929. *Courtesy Minnesota Historical Society*

George Long relaxing on Five Mile Lake, a favorite spot near today's Corporate Headquarters in Federal Way. Circa 1925. *Courtesy Mrs. Carol Feige*

LUMBER SPECIFICATIONS
can now be WRITTEN with CONFIDENCE

Species

and

Grade

L UMBER now takes its place among materials which can be specified with the certainty that there will be no substitution of inferior quality.

It is 4-Square Lumber—*packaged* lumber—*finest* quality lumber.

Every package is labeled and marked with the *species* and *grade* it contains. It is delivered from the mill—to the dealer—to the job in *original* packages for *identification* and *protection*.

It is *more* than packaged lumber. It is *guaranteed* lumber—the finest money can buy. Every operation from tree to job is controlled by precision standards. It assures finished results which measure up to your requirements.

Every piece of 4-Square Lumber is cut to *exact* length and *trimmed square at both ends*. Thus time is saved on the job—an added advantage.

Progressive lumber dealers now have—or can get —the items listed on this page to fill your specifications for 4-Square Lumber.

Finishing Lumber

Bevel Siding · Drop Siding

Colonial Siding

Softwood Flooring

Ceiling and Partition

Shelving · Stepping

Casing · Base · Mouldings

WEYERHAEUSER FOREST PRODUCTS
ST. PAUL, MINNESOTA
Weyerhaeuser Sales Co., *Distributors*, Spokane, Washington
District Offices: Minneapolis, Kansas City, Chicago, Toledo,
Pittsburgh, Philadelphia, New York

4 SQUARE LUMBER
Species and Grade are Marked and Guaranteed
TRIMMED SQUARE ·· PACKAGED ·· READY TO USE ·· GUARANTEED

The sort of 1929 advertisement that bothered George Long. Despite his opposition, 4-Square Lumber became a mainstay of Weyerhaeuser Sales Company campaigns. Whether it was a success in a promotional sense or simply an added expense was continually debated. It died a natural death in the post–World War II era.

Another of George Long's office poses. In truth, he was about as uncomfortable using the phone as he appears. Important personnel decisions and the onset of the Great Depression probably added to his worried expression in this 1929 photo.

Grandfather, father, and two-year-old—
three generations of George S. Long at the Prospect Hill home, 1929.

commencing our observations there," and then go over to the coast "in the general all around scheme of investigating conditions in the East for the marketing of fir, etc." Hornby promised to be at Long's beck and call.

In the meantime, F.E. closed the deal with George Thompson, advancing $100,000 from the F. Weyerhaeuser Company (Minnesota), a family holding company, with the understanding that the Weyerhaeuser Timber Company would repay that amount as soon as possible. There was always the possibility for confusion in such instances, operating as they did out of two or more offices. F.E. appreciated the danger: "I am very fearful of crossing wires in a way that may prove disagreeable to you, and I trust in such an emergency you will be frank enough to tell me so that it will not be repeated." Matters were further complicated by the current organizational arrangements. As Long explained, "We look upon the Tacoma office as the headquarter office of the Weyerhaeuser Lumber Company, and we treat the Weyerhaeuser mills at Everett as a branch . . . so that all of the investments pertaining to the Thompson Yards will be handled through the books of the company in Tacoma."

The new mill was in operation, experiencing the expected difficulties. Long was trying to help procure the necessary raw material. To Boner he advised, "It looks a little as though it is going to be a good time to make some log buys." They really had no choice.

There remained a few who felt that Long was exceeding his authority in terms of the manufacturing effort. In fact, they had never been convinced that the Weyerhaeuser Timber Company should be anything other than a dealer in timberlands. The timber had been the investment opportunity, and the sooner that investment was realized, the better. The most expedient way to obtain a return was to sell the timber, not to invest in manufacturing plants. Although Long was acutely aware of this attitude, he hated to see the timber resource wasted, arguing that proper utilization demanded efficient manufacturing facilities. Snoqualmie Falls would be an integrated model of Long's vision, early to be recognized as such.

Long left for the East on May 24, going initially to Washington, D.C., as planned. He spent the entire month of June away from his desk, stopping in St. Paul on the return trip, primarily to meet with Thompson. In addition to Hornby, Long was accompanied at times by Rudolph and J. P. Weyerhaeuser, the latter's son Frederick K., and F. H. Thatcher. While Long found the tour instructive, he realized it was only an introduction, "that the subject [of Atlantic Coast terminals] was so big it would require still further investigation and a great deal of thought before any definite policy was outlined." As for selling Douglas-fir lumber in eastern markets, he admitted that would be "a good-sized problem." Still, once the product became known and accepted, business would likely "come with a rush, and it will be the old question of the one having the best facilities who will do the most business."

Long was happy to return home in July, even though his absence of nearly six weeks meant that he was "buried up with work." He was eager to visit the new mill at Everett and also to discuss Snoqualmie plans with Warren. The manager doubtless understood no major decisions were likely, for Long had sent him an unusually pessimistic assessment: "My observation of the lumber situation is that it has never been as it is now on the Pacific Coast and that the demoralization is just about as bad elsewhere."

The war, though still distant, was having its effects, some of which were a bit unusual. Americans wanted to help the sufferers, and at the moment the most popular of the war's victims were the Belgians. When in Washington, D.C., Long encountered a few politicians who were proposing the resettlement of displaced Belgians on cutover lands. Long did his best to discourage such plans, later suggesting that what many "wanted to do was to separate the Belgian from what little money he has." His own position regarding settlers and the cutover was unchanging: "Whenever the settler arrives and takes up this land, he must do so on a basis that would make him a good profit." And west of the Cascades, the chances of that happening were slim at best.

Although the new mill at Everett was up and running, there was always room for improvement. Boner suggested that they would benefit from another resawing facility. Long vetoed that plan, saying they should wait "until we really know just what is the best thing to do," He thought that perhaps Boner was paying too much attention to the plant and not enough to those who ran the machines. As Long reminded, success depended on having good lumbermen on the job. He then offered a general observation: "I am not so much stuck on making a whole lot of lumber as I am on making it right, grading it right, shipping it right, and doing everything right after the log enters the mill for that is what counts and if we do not reach the full opportunities that can be developed for us, we are losing the most advantageous benefit of a brand new up-to-date plant." Boner got the message. He would be quiet for a time about further improvements.

At Snoqualmie they could still plan for whatever would provide the maximum advantage. Long urged Warren to use the time afforded by the delay in construction to collect information and to think about "exactly what kind of a mill we should put up." He then stated his belief "that the average mill man out here has not yet commenced to thoroughly understand the very best way to handle his good timber." Flooring had been the principal product from the higher quality logs, but Long doubted that the market could absorb all of the flooring being produced. While he wasn't certain in his own mind about much that went on in the mills, that didn't stop him from raising concerns. He was convinced of one thing: "This habit which has grown on the mill men of recent years to put in the resaws is a habit stimulated more by the desire to get increased cut than anything else." That may be all right in the case of common lumber, "but in good lumber the main point is to have it cut right." He wanted Warren to accompany him someday on a tour of the Everett mill to explore "the idea of having the mill adapted to quality cutting in preference to quantity cutting. Possibly a compromise between these two extremes is really what we need."

Although everything was in abeyance, Long found it impossible to keep from worrying over the Snoqualmie mill construction plans, in part because it required such critical decisions, but also because he had no practical engineering experience. He had long recognized this as a serious deficiency, and in fact thought for a time that George, Jr., might acquire those skills he lacked. With that specifically in mind, the previous summer Long had inquired of an old Eau Claire friend, John S. Owen, whose sons had attended Cornell University, as to the "practical results" of their education. Long left no doubt about what he had in mind: "I think my boy would want to take a course that would fit him for the practical end of lumbering, more or less, like mechanical and electrical engineering . . . as well as to take a liberal allotment of classics and literature generally." George, Jr., entered Cornell in the fall of 1915, but the war would interrupt and he would never return to school. Through the years he would be of large help to his father, but it would be more as a chauffeur, errand runner, companion, and confidant. Professional assistance would have to come from other sources.

At the moment, Long had a professional engineer in mind, Al Onstad, the individual who had assisted with the Everett mill construction. Onstad had been employed by Pracna for the past year and a half, but with the business conditions as they were, new mill construction had ceased, and Pracna, on the verge of letting Onstad go, inquired whether there was any possibility of Weyerhaeuser hiring him as a consulting engineer. Long was interested. As he explained to F.E., "We will need someone out here before a great while possibly in connection with the Snoqualmie Falls plant . . . so that I am a little bit inclined to think it might well be enough to retain Mr. Onstad, if we can do so on a reasonable basis."

F.E., though he didn't disagree, was feeling a bit testy for several reasons, chief among which was what he perceived to be a lack of overall direction to the many Weyerhaeuser investments. It was difficult if not impossible for F.E. to stay informed on matters generally, but still he tried, and central to his effort was getting all operations to follow uniform procedures in record keeping. He dispatched his accountant, Frank B. Poole, to oversee, and Poole's annual inspections were at best tolerated.

It so happened that the Everett office had recently chosen to ignore some of Poole's recommendations, and F.E. wanted an immediate explanation. "You appreciate of course," he wrote to Long, "that we are endeavoring to unify our books so far as possible, and if any good reason exists for discarding any books furnished by this office it appears to me that Mr. Poole should have been given an opportunity at least to explain the reasons why they were installed." The facts were, as seen from St. Paul, that unless Poole had some authority, "his usefulness as an auditor would be very seriously impaired." Clearly some corrective action was expected.

Long understood what was difficult for F.E. to understand—that individual operators kept their books in a way that seemed most useful to them—and he tried to make the point, politely, of course. "Their explanation," meaning Everett's, "was that by their new books they were able to accomplish the same results with a great deal less duplication of work and they think also without being quite so confusing." But Long realized his justifications would not satisfy. He assured F.E. that in the future whenever they disagreed with one of Poole's suggestions, they would "take it up with him direct and fully and carefully before trying the innovation of substituting something else and I think we will have no repetition of an instance of this kind in the future."

F.E., however, wasn't about to let the matter drop without some assurance that Long treated it seriously. While he might agree that there were "a thousand ways of accomplishing the same results in accounting," he also wanted to make clear that Poole's suggestions were anything but whimsical. In fact, they had "made a very careful study of the whole question of bookkeeping," and the methods agreed upon had been determined "very largely by Mr. Hugo Schlenk of the Northern Lumber Company, Mr. C. A. Barton then of the Northland Pine Company, Mr. Charles Esplin and Mr. Poole." These four experienced men had concluded that there were important benefits to be gained by having "practically the same system of accounting at each plant."

Actually, accounting procedure was only a fraction of a larger, more general concern that had been troubling F.E. for a long time, brought to a head by the passing of his father. How does one achieve some overall advantage from the common ownership and management of separate operations? How do you keep separate but similar operations from competing with each other? These crucial questions had no easy answers. To F.E., the resolution seemed to lie in the direction of consolidation: "We have not arrived at any very definite conclusion, but I am very sure that in an organization so large as ours is growing to be we must have men who are particularly gifted to outline the general policy for all of our companies in their different departments."

It seemed a logical enough approach, at least from F.E.'s vantage. Long wasn't so sure. The effort to consolidate would constitute much of the Weyerhaeuser story in the years to come, but for every F.E. there were others who saw larger benefits to be derived from autonomy and decentralization. Anyway, St. Paul didn't always know best, and intramural competition wasn't always a bad thing.

George Long had also been thinking about sales. On the last day of September, he assessed the situation as he saw it. "The great big question," he wrote to Boner, "is not much different from what it has been in the past, i.e., the proper marketing of lumber, but it has more stress these times than usual and we cannot give too much thought and we cannot make plans on too broad a scale."

Boner had scheduled a trip east to discuss the sales situation with Thompson. Long now wrote to Thompson, in part to prepare him for Boner's visit. Thompson had recently complained that some of his orders had not been promptly filled, and Long facetiously suggested that Thompson quote Boner's favorite expression back to him: " 'What you want, when you want it, all the time'—and tell him to live up to his maxim." On a more serious note, Long outlined the difficulties confronting wholesalers and retailers, and the problems they had in understanding one another. In this instance, however, Long was confident that Thompson would be able to "line Mr. Boner up where he will work with you very nicely in that direction and I am equally sure that you are going to do everything you can to help Mr. Boner out." The relationship never proved to be quite this happy, but Long was operating in character, appealing to and expecting the best possible.

A major event on the agenda in the fall of 1915 was a joint session of the Western Forestry and Conservation Association and the Pacific Logging Congress in San Francisco from October 19 through 21. As it turned out, Long was able to attend only a portion of the meeting, much to Allen's dismay. Still, he participated behind the scenes, reading and commenting on various papers, and he thoroughly enjoyed San Francisco. He had taken his daughters with him and spent considerable time "with them on the Fair Grounds and at other places which they had never seen." He later described it as being "the finest trip we ever had anywhere."

Maybe things were just looking up generally. The company was successful in several tax suits. In Pierce County, Washington, the timber taxes were reduced by nearly $12,000. Similar results had been obtained in Oregon's Clackamas and Klamath counties. Court actions had drawbacks, of course, as neither Long nor the Weyerhaeusers ever wanted to bring unnecessary attention to themselves and their operations. At the same time, however, Long was increasingly convinced that they needed to create " a healthy sentiment in the direction that we won't stand for too much abuse." But the major cause for optimism was an improvement in the business climate.

Just as Long was leaving San Francisco, Captain White had pulled him aside to report on recent news from his southern mills. It seemed that "they were swamped with orders," and were getting $1.50 per thousand more for lumber than a month previous. As Long indicated to Boner, "It begins to look like a boom." His advice, doubtless unnecessary, was simply to let their lumber dry, "so that we will have something to give the trade in their spring buying that will bring us more money at the mill and more money in underweights." It was the first encouraging word in many months.

Others had been studying the industry and Weyerhaeuser, among them Austin Cary of the Forest Service. Allen found Cary's draft of a report on Pacific Northwest lumbering interesting, noting that it included a good deal about the Weyerhaeuser Timber Company, mostly very favorable. "I don't know how correct his figures are," said Allen, "but certainly no one can complain of his conclusion." Allen quoted Cary as saying that the Timber Company was "competent and strong, yet never greedy or unfair," and added that Cary "holds you up as an example of the large scale unit at its best as an influence for good."

Cary's work was destined for inclusion in a joint report of the Federal Trade Commission and the Forest Service on the lumber industry. He continued to share his efforts with Allen, and Allen continued to try to be helpful. As he wrote to Long, Judge Flewelling, president of Western Forestry and Conservation, "thinks this forth-coming Government report is a unique opportunity for good or evil, such as may never be repeated and which is coming at a most timely moment therefore that it is our duty to be on the job to do our best, be this much or little." Allen proposed a trip east, though he knew that it would be expensive and that Long might have reservations, fiscal and otherwise. Allen suggested that he might be able to arrange to participate in the Federal Trade Commission–Forest Service business as well as hearings on the Weeks law, also scheduled at about the same time. The Weeks Act of 1911 was a basic conservation measure, authorizing federal purchase of lands protecting the headwaters of navigable streams and extending state authority for programs aimed at conserving timber and water resources. It also provided for $10,000 a year in matching funds to states whose conservation programs were deemed worthy of support. These provisions would eventually be broadened by the Clarke–McNary Act of 1924, and the hearings Allen notes marked the early stage of the revision process.

Long had recently met with Cary, after which he concluded that the task was more than the young man could manage, at least without a fair amount of assistance. He was quick to approve Allen's eastern trip. Long realized that his lieutenant had the rare ability to combine "logical thought, full grasp of the essentials and a high art of feeding to those who are to digest it, the food in such a way that it will be palatable and every course pronounced superior to the one previously partaken of." Long knew what he liked to read, and he thoroughly enjoyed reading what Allen wrote. That's what brought them together in the first place.

Change was in the wind at Snoqualmie. Warren had been negotiating with Pracna to design and oversee the construction of the sawmill, but there was a snag. Pracna's proposal included a monthly salary of $1,400 for a year and a half. Long responded that if he agreed to such terms, "it will be the most astonishing thing that I have ever O. K.'ed." Long hadn't felt quite right about taking engineer Onstad away from Pracna, but he now decided otherwise. After Onstad phoned Long on the morning of December 14, the two got together that same afternoon. As Long subsequently reported to Warren, "[We] have practically made arrangements with him to enter into the employment of the Weyerhaeuser Timber Company for the year 1916." Long, Warren, and Onstad met the following day to discuss plans for Snoqualmie Falls.

EIGHTEEN

A Costly Boom

LONG CONTINUED TO HAVE HIS DOUBTS ABOUT GEORGE Thompson, but with the establishment of Thompson Yards, Inc., there was no turning back. Weyerhaeuser and Thompson were bound together, for better or worse.

Thompson was something of a high roller who had large expectations and limited business acumen. He enjoyed seeing the flags that dotted his map, each one indicating another outlet of Thompson Yards, Inc. And regardless of the circumstances, he was incessantly thinking about adding more flags. Long soon understood that Thompson required careful supervision, but he wasn't yet sure how much.

The Weyerhaeuser notion of desirable expansion was always based on a presumed ability to compete—that is, to make a profit. Heretofore, the so-called natural market for Douglas-fir lumber had been in the Dakotas, farther removed as they were "from [southern] yellow pine and not near enough to the few white pine mills to expect any opposition." The time had come for Pacific Northwest producers to consider new areas, specifically northern Iowa and southern Minnesota. But first they must do their homework, so as to be able to move purposefully when conditions were right. And they should not expect immediate success. Long accordingly advised, "I would not object to breaking into advantageous points at any time, even if yards so located would have to do some missionary work in pushing fir."

F. E. Weyerhaeuser was "convinced that Mr. Thompson's method of making lumber sales at retail is the only proper system to follow." Thus he supported Thompson's plans to expand operations into the Twin Cities area. Initially he could take over the retail department of the Northland Pine Company, perhaps to include the Midway Yard, the large wholesale plant located between Minneapolis and St. Paul. F. E. anticipated that the Weyerhaeuser Lumber Company, which was to say George Long, might be hesitant about "making a large investment in a channel which would market largely if not exclusively, northern pine products for some years to come." At the same time, he reminded Long that they were "carrying this investment very largely for the future benefit of the Weyerhaeuser Lumber Company." Regardless, he hoped that Long would give the matter some thought and that he would attend the annual meeting of the Thompson Yards, Inc.

Meanwhile, Long tried to keep up with the progress of the Trade Commission's hearings in Washington, D.C. The general manager of the Bridal Veil Lumbering Company, E. B. Hazen, had attended, and willingly shared his reactions. One frustration that surfaced many times involved the price of stumpage. Hazen complained that every time lumbermen talked about "losing money in operating, etc., the question always came up by the Commission (it seemed to me sometimes in a sort of cynical tone of voice), 'At what are you figuring your stumpage?' " Hazen tried his best to provide an answer, for it "seemed to be uppermost in their minds and of course we all know that this is the meat of the whole thing."

Hazen's method of calculating stumpage prices was hardly unusual. "The value of the trees (or the stumpage, as we call it) is simply the difference between the cost of producing these manufactured products and the return realized by the manufacturer from the sale of them." In short, stumpage was a residual, computed by figuring backward. For example, if the product cost $12 per thousand feet to manufacture and it sold for $15, the $3 difference would be the stumpage figure. In recent years, however, some West Coast operators had been selling below cost, so stumpage actually had a negative value; it was simply being "given away." Many operators, however, thought first in terms of cash flow, and too many never got beyond that.

It proved difficult for those on the Commission to see things as the lumbermen saw them. Whatever value one might place on vision and courage, those initial investors got quite a bargain, buying lands at between $1.25 and $2.50 an acre, even though "they assumed risk, taxation and other burdens in taking them over." Hazen argued that in almost every instance, the first owner, "this pioneer," had long since sold out, "collected his subsidy." And it was "the second owner [who] solidified tracts to hold against the day when there would be a demand and consequent increased value:

> Most of the privately owned stumpage is today in the hands of those who paid this premium—this earned increment to the pioneer. . . . These operating owners are the captains of industry in the second largest producing business in this country. These men have legitimate investments which must be given consideration in the determination of the cost of lumber. These investments are not excessively speculative. They do not represent inflated values. They merely include the subsidy earned by and paid to the first owner, plus the natural, added value coming to all property when the country holding it emerges from a wilderness into civilization.

The confusion over stumpage costs or prices would continue. But as Long contended: "The mill, as a mill, should show a profit, and the logging operation as a logging operation should show a profit sufficient at least to take care of the capital invested and the wear and tear of the machinery, and let stumpage have what is left." The problem was basic—the location of stumpage, "situated so remote from the people who use it."

One question was resolved. Long hired engineer Onstad at an annual salary of $4,000, "to give such services as we may call on . . . in connection with mill designing, mill development or any other class of technical work . . . that may pertain to the business of the Weyerhaeuser Timber Company, or any of its associated or allied interests." At the moment, Onstad was about to begin work for the Snoqualmie Falls Lumber Company.

Long left for St. Paul in mid-January 1916, to attend the annual meeting of Thompson Yards, Inc. Not everything went smoothly. Indeed, Long returned to Tacoma convinced that George Thompson should not get involved in Twin Cities retailing, and said as much in a letter to Charlie Weyerhaeuser: "Mr.

Thompson has problems enough of his own in his outlying country yards . . . without mixing him up, at the present time, with the detail and responsibility of looking after the local distributing in Minneapolis and St. Paul." Clearly, Long was beginning to question Thompson's competence.

There was additional cause for worry. Thompson had gone ahead and purchased some more yards, apparently with F.E.'s approval. Since it was after the fact, Long had to be careful how he complained, but he was more than a little annoyed by what seemed an unnecessarily hasty decision. They had employed John Kendall for the precise purpose of "passing upon and investigating any location where any of us might desire to buy yards," but Thompson had ignored Kendall in this instance. The purchases may have been wise or otherwise, but one thing was clear: Thompson had not followed procedure. F.E. could not argue that point.

When Long arrived home following his St. Paul trip, he found western Washington in the midst of an old-fashioned Wisconsin winter, but with one important difference: Washington logging wasn't made easier by the snowfall. On the contrary, woods operations came to a halt and mills were forced to shut down for lack of logs. By the end of January, Boner had already closed the old mill at Everett and was about to do the same with Mill B. And as the snow continued, prices began to rise, $1 on both logs and lumber. It was more than a little frustrating.

Long, as usual, counseled patience. The enforced slowdown might be just as well, encouraging other operators to do the same thing, and also to test the waters before raising prices too high. But prices had been demoralized for such a long time that the temptation to raise them was nearly impossible to resist. Warren had recently returned from a tour of his old southern territory, a return delayed by several days, the train being snowbound in the Cascades. He reported all of the yellow-pine lumbermen as being "very optimistic, with one exception." That one had published a price list so much higher than anyone else that he was not getting any business.

In mid-February, Long proclaimed winter finally over, and as for business, it continued to improve, slowly. Although they had advanced prices $2 to $3 per thousand on an average, they were still not enjoying "the full taste of the prosperity that has come to the southern lumbermen." One of the reasons was that the export trade had gone sour, Long assigning blame to "the very high rates that ships can command."

Long couldn't do a thing about shipping rates, but he resolved to do what he could about Thompson's hyperactivity in acquiring retail yards. This time, Long did not hesitate in sharing his concerns with F.E.: "I do not want to see him grow so fast that he cannot give to each particular part of his business all of the attention that it requires to make it a success." He wrote much the same directly to Thompson: "I am perfectly willing to grow and grow fast, if we are quite sure that we do not get overloaded." He concluded his lecture with the admonition that "this is an opportune time to sell a yard, and possibly a better time to sell than it is to buy."

Long might have saved the paper; Thompson was on his way to Tacoma. The primary purpose for the visit was to propose a grand scheme to purchase some fifty retail operations in the Dakotas, the so-called Rogers Yards. Thompson found Long less than enthusiastic. Long later described the meeting in great detail for F.E. Perhaps what bothered him most was that Thompson clearly had not done his homework. He had no real strategic understanding of the situation, no appreciation for the diplomacy required when dealing with four different railways. But the decisive difference had to do with geography; Long contended that they ought to be focusing on points farther south, "points on the Northern Pacific and on the Milwaukee in South Dakota, and as fast as the opportunity presents itself into southern Minnesota." Thus, Long opposed purchase of the Rogers Yards. F.E. immediately sent a letter to Thompson, instructing him "to advise Mr. Rogers that we do not care to purchase his string of yards at the present time."

Although Long successfully squelched this scheme, he didn't squelch the scheming. Back in St. Paul, plans were going forward to make "an earnest attempt to hereafter distribute the products of these [Weyerhaeuser-affiliated] mills along lines of higher efficiency and more careful economy than in the past." Someone had proposed a name for the new organization, the Weyerhaeuser Sales Agency and Weyerhaeuser Products. Harold J. Richardson, representing the Laird–Norton interests, sought

Long's opinion: "We will be very glad if you will kindly write, offering any suggestions or modifications which may seem to fit." Long's reply was tepid, but he had no objections, at least none that he cared to offer.

Long was also disappointed with the slow improvement of the lumber business. In early March he had written to a Louisiana friend, "We are all looking to the yellow pine people as the barometer of the lumber market, for we feel sure that if they get going in good shape, we will trail along and do the best we can." It would have been easier to be encouraged, of course, had it not been for the war, which dragged on, destroying so many and so much. Even if some seemed to reap benefits, there was little if any joy in the results. Furthermore, as Long observed, the "chaotic condition of everything" made safe predictions impossible. Like businessmen generally, he longed for some stability.

Still, even George Long had to admit that despite all of the reasons for concern and worry, the lumber business was good and getting better, and it would soon be reflected in the market for timber just as surely as in lumber sales. He also allowed that, for the moment anyway, President Wilson seemed to be steering the ship of state satisfactorily. Although he would have preferred a Republican in the White House, he admitted that he "never believed that the Democratic Party had as strong a national character in it as the President has demonstrated."

Though the purchase and sale of timberlands attracted the most attention, the other side of the coin, logged-off lands, continued to be of interest. Forest Service friend Austin Cary had recently inquired whether large property owners such as the Weyerhaeuser Timber Company "might be willing to turn this property over to the government or state *gratis*." Long contended that there was a more basic consideration. "In this matter," he explained, "I have always entertained the idea that in Western Washington and in many parts of Western Oregon, the logged off area, to a very large extent, would ultimately be found to be better fitted for reforestation than for agricultural or grazing purposes." If reforestation were indeed the objective, Long thought that property owners "would be inclined to make a very low price on the land," even if they were not quite ready to deed it over *gratis*. What encouraged this approach was taxes, pure and simple: If private owners could not afford to reforest, then perhaps the government, free of the burden of taxes, could. The possibility of private reforestation still seemed remote. "Again, it might be entirely possible for the state to enact such laws that would enable individuals to do this, and I am sure if we had such laws in our own state, our company would give very careful attention to reforestation, but today it is a financial impossibility." And that was that.

Allen had spent most of March in Washington, D.C., trying mightily to influence the report on the lumber industry. He was uncertain whether he had been successful, but he was convinced of the timeliness of the effort. As he reported to Long, "It happened to be a most propitious time to get our ideas started right with the Trade Commission and also to help Greeley with the preliminary report on the industry by the Forest Service." Bill Greeley's work required the approval of Chief Forester Graves, and even then, the Trade Commission might refuse to accept the Service's final report, but three things seemed likely to be recommended: (1) the right to curtail and make some price agreements under government regulation; (2) the right to circulate price information; and (3) the right to cooperate in the area of export trade.

Allen now needed to know just how far they should go in the name of their Western Forestry and Conservation Association. He feared that without the association's participation, the discussion from the industry side might narrow to "helping a few manufacturers to sell a few boards at a higher price." He thought that the question was whether the association wanted to leave its case in the hands of the manufacturers and their counsel or continue to exploit the reputation it had gained of speaking for the industry in a broad-minded way.

Although Long had his doubts about the outcome, he advised Allen to return to Washington, D.C., and do his best. "I think we better make the most of our opportunities and follow up the work that you have already accomplished," he advised. Then he warned about two probable outcomes "of the work of commissions like the Trades Commission." First, "nine times out of ten," they wouldn't recommend anything helpful to the industry unless it was good politics. And second, whatever the recommendations were, they would be so delayed that "before they get to the public, conditions may have changed once or twice, and the public mind is not in an attentive mood and very little, if anything, is accomplished."

The board of the Western Forestry and Conservation Association met in Spokane on May 3, and all supported a brief that Allen had prepared for presentation at the Federal Trade Commission hearings. Long noted that there was one issue to which everyone reacted, the suggestion that the government should regulate the price of lumber. In Long's words, "We all felt that this would be a dangerous experiment," adding that "none of us were yet in favor of this paternalism." The next morning, Long sent Allen a follow-up note, urging that he "emphasize and reiterate the absolute necessity for the propaganda of reforestation" at every opportunity. Long didn't care who did it, only that it was done. Public or private, it didn't matter. And he closed, "My own opinion is that it is the function of the State or Nation, however, and I think it cannot be emphasized too much or too often."

Meanwhile Snoqualmie Falls was a very busy place. Warren ordered the first electrically powered donkey engine, and reported that work on the railroad was progressing, "clearing the right of way from the slough up to the timber." They were also pouring the foundation for the boarding house and clearing ground for ten cottages. Because of its isolation, Snoqualmie Falls was destined to become a company town. Long had misgivings about this, some of which he shared with Warren. He urged that there be variation in the cottages, explaining that he had recently seen "a bunch of all alike houses, and it really looked quite common place," when it would have been easy to have provided "more individuality."

Long wanted to be kept abreast of the mill plans, but he needed some visual assistance. Thus he requested that Warren have Onstad "make out for me a very simple crude sheet, showing the area of our mill site . . . and then have him roughly sketch on this sheet the location of his mill, dry kiln, timber dock, planing mill storage shed and likewise his sorting chains." With this, Long thought it would help "get out of my system some ideas I have heretofore had about the way I would like to see your lumber handled into your yard, etc."

Then there was a planned tour of the Thompson Yards,

to get "a bird's eye view." Long pretended that he looked upon the trip as something of a vacation, but Thompson knew it would be considerably more than that. He had tried to reassure Long, forwarding a report made by an independent inspector, one Mr. Jones. Long wasn't reassured, as his response disclosed: "Assuming, therefore, that this report is a true reflection of the public sentiment in the vicinity of your yards, it is, of course, quite complimentary." This, however, was clearly an assumption he wasn't ready to make.

Long had no such doubts when it came to Louis S. Case, in charge of retail sales for the Northland Pine Company in the Twin Cities region, and he encouraged Boner to cooperate fully with Case. He had a few misgivings, however, one of which concerned the cars-in-transit scheme of entering into eastern competition. This involved allowing cars of lumber to leave the plant unsold, trusting that the contents would be purchased piecemeal along the way, not unlike the old method of auctioning off portions of the Mississippi River lumber rafts as they floated downstream from port to port. Although cars-in-transit could be an option for selling Everett lumber, Long thought they had best wait until there were ample facilities at the Minnesota end "for taking care of anything that might not be disposed of."

There was a slight decline in log and lumber prices in late spring. Boner seemed unworried, although he admitted that they were not quite shipping their cut, having accumulated "five or six million feet." The Puget Sound loggers met to discuss the situation on June 20. Boner thought they were like the lumbermen, "more scared than hurt," but the hurt proved to be sufficient. Joseph Irving of the Sultan Railway and Timber Company reported that it had been a "largely attended and enthusiastic meeting," and the assembled unanimously voted to begin closing down the camps and to keep them closed through July.

Long agreed with Boner that the circumstances weren't all that serious. Still, they had to operate with an awareness of the industry's position: "I presume it is a time of year and a general all around atmospheric condition when it is a little hard to be able to determine just what is the best thing to do." And he reminded, unnecessarily, "that everybody in the country is watching what the Weyerhaeuser Lumber Company does, and if we shade prices $1.00, they will shade them $1.50 and so forth and so on."

Of greater immediate worry was the deteriorating labor situation. As Long noted for the benefit of Horace Irvine, "There are more or less labor troubles out here, and men are getting a little scarcer and a little bit more restless every day, so that there are some reasons, other than market conditions, that seem to make a shut down rather desirable." Troubles had begun, at least in an obvious way, on May Day, with the appearance of a few pickets. While the mills kept operating, there was tension in the air and tempers became short. What to do was a puzzlement.

The summer of 1916 was unusually wet. The loggers, however, didn't mind; they were sitting at home. Long couldn't help but look forward to the resumption of logging, but he wasn't anxious. Indeed, although he had initially questioned any stoppage, by July 20 he was instructing the Cherry Valley Timber Company manager to plan on "a little lighter output, provided it can be done, in your judgment, economically." August 1 was the date circled for starting up, but Long was willing to accept an additional delay, provided that work resumed sometime in August. A later startup would mean that the Everett mill would be forced to shut down.

As for marketing the mill's output, both Long and Boner continued to fret. Business had improved—of that there was no question—but the arrangements with Thompson Yards still troubled. F.E. was sensitive to these concerns and tried his utmost to provide assurances, expressing confidence that "Mr. Thompson is giving our affiliated interests the best service he is capable of." He then outlined what was to be his major program for the years ahead, a single sales agency serving all of the affiliated Weyerhaeuser mills:

> I trust you and Mr. Boner will not become uneasy over the Weyerhaeuser Sales Company activities even if they do not bring you the results in actual sales that you expect. We have a many sided problem here, and you may rest assured that Mr. [Louis S.] Case and this office will always endeavor to serve all companies interested to the best of our ability and along lines that are absolutely fair as between the different corpora-

tions. I am very strongly of the opinion that the Weyerhaeuser Timber Company before many years, on account of its manufacturing operations which must necessarily increase rapidly, will have to cover practically the whole northern half of the United States with an efficient sales organization. Accordingly everything that we are doing in the way of building up a sales organization is with the ultimate purpose of having it become a Weyerhaeuser Timber Company organization.

Long wasn't so sure. Like other managers, he was continually weighing the relative advantages and disadvantages of centralization as opposed to decentralization. If at the moment he seemed to favor centralization, he also knew there were dangers inherent in a two-headed operation—that is, quartered in both Tacoma and St. Paul. Still, he wrote F.E. on July 31, there seemed every reason to launch "an organization that will meet the issue, whatever it is, of disposing of our lumber products. . . . I am more and more convinced that the Weyerhaeuser Timber Company especially will continue to grow in a very steady manner in the direction of manufacturing lumber, and I think we ought to get ready for it just as fast as we can."

But could a central management balance the various interests? Long had already heard hints of unhappiness from a couple of the Idaho companies regarding unfair competition. Could that be avoided? Still, the public would assume that "Thompson Yards, Weyerhaeuser Timber Company, Weyerhaeuser Lumber Company and all of the Idaho companies [were] one organization." Though they were separate in fact, "We have the name of being one . . . and I keep harking back in my own mind all the while to the many apparent advantages of really making them all one." And, after noting that there might be unforeseen difficulties in any large-scale merger, he added, "I am very much inclined to think that if this could be brought about harmoniously among our people generally that it would be a very wise and economic performance."

Allen obtained an advance copy of the Forest Service report in mid-August, Chief Forester Graves soliciting comments from Long and others within Allen's circle. Allen thought that "unless there is some lively criticism," it seemed unlikely that any further conferences would be necessary. Such was not the case, however, with the Trade Commission's efforts, Allen holding himself in readiness to return to Washington, D.C., for such a meeting. Long agreed that Allen should be on hand, adding, "I think anything that we could do that would have a tendency to improve the tax question would be about as important as any problem that could be passed up to the Trade Commission."

As for the Forest Service report, Long thought he saw "a vast amount [of Allen] in Mr. Greeley's conclusions and comments . . . [and] taken as a whole, I think it is a masterpiece and deserves and is worthy of the highest praise." He did have one complaint: "Somewhere it says that the Weyerhaeuser Timber Company has ninety-six billion feet of timber and their annual tax burden is one million dollars." While not wishing "to attack the reliability of a Government report on stumpage," Long allowed, "If there ever was a case where our earthly possessions have been over-estimated, it is in the Government's report." He also wondered why "our name is the only name of a lumber concern used in the whole report."

As promised, Will Parry of the Trade Commission paid a visit to the Pacific Northwest, and Allen spent most of September 2 escorting him about, "much flattered by his attitude." Wilson Compton would arrive a week or so later, Allen indicating that Parry wanted the professor "to meet lumbermen personally and informally, to see their mills and camps and generally get an inside and sympathetic viewpoint." Thus Allen inquired of Long, "Who should he see?" And then he offered his own opinion: "He should meet both kinds, the narrow and difficult lumbermen as well as the broad and progressive, but I want to be sure he gets the latter."

Allen would have to do the hosting because Long was off to other parts, beginning a five-week absence from the office. The time was largely devoted to touring the retail yards in the company of Thompson and Kendall, but it would conclude with a visit to Indianapolis, where "Mrs. Long and I are trying to renew old acquaintances and join in the State Centennial celebration of Indiana's hundredth birthday."

The development of a sales organization was proceeding rapidly, perhaps too rapidly from the Tacoma vantage. F.E. was leading the effort, but the mere fact that he was his father's son failed to make him immune from criticism. Boner, for example, wondered what would become of Everett's own salesmen. Were they soon to be odd men out? F.E. tried to reassure Boner that he had no intention of decimating existing arrangements, at least not in the near term. His purpose, as he explained it, was simply "to get all the information I can in regard to our whole sales problem, and to work out the most efficient system we can that thoroughly complies with our federal and state laws." There would, of course, "be some crossing of wires and perhaps unfortunate individual incidents during the time that we are re-districting the territory to be covered by traveling salesmen, but I trust you will bear with us patiently during this period, as I am satisfied that finally you will be able to sell your lumber more intelligently and more directly to the consumer than has been possible in the past." While Everett remained free to sell its lumber independently, "where you can find that you can do it to better advantage . . . as a matter of general policy we are extremely anxious to cut out as many middlemen as is possible in the marketing of your product."

Although F.E.'s intentions seemed reasonable enough, "cutting out the middlemen" wasn't going to be easily managed. Some of those so-called middlemen were actually salesmen representing various plants in the affiliated companies, and what seemed like duplication and unnecessary competition from a desk in St. Paul could be considered a healthy association from Boner's perspective. Boner wasn't yet convinced that Everett would be as well served by someone whose loyalties were elsewhere or at best divided, or whose understanding of Douglas-fir lumber was limited. These were uncertainties that endured.

Throughout, Long continued to wrestle with his doubts concerning George Thompson. There was that seemingly irresistible urge of Thompson's to buy everything up for sale, and Long urged F.E. to "assume the responsibility of passing upon the purchase of individual yards." F.E. wasn't enthusiastic, but agreed to "undertake it for a time," until another solution could be arranged. And then he added, almost without pausing for a breath, "I imagine you already suspect that Mr. Thompson has something new up his sleeve, and I shall not disappoint you. He has an opportunity of buying a yard at Mandan on the right-of-way of the Northern Pacific." F.E. saw nothing wrong with the plan and had authorized Thompson to go ahead, providing further investigation showed "the property to be as desirable as he thinks." George Long hoped that those in St. Paul would tighten the reins, but he had yet to see any evidence of that.

The fact was, selling lumber in the fall of 1916 wasn't much of a chore. Other problems, however, demanded attention. Labor unrest in Everett festered, with no solution in sight. Although these disturbances were common throughout many western communities, they reached a tragic climax in Everett on Sunday, November 5. The IWW leadership had planned for a show of strength on that day, but others opposed, and the confrontation turned violent. When the *Verona* passenger vessel, loaded with some 250 Wobblies, was about to tie up at the Everett dock, gunfire erupted. The source of the first shot was forever obscured in the confusion of the moment, but death and injury resulted. It was momentary madness.

Boner wanted no part of the affair. Indeed, he took great exception to a suggestion attributed to H. C. Gill, mayor of Seattle, that the Weyerhaeuser Lumber Company had played a role. In a letter to the mayor, Boner vehemently disclaimed any such involvement.

Long wasn't exactly happy that Boner had challenged Mayor Gill. While he agreed that the mayor "ought to have been placed right in the matter," he also feared that if Gill were "bombarded much more, and to an extent that may make him turn and fight, he will take hold of any weapon that he can put his hand on." In short, one didn't want to increase the enmity. Thus, Long offered his own opinion as to "a better way to handle such things": go directly to Gill and talk the situation over with him, "for then you will be able to size him up a little better and he can size you up, and if he has got any human instinct at all of fairness, it will come to the front better in a personal interview than in any other way." That was Long's method—to use the opportunity to get to know the person— and for Long, more often than not, it had proved effective.

The violence in Everett, as sad as it was, could not compare with what continued to rage across the battlefields of Europe. The end seemed as distant as ever, and the longer the war lasted, the more likely it would spread. Thus far, those in the Pacific Northwest had been distant observers, but there were indications of change. One such was evidenced by an

October letter from an English broker expressing an interest in spruce. "My friends in Liverpool are in urgent need of this particular kind of material being wanted for the manufacturing of our Air Craft," he wrote, adding that if even a small lot could be provided, "business mutually advantageous would result at the close of this War." Neither Long nor his colleagues could imagine what this would come to mean.

At the moment, Long was busily preparing for another St. Paul meeting with F.E., the most obvious result of which would be an end to the confusion surrounding the western designations. Henceforth, they would return to the old arrangement whereby the Weyerhaeuser Timber Company constituted a single organization. Boner was accordingly advised that he would shortly become manager of the Everett branch of the Timber Company, the Weyerhaeuser Lumber Company ceasing to be; and the holdings in Oregon would no longer be called the Weyerhaeuser Land Company, but would simply become the responsibility of an agent, a representative of the Weyerhaeuser Timber Company. And it would remain the Weyerhaeuser Timber Company for nearly half a century.

NINETEEN

War Comes to the Northwest

THE GOOD TIMES THAT HAD BEGUN IN 1916 CONTINUED into the new year, but there were causes for worry, chiefly concerning the worsening labor situation. Boner might occasionally complain about "disloyalty," but he knew the problem was deeper than that. He foresaw important battles in the spring, the foe being the IWW membership. Now was the time to prepare. On February 17, Boner sent an open letter to the employees of Mill A, who, he maintained, were receiving "the very highest wages of any saw mill and have made the Weyerhaeuser Mill the best mill on the Pacific Coast for the working man." The fact remained that an increase in wages was overdue. The cost of living had been rising, and there was no denying the profits that the Everett branch had recently earned. Thus Boner forewarned Long to expect wage increases, though the forewarning was merely a formality.

In the meantime, what was Long to do with his mounting cash surplus? The answer that came immediately to mind was to purchase the Thompson Yard indebtedness back from the Timber Securities Company, the recently reorganized Weyerhaeuser family holding company in St. Paul. Long advised F. E. of these intentions in a January 11 letter. "We are piling up some money," he wrote, and "if it is agreeable all around, I think we better pay off $300,000.00 of this [$500,000.00] indebtedness, and if you approve of this, we will send you funds from this end for that purpose."

In reply, F. E. said that though he preferred "to continue carrying the Thompson Yards account . . . you are at entire liberty to take up any advances . . . made by us at any time you

203

see fit." F.E. had big plans, and not only on the sales front. His initial intention was to send a letter outlining his program "to most of our active stockholders," but suddenly he had some second thoughts. Having just organized the Weyerhaeuser Sales Company, was he moving too fast, thereby giving "a wrong impression"? He sought Long's advice, hoping for a conference during the annual meeting. The details, of course, were up to Long, F.E. advising that two days should be sufficient, and "in case it proves uninteresting, we can spend our extra time at Snoqualmie Falls or Everett."

J. P. Weyerhaeuser appended a handwritten paragraph at the end of his brother's letter, suggesting that they delve into the particulars of F.E.'s plan, especially as concerned the relationship of the Sales Company and the mills. Long demurred. The first objective, he argued, should be to "get our crowd fully committed to the idea of a broad treatment of the big problems that underlie our properties," after which the details could "be handled by committees who are looking after the departments that they are going to supervise." He further explained, "Human nature is rather a peculiar thing, and once in a while you run across a temperament or an individual who picks up a little detail and makes a great big thing out of it, when as a matter of fact, it is insignificant, compared with the larger problem behind it."

It was a good thing that Long gave serious thought to the matter, because he was immediately called upon to discuss it with F. C. Denkmann. Over the years, the Denkmanns had been relatively conservative in their general outlook, far more interested than most of the Weyerhaeusers in enjoying the fruits of present labors. At the moment, Fred Denkmann was appalled at the extent of the investment already made in the retail yard business through Thompson Yards. On the one hand, Long could certainly sympathize. On the other, as he reminded Denkmann, they simply had to confront the problem of matching "[our] ambitions in the way of manufacturing" with an ability to market profitably. In this equation, the Thompson Yards involvement was an experiment, "more to test out the real advantages that may accrue by distributing lumber in that way." Denkmann no doubt shook his head at that explanation. For an experiment, Thompson Yards seemed awfully expensive.

While the others were preoccupied with retail opportunities across the northern tier of states, Long was continuing his investigation of Atlantic Coast properties, dispatching one John Richards to study "the whole eastern terminal situation." Although these were chaotic days, and it was unlikely they would be making any commitments, Long nonetheless thought it was a good time to be looking about. That very morning he had received a telegram from Richards regarding an option on a Baltimore site, some forty-six acres that could be acquired for about $30,000 and developed for another $10,000. Long wired back: "Take option in your name. While in Baltimore would investigate all possible good locations that have railway facilities for shipment and good water regardless of immediate proximity to Baltimore." Long considered the district a prime location, but saw no need to be in the city itself. "The main point after all is to get plenty of room," he instructed, "and a good location for rail connection and water shipments, even more important than it is to get down into the heart of things."

Long's optimism about the future in Baltimore failed to assuage his general discouragement, primarily because of the "chaotic conditions" the world over in the spring of 1917. These uncertainties made him unusually pessimistic when he thought about timberland. He could not foresee, for example, any significant increase in stumpage values "for a great many years to come." In the first place, there was so much of it, and, given current taxation policies, there was every incentive to cut, to liquidate, creating a continuing glut on the market. "The lumber business," he concluded, "seems to be one where we get two or three good years, and then seven or eight lean ones, and my own personal view is that this situation is likely to occur for a long time to come."

In early April the directors of the various Weyerhaeuser operations received a letter of clear importance. Although signed by George Long, the message originated with F. E. Weyerhaeuser. The letter began by noting the cooperation obtained by the Weyerhaeuser Sales Company, which "has undertaken . . . the sale of the lumber in the common markets of the country of practically all companies where the stock ownership is more or less affiliated." There was then brief discussion of some of the changing conditions facing the industry: "Lumber substitutes are now being used to an extent that was not dreamed of in the past, and it is more than probable that the task of getting satisfactory returns from stumpage holdings is a more serious and doubtful question than it has ever been before."

The solution to such problems was deemed obvious: "In many other directions, the advantages and economies of joint action seem so manifest, that there should be no hesitation in acting unitedly instead of separately." The general recommendation that followed was that they should create "in the ranks of the so-called affiliated companies, an organization that will handle all problems of logging, of manufacturing, of insurance, of taxation, of wages, of the utilization of by-products, of conducting the manufacture of our lumber in such ways as will minimize competition among ourselves, and in many other ways improve the conditions under which we have been working in the past." Thus it was suggested that the directors meet in Tacoma on Thursday, April 26, and devote two or three days to the discussion of this matter. The letter closed as follows: "No more important conference has ever been held, pertaining to our own affairs, and it is earnestly hoped that we can count upon your presence."

In April 1917, however, cooperation and service in a far different sense was in the offing. With war formally declared on April 6, George, Jr., telegraphed his father seeking advice. Young George expressed an interest in returning to Tacoma, where he might enter the armed service alongside his former companions. George, Sr., preferred that his son stay in school for the time being, observing that when "a company or regiment [was] organized from the students at Cornell, it would include a mighty fine bunch of fellows." Although they couldn't foresee the details, both Longs would go to war, and in many ways the one wearing the uniform would have the easier time of it.

The war came quickly to George Long, through his appointment to the lumber committee of the Council of National Defense. He learned of this from reading the *Tacoma Tribune* April 25. No one as yet knew how the committee would function, but, E. T. Allen advised, "Two things were clear—that it would have power to do harm to the industry even if it did no good, and that in any event Washington would soon be swarming with schemers trying to put various lumber projects across."

At the behest of the lumber committee, Long departed for the nation's capital. He didn't, however, neglect Timber Company business, arranging to meet any available Weyerhaeuser brothers for a tour of Baltimore port sites. As it turned out, Long spent most of May on the East Coast. He did meet his group in Baltimore, where one and all agreed that they should make the purchase, provided the city imposed no unreasonable requirements.

Long stopped in St. Paul on his homeward trek, and there he received word from Boner that the market was "running away and going by leaps and bounds." Similar reports awaited him in Tacoma, among which was a letter from Mark Draham, of the Mud Bay Logging Company subsidiary. Knowing that Long would oppose an increase in log prices, Draham endeavored to answer the opposition in advance. "Everything we buy is constantly going up," he reminded. "It looks to me as though, if the loggers keep their feet on the ground at all, we will be left alone, as everybody appears to be in the air." And he closed, "What do you think about raising logs another dollar about July the first?"

Opposition was akin to stemming the tide. "The whole country has gone just a little bit hysterical as to prices, and the lumbermen have caught the fever," Long wrote in answer to Draham. Still, if lumber prices held at current levels, loggers deserved a higher price. He did advise "to go a little slow," in this instance, to lead by following.

Long was only beginning to understand how much time he would have to devote to his new responsibilities on the national lumber committee. He and J. W. Gregory were the committee's West Coast representatives, but all of the Weyerhaeuser regional organizations were involved one way or another. Long, for example, began working with the West Coast Lumber Manufacturers' Association to fix a schedule of prices and specification sizes for ship building and also to supervise the purchase of lumber needed for the new American Lake cantonment, soon to become Camp Lewis.

Another subject clearly of importance was spruce for

airplane construction. This use of spruce was unfamiliar to the region's lumbermen, and spruce producers were few. Long and Allen, from his Washington, D.C. office, encouraged spruce users and suppliers to get acquainted. In Long's words, "I am trying to impress upon the dealers out here the necessity of sending a bright, active lumberman to the aeroplane people, who knows all about spruce, and to get the aeroplane people to send one of their men out to the mills." In this way, he was convinced that the government would get better service and better prices than if the need was met by "some jobber . . . [who would] jump in and buy up the stock and sell it to the Government." But time was the enemy.

Indeed, events seemed to be racing, creating a blur. Long noted that the "cantonments for Tacoma evidently are going to be built quickly," and that if there were "no slips betwixt the cup and the lip, the Local Committee out here will have the handling of placing the orders." This was a very sensitive business, as Long well understood, "not an honor that I am looking for, nor a responsibility that I care for, but if it is up to the Committee, I think it will be handled all right." As he quickly learned, however, it wasn't always possible to keep patriotism and the scramble for profits separate.

The local committee met on June 22, and after voting to call themselves the Fir Emergency Committee, decided that their expenses would be paid by means of a 2 percent assessment on the invoices of those members supplying lumber for the construction of the cantonment. Most questions weren't so quickly resolved. The freight rate, for example, had long been a serious point of contention, and Long feared, as he wrote in a letter to one R. D. Brown, that the government might instruct that "we should send the orders only to the points that take the lowest freight rate. . . . [This] would not be practical, nor in harmony with the idea of scattering these orders to the membership of the West Coast Lumbermen's Association, nor in harmony with giving some of the orders to mills who are not in the Association." As in many crises, efficiency took a back seat to democracy.

Another concern was the manner of assigning orders. J. H. Bloedel had recommended that since the annual dues to the association reflected output, the distribution of orders "should be made in harmony with the output of the mills, as reflected by the dues which they pay into the Association." Although there was no formal vote on this suggestion, Long assumed that "it was taken for granted that this would be both wise and equitable."

As in all previous wars, the initial contribution of the industry to the effort was manpower. And as more men left the woods and mills for training camps, they became increasingly difficult to replace. Long could only sympathize with his managers—sympathize and try to encourage. The declining numbers of workers available naturally affected attitudes. The Puget Sound loggers attempted to inaugurate an employer policy regarding hours and wages, but Long would not allow his units to participate. He advised T. M. Williams of the Cherry Valley Timber Company that the reason was not because "we are disposed to be reckless in the matter of wages, and not act in harmony with the things that are best for loggers to do, but because we are of the impression that an agreement of this kind is unnecessary and not politic."

The whole situation was a conundrum, as Long probably stated, to both employers and employees. The instigators, it was assumed, were the members of the IWW, and their influence was felt first in the woods and later in the mills. By mid-July, Boner found himself in the middle of real problems at Everett.

It was easy to label all troublemakers as Wobblies, and to blame every difficulty on them. The fact was that there were legitimate causes for complaint, but these were often buried in the extremity of the IWW solution. In the midst of the battle, it became a matter of "them" versus "us." Boner felt helpless as he watched both of his mills close down. Still, there was plenty to do at the loading docks, and those jobs were publicized. "Quite a number of the men wanted to work," Boner informed Long, "but they seemed to be afraid to do so," even though there had been no confrontations, not even any trouble, and, indeed, no picketing. He continued: "Some of them came down to ask when we would start up again and I told them, 'When their yellow streak turned red.' "

Where would it all end? Boner despaired and saw nothing but gloom. "In my opinion, we are into industrial warfare and we will remain closed down until the good honest working men, if there are any left, will drive the IWW's from their midst, and in my opinion, it will take a month or longer." Looking on the brighter side, if they somehow managed to continue to ship lumber, the strike could be a benefit. Everett did have lots of lumber piled about.

Boner's labor problems at Everett hardly compared with those in the Grays Harbor mills and logging camps. The International Union of Timber Workers announced their demands on July 6 at Aberdeen: an eight-hour day in the mills with a minimum wage of $3.00; a nine-hour day in the woods with a minimum wage of $3.50; thirty days' notice for wage changes; no loss of time as a result of breakdowns; no discrimination against organized labor; freedom of speech; no tampering with the mails; and time-and-a-half for overtime and Sundays.

Both sides made much noise about the detrimental effects of a shutdown. The debate frequently returned to questions of patriotism. If it was unpatriotic to stop production by striking, so too was it unpatriotic for operators to deny reasonable demands, thereby prolonging strikes. Nevertheless, by July 18, many operations shut down, with an estimated 17,000 men on strike, including 500 employees of Weyerhaeuser's Everett mill.

The scene was anything but quiet at Camp Lewis, where nearly two thousand men were constructing an army post from scratch. George Long was learning about Army red tape as he worked with Major David L. Stone to meet the building requirements. The blueprints had been made on the East Coast, and the specifications frequently needed to be modified to meet western realities. Fortunately, Major Stone was far more interested in results than blueprints. The modifications were approved, and Camp Lewis began to take form.

Though work was proceeding, the general strike was having its effects. In his capacity as a member of the Fir Emergency Committee, George Long called a meeting on July 25 that included government representatives and a group of West Coast lumber manufacturers. Obviously the old formula for placing orders could no longer be followed. "The war department must have lumber," Long reminded, and "the only possible way to do this was to take the orders away from those who could not handle them and give them to those who could."

As the summer of 1917 wore on, pressures to settle the

strike mounted, but heels dug in seemed immovable. Boner sadly watched his Everett work crews scatter, many doubtless lost forever. Although he looked forward to hearing the whistle blow, he didn't expect the strike to be settled soon. In one sense the prolonged strike was just as well, as prices were none too strong "and we must maintain a good market if we are going to operate under an advanced cost." Further, the first move was up to the loggers, there being little incentive for lumbermen to end the strike if there were no logs to saw. In Boner's "private opinion," they would eventually compromise on an eight-hour day: the only question was the terms and conditions.

The battle was not simply between labor and management. In many places, it was also a contest between the American Federation of Labor organizers and members of the IWW. Some operators saw the divisions within labor as an advantage, assuming that all could be discredited on the basis of the IWW performance. Was the length of the strike increasing the IWW's power? That possibility was what Boner feared most.

Allen had read the IWW literature and researched its leaders, and nothing he had learned was reassuring. These were not the usual adversaries. Their concerns with working conditions and wages were dwarfed by the conviction that it was the system itself that must be changed. They wanted a world turned upside down. No wonder they struck fear into the likes of George Long. "They are a bad lot every way," Long observed in a July 31 letter to Allen, "and very shrewd and very inconsistent and have a potent influence for disturbance, not only in local labor matters, but in all matters pertaining to the stability of anything that the world has built into a stable form by centuries of evolution."

That was unusual vehemence, clear evidence of the extent of Long's fears. He even wondered whether the time had come for government intervention. One thing was clear: the times were ripe to wed feelings of patriotism to opposition of the IWW. And so it would be. But Long was still able to see beyond the moment, trusting that when the "hurly-burly is over," they would get on with "some deep seated instructive work." The industry had labor problems, and if nothing was done to correct them, these troublesome conditions seemed destined to return.

By no means was all of the blame for the strike to be placed on labor. The *Tacoma Times* of August 13 featured an article titled "Business Slackers," which reported the wartime need for lumber accompanied by a request from Secretary of War Newton D. Baker that millmen reopen their plants, "as a patriotic duty" on an eight-hour basis, with time-and-a-half for overtime. Washington Governor Ernest Lister implied that lumbermen had exploited the IWW as a straw man. "I know and I believe the lumber operators know," the governor asserted, "that there are hundreds and thousands of their striking employees who never were, are not now, and never will be connected with the I. W. W."

The stalemate continued. On September 17, Dr. Henry Suzzallo, chairman of the state Council of Defense, and Henry M. White, a federal mediator, announced their intention to reopen negotiations. The extremists on either side, however, weren't interested. The IWW leadership threatened to "strike on the job"—that is, to return and work eight hours, period. F. S. Grammer, chairman of the Lumbermen's Protective League, countered by promising that any employee who used such subterfuge would do so only once. Nonetheless, by the end of September, enough had tired of the impasse that some mills, including St. Paul and Tacoma, were able to resume operations on the old ten-hour basis. Clearly, it was a strike—like most strikes—with no winner.

Long agonized over the labor situation for more reasons than he could count. "I am mixed up in the question from a half dozen different angles," he wrote to Horace Irvine in early August. His responsibilities at Camp Lewis had "multiplied a good deal since the mills shut down. The men even refuse to load lumber on ten hours per day work, when the lumber is destined for ship yards or to the cantonment building." It was frustrating for all concerned, but the company was nevertheless providing between thirty and forty carloads a day, and, Long predicted, "if we can keep the lumber moving for another week at this rate, we will be out of the woods."

Nor was that all: "Next comes the shipbuilding stunt, and then the aeroplane stunt, there being a Government Commission on its way out here now to try and increase the quantity of spruce lumber about 200 percent over that which has ever heretofore been produced, so we are up against some knotty problems." (No pun was intended; George Long, for all of his rhetorical failings, didn't deal in puns.) The fact was, however, that wooden ships were fast becoming relics. Such was not the case with airplanes and the associated construction requirements for spruce. In early October, Long announced that the government was prepared to purchase upward of 100 million feet of spruce "at a price ranging from $50 to $100 a thousand." This large demand would create correspondingly large problems.

In mid-August, Long briefly left his wartime responsibilities to attend a Sales Company meeting in Spokane. The results were important, but hardly unexpected. Louis Case proposed that the Weyerhaeuser Sales Company be charged with representing "all of the various affiliations in their sales and solicitations for orders," this to include the Eastern Lumber Company sales force and also the old Northern Idaho Sales Agency. He further recommended that they enter the eastern trade with a single price, "whether the buyer was a retailer, an industrial institution or a wholesaler." Long questioned the advisability of effectively eliminating the wholesaler, but the consensus was to proceed on that basis. The outcome certified the Sales Company as the sole sales representative for all of the mills, from northern Minnesota, to the Inland Empire, to the Pacific Northwest. It was an important decision with far-reaching ramifications.

George Long reported these developments in a letter to F.E., knowing that he would be pleased. Long also told of recent efforts to end the strike. The federal negotiator had assumed that if the IWW threat were eliminated, industry leaders "might be persuaded to go to the eight hour basis." The negotiator assumed too much. The other key official was Dr. Suzzallo, president of the University of Washington and chairman of the state Council of Defense. Long genuinely liked Suzzallo, describing him as "a great big fellow," who had "waded into our strike with all of the vigor, enthusiasm and fairmindedness that ought to be characteristic of a man who was trying to solve a difficult problem."

It was Suzzallo who finally succeeded in getting the two sides together for negotiation. No agreement was reached, but the labor representatives dropped their demand for ten hours' pay for eight hours' work; rather they wanted recognition of the eight-hour day, "let the wage fall where it might." The Board finally suggested the following:

> That labor return to work on the basis of ten hours work, at the present wage, until April 1st, 1918, on the same basis of prices that were in existence when the strike occurred. Second, that on April 1st, 1918, the mills and camps go to work on the basis of eight hours a day in mills and nine hours a day in the camps, on the basis of the wages that are now in existence for ten hours work. Third, that committees be appointed to arbitrate and adjust all such questions as sanitary regulations in camps, and wage propositions. Fourth, that the industry in the meantime bring what pressure they could to bear to bring about nationally an eight hour day through all of the lumber producing districts.

Still there was no agreement. Management argued that they had no idea what conditions would be on April 1; and labor couldn't accept some indefinite promise.

For all of the problems, cantonment construction went forward, even if a major share of the lumber came from Oregon, as yet largely unaffected by the strike. There was a visit to western Washington by the so-called Airplane Commission, consisting of prominent officers from France, Italy, England, and the United States, "delegated to take charge of the assembling of the spruce material that is needed." Accompanied by Allen, the tour had a glamorous aspect; motion picture crews followed these knights of the air as they proceeded from Seattle down the coast to Portland, making "a great plea for spruce, spruce, spruce, and that the fate of the war is to be settled by the activities of the airplane fleet and that vast quantities of spruce are immediately needed for that purpose." Despite the pleas, however, the spruce districts were strike-bound: there were no spruce logs being cut and no spruce lumber being sawed.

By mid-September, operations in both woods and mills began to start up, but conditions were far from normal. The war caused a shortage of men, and those who remained were scattered, particularly the loggers, who tended to be, in Long's words, "generally of a more roving type." In any event, it was clear that logs would be in short supply, and given that fact, prices began to rise. Long did his best to slow things down, allowing, in a September 19, 1917, letter to H. C. Clair: "History makes itself pretty fast these days, and what we may have thought was a reasonable price a few weeks or months ago, ceases to be a reasonable price under abnormal conditions." But Long argued that they were still making a reasonable profit on their logs, and they should be satisfied with reasonable.

In the midst of these concerns, St. Paul called. F. E. Weyerhaeuser wanted to reconvene the advisory committee, that inclusive group formed to consider problems and opportunities common to the affiliated companies. There was no saying "no" to the invitation, so Long set off, momentarily leaving behind his Pacific Northwest war responsibilities. It was something of a relief.

Although a good deal was accomplished at St. Paul, not all the results pleased Long. For example, George Lindsay, chairman of the advertising subcommittee, announced plans that struck Long as excessive. "They have worked up quite an idea about advertising lumber," Long informed Boner at Everett, adding that he was in full accord with most of the details, but not all. For instance, he questioned the plan "to mark each and every grade on each and every piece of lumber that is made," and the "propaganda then is to advertise our lumber and that we guarantee these grades." Long wondered just what the effect of such a program of grade marking and guarantees would have on the average retailer. Would they not be forced to depend exclusively on our stock, and when we failed to supply their needs, what then? Long would have preferred that they begin by simply trademarking and advertising, forgetting about grade marking, "not so much on account of the expense and difficulty, which will be colossal, but more especially [because of] how it will be accepted by the retailers."

At Snoqualmie Falls, Warren and his crew probably wished they were closer to worrying about lumber sales, but at the moment it was a lack of labor and delays in the delivery of machinery that frustrated. As the weeks passed, the logs accumulated. Given the lively market, Warren wondered whether they might sell some of the logs that were piling up. Long had no objection, observing that at present prices they "could make just about as much money selling logs as we could to make lumber."

Great changes in Pacific Northwest forests were imminent. A hint of these was provided in an October 23 letter that Long sent to six of his state colleagues in which he announced that Colonel Brice P. Disque had been authorized by the Secretary of War "to take whatever steps are necessary to get out what the Government needs for aircraft spruce lumber." This meant, or so Long surmised, "some kind of a mutual harmonious solving of the labor problem," and it was for this purpose that he called for discussions on October 25 at the Benson Hotel in Portland: "No more important conference has yet been called and I urge you to drop everything and be present."

For Boner's benefit, Long predicted movement on the labor question. "It is among the possibilities," he advised, "that the Government will organize four or five regiments of men to put in the logging camps in the spruce regions" to log, and if this were done it "undoubtedly will be an eight hour stunt." No one needed to explain the ramifications of that.

The Portland meeting was complicated by an unexpected coincidence. The Fir Exploitation Company, including "the leading manufacturers of fir, engaged in export business," had scheduled a meeting at the same time and place. Disque had agreed to meet informally with Long's contingent of loggers, and the colonel held to that agreement. Whether the discussions were fruitful depended on one's expectations. Disque allowed that "he had learned some things that he ought not to do, as well as those that he ought to do."

Colonel Disque immediately immersed himself in regional problems. One of his first public statements concerned the inevitability of an eight-hour workday, and Long took exception to that, not from a consideration of its correctness, but of its timeliness. "Very often," he reminded the colonel, "the most worthy cause is misdirected on account of some slip of the tongue or failure to meet a situation that develops unexpectedly." Then he added, "It is the man who can give a satisfactory answer quick off the bat that carries conviction, and in order to do this he has to be pretty well acquainted with the possible questions that may be fired at him." In other words, one "has to know the territory." The colonel was right about the eight-hour day, however, "and the only question that we have to confront is when." Weyerhaeuser was willing, Long assured, "whenever the industry as a whole is ready for it."

Spruce procurement was proving an awful headache. The species grew only in a few coastal areas, and even there its density seldom amounted to a quarter of the timber. Since spruce commonly was found in rough terrain, intermixed with Douglas-fir, cedar, and hemlock, the removal of only the spruce was difficult at best and impossible in terms of profit. And the government wanted only high-quality spruce. As Long explained it, "If the spruce grew within the immediate environment of all of the mills out here, and if it grew in solid forests as the fir does then the question would be a very simple one and a very easy one, but there has never been put up to the lumber industry, at any time anywhere as hard a problem as to speed up the spruce production at once some 200 or 300 per cent above the normal and natural production of the lumber." To this he added, "Those who criticise the action of the lumbermen are not familiar with the facts."

Long was soon on his way back to Washington, D.C., immediately following Thanksgiving, for another hearing of the Federal Trade Commission. He sighed before leaving, "There is no telling what they may want us to do." He had some parting words for Colonel Disque. He wasn't going, he assured him, "to butt in, but to attend a hearing" whose topic pertained not just to the Pacific Northwest, "but to lumber all over the United States and especially the Yellow Pine lumber district of the south." Long predicted that the Federal Trade Commission would encourage lumbermen to "behave themselves, as to prices, and I think they are right." Regarding their Pacific Northwest concerns, Long repeated his view that it would be "a wise thing to have all of the Government requirements for ship building and aircraft material handled through your office and under your supervision," once more promising the colonel that he was "not going to work at cross purposes with your functions out here, but with them in every way."

The FTC hearing was not as thorough as George Long had expected—"not very profound" was the way he described it. There was a consensus that "the pine people of the South and the fir people of the West," in cooperation with government officials, should agree to a schedule of prices. Such agreement, however, was assumed to be temporary. In any event, Long predicted that sooner or later they were bound to have a careful scrutiny on the part of the Federal Trade Commission as to costs, "and no doubt the whole subject will be in the lime light at some time in the future."

Long didn't get back to his Tacoma desk until December 19. There he found a number of complaints from spruce producers, most of which had to do with the differing specifications demanded by the individual airplane manufacturers. One involved a Curtiss aircraft representative who had spent a month visiting the mills, and "the longer he stayed the worse the specifications were that he wanted to be made official." Long sympathized, agreeing that smoothing out "all the wrinkles that are wrong would be a bigger job than to lick the Germans, so about the only practical thing to do is to catch as catch can and do all you can and in nothing does this apply with more force than the getting out of aircraft spruce and aircraft fir."

Another rankling aspect of the spruce program was that opening up tracts for the sake of a tree here and there heightened the fire danger substantially. But there seemed to be no choice. Only a few trees met the standards, and there was simply no market for all of the adjoining timber. Long informed George C. Joy of the Washington Forest Fire Association of the increased hazards: "I do not know just how much grief there is going to be in this game, but I am inclined to think that we will have quite a little inquiry along the same direction, and I think I will want you to look after it for us." There was little alternative to letting "the Government take what they want," but it was hoped that they would "surround the operation with the right kind of precaution so that we will not have a horrible fire hazard on our hands and if in cutting the timber they do not destroy a whole lot of other property." And he advised Colonel Disque along similar lines.

Wrestling with such matters was hardly conducive to instilling a Christmas spirit. Long was sure that he could find pleasanter ways to spend the holidays. He could, for example, have tried to make his way up to Snoqualmie Falls for a visit with Warren at the new plant, which had opened during his stay on the East Coast. The November start-up had gone well enough and things had soon settled into a fair routine. Meanwhile, work progressed on the cedar mill as well as on storage sheds and houses.

Snoqualmie was a busy place, and George Long had every reason to be interested. After all, this was his creation. Everett had been an operating plant at the time of purchase, and however much it had since improved, it would always bear the mark of those who had gone before. Not so Snoqualmie Falls. That plant was there because of Long's vision and dealings. More than thirty years later, another Weyerhaeuser operating head, J. P. Weyerhaeuser, Jr., led a shareholders' tour to Snoqualmie, "our oldest integrated plant," as he then described it. This was "the first place where a well-blocked, prime, big body of timber for operation came together without any obligations to others." And he paid the credit due: "Mr. Long built a monument." Snoqualmie may not have seemed monumental in late 1917, but it showed plenty of promise.

TWENTY

War Is Wearing

PERHAPS TURNING THE CALENDAR TO 1918 CAUSED Long to count up his years, and admit to the possibility that he might be getting old. He didn't feel old, at least not physically, but now and then he wondered if his thinking was in keeping with the times, or if he had become overly conservative. Occasionally it may have seemed so to others. One who doubtless felt that way was George Lindsay, enamored with his own creation, the advertising campaign.

While Long was in St. Paul the previous fall, Lindsay had presented him with a copy of his advertising plan, which would subsequently be approved unanimously by the advisory committee. Long read it on the way home, but he took his time responding. Lindsay was likable enough, but he did tend to involve himself in matters beyond his grasp, or so it seemed to Long. Thus when he finally responded, on January 2, 1918, the words did not come easily. "It certainly is a very forceful document in every particular," Long began, "and I do not see how the question could be presented more ably than you have presented it." One can imagine Lindsay reading this, and taking a deep breath in preparation for what was to follow. "I fear that I am getting old, for the unanimous vote that was given by all of those present to go ahead with the campaign as outlined by you is one, where if I had been present, there would have been one discordant note, for I would not have voted for it, and consequently the true logic of the situation is that I am wrong and the crowd is absolutely right, and continuing the logic a little further means that old age is creeping up on me and dulling my judgment and suppressing my aggressiveness, etc., all of which means that there comes a time when a man is hardly up to date, especially when new things have to be done in a new way."

Though Long was not convinced by Lindsay's proposal, and certainly not as concerned grade marking, he nonetheless promised to cooperate, in his words, "to enter heartily into this game and try and work it out and pull just as strongly for it as if I was cordially in favor of it." But that was easier said than done, and the debate would continue on and on.

Possibly it was the war that forced Long to think about aging. Responding to a note from Lafe Heath about a tract of timber for sale, Long seemed less interested in doing business than renewing a friendship. "It seems like an awfully long while since I have seen you," he wrote in reply, "and I don't know of anyone I would rather sit down and have an hour's chat with." Then he made passing reference to the moment, observing that if Lafe had been "about seventy-five years younger," he might well be on his way to France "at the head of that Forestry Regiment, but old fellows like us have to sit back and give advice." As it was, George Long, Jr., was the one on his way to France, a member of the 20th Engineers (Forestry).

Meanwhile, his father grappled with thankless responsibilities. Long had agreed to chair the West Coast Lumber Emergency Bureau, and it was in this capacity that he called a January 11 meeting in Seattle of all the Washington and Oregon loggers west of the Cascades. The purpose was clearly stated: to explain in detail the nature of the wartime requirements, which "will not be met unless all branches of the industry work together harmoniously in getting out quickly that which the Government needs most." Additionally, there was, as always, the question of prices.

The meeting with the western loggers went about as expected. More than a few wanted to raise prices as high as the market would bear. But though the government might not complain about an increase in log prices, notice would surely be taken when lumber prices subsequently rose. Thus Long preached against yielding to temptation, urging one and all to hold the line, advising that they go slow, at least until they had "some clean-cut plan of handling Government business."

There would be problems aplenty. Representatives of the Warren Spruce Company and the Grant Smith & Porter

Brothers were soon pestering Long, both claiming to have contracted with the government to procure spruce, and both of whom now requested permission to enter Weyerhaeuser lands for that purpose. Long sought Colonel Disque's confirmation of the facts, including the details pertaining to "the way the debris and other materials that are hauled in the woods in getting out spruce will be handled." The request served a double purpose. If Disque had included such details in the contract, they should be shared; if he had not, Long's letter reminded him of the importance of doing so.

As matters evolved, Long would make his own contracts, including stipulations for fire safety. (It should be noted that a new word, "rive," was coming into common usage. This described the activity of splitting Sitka spruce logs lengthwise to check the straightness of the grain, with only the straightest being accepted for airplane stock.) The general "terms and conditions" were (1) that the timber suitable for riving within the tract would have to be cut and delivered to the government within twelve months; (2) that the selection of suitable trees would be made jointly with a Weyerhaeuser representative, that an estimate would then be made, Weyerhaeuser accepting the "price which the Government has named for the trees so selected, i.e., $7.50 per thousand feet and the timber to be estimated and paid for before any cutting is done"; (3) that the work be done with the approval of Weyerhaeuser representatives so as to minimize the fire hazard; and (4) that any timber damaged in the course of the operation "will be paid for by you at the current prices that may be established by the Government." The work would go forward, but insofar as possible the terms would be George Long's.

A somewhat more cheerful aspect of the times was the arrival of soldiers, soon to be loggers. In mid-January, sixty troops reported at Snoqualmie Falls, where Cutler Lewis, the logging superintendent, "seemed to be mighty well pleased with the appearance of the men." At least in numbers, Snoqualmie now had a full crew. Experience was quite another matter, of course. Long counseled patience, observing that Snoqualmie may not have gotten the pick of the crop, other camps having been earlier supplied. Nonetheless, he was certain that "the experiment is worth trying."

Warren acknowledged that some of the soldiers had plenty to learn about woods work, but he also noted that their spirit was "fine and it is spreading to our other men." Warren liked Lieutenant Eckels, the officer in charge, describing him as "a fine young fellow and very anxious to make good," who was "highly pleased with camp conditions."

For the moment Boner had other things on his mind; he was beating an old drum, complaining about his treatment at the hands of George Thompson. Long tried to be sympathetic, but it wasn't easy, what with so much else to worry about, and Boner soon found himself with more pressing concerns. Although the soldier-loggers had helped to alleviate the manpower shortage, there were other serious problems. The Wobblies continued to be the principal agitators.

Colonel Disque had sought some reasonable way out of the "psychological" dilemma, as he described it, with the IWW, and his solution was largely based on the belief that they should and could all "get together." Thus it was that the Loyal Legion of Loggers and Lumbermen (4-L) came into being, "a 'union,' which might bring employer, employee, and the military oversight together."

George Long supported the colonel's efforts. He agreed entirely that the owner-managers had been seriously deficient in their discussion of labor problems. There had been no means of communication, no alternative to the IWW, and the Loyal Legion now seemed to provide that. Still, Disque's job was "herculean," made all the more difficult by the lack of a dependable individual to manage the woods end of the responsibility. And, as Long noted, "The whole problem is getting the logs to the mills, and after that the task is easy." Well, maybe not easy, but possible.

Although the Loyal Legion would come to be maligned as a "company union," initially it attracted solid support from labor. Boner, for example, reported that by early February 1918, the 4-L had succeeded in organizing the Everett mills "practically 100 percent," and had appointed one of his own office staff as its local secretary. Within six months of its December 1917 start-up, membership totaled more than 80,000.

For 4-L organizational purposes, the West Coast region was divided into seven districts, and officers visited every camp and mill, organizing "locals" and leaving a secretary behind. The issues that had provided the IWW platform—the eight-hour day and working conditions—were largely assumed by the Loyal Legion. There is little doubt that the vast majority of loggers and millmen preferred the label of loyalty, all other things being equal.

Each secretary soon received a bulletin from Lieutenant M. E. Crumpacker of Disque's Portland office. The bulletin emphasized the colonel's commitment to solving the problem of hours of labor in the mills and camps, and his wish to do so "in a spirit of utmost fairness to everyone concerned and, for this reason, [he] asks the advice of the members of the Loyal Legion." Specifically, Disque requested that local secretaries lead "a frank and open discussion . . . [of] a proposition something like this: . . . As the hours of daylight grow longer, providing additional time to make weapons against the Kaiser, would it be agreeable to the men to operate on a basis of an eight hour day, and devote ten hours actual work when daylight conditions permitted, with the understanding that eighty percent of the present wages be paid for the eight hour basic day, and time and a half paid for the extra hours."

Applying such a formula, a laborer currently earning $5 per day for ten hours would receive 80 percent of that, or $4, for the first eight hours worked. For work beyond eight hours he would receive 75 cents an hour, or a total pay of $5.50 for a ten-hour day, an increase of 10 percent.

In closing, Lieutenant Crumpacker reminded one and all of the cause: "It is impossible to attach too much importance to this request from Colonel Disque for your patriotic co-operation and the patriotic co-operation of every member of the Legion." The recipients of this bulletin hardly wished to be unpatriotic, but many must have wondered where it would end. As management and labor alike were beginning to learn, the old rules no longer seemed to apply. Everywhere one turned, one was likely to see a uniform or to face new regulations, reinforced always by the torch, or the club, of patriotism.

In early February 1918, George Long experienced an attack on his own patriotism—an attack that had to hurt deeply. A few detractors, led by ex-governor Oswald West of Oregon and recently relieved Major Charles R. Sligh, implied that Long had used his influence to benefit the Weyerhaeuser Timber Company to the detriment, of course, of the war effort. What their motives could have been is unclear, but Sligh's letter to Secretary of War Newton D. Baker offers telling evidence: "A baneful influence has been exerted by the representative of the German Weyerhaeuser lumber interests . . ." Sadly, for many Americans of the period a German name alone was reason for suspicion. Since the Portland press had been among the first to publicize the charges, Long hurried a letter off to publisher friend George Cornwall, knowing that Cornwall and his *Timbermen* were bound to get involved in any war of words. "In the whirligigs that are flying around in every part of the country, resulting from war activities, I of course am getting my little share," Long wrote, and proceeded to review his involvement in war work: member of the National Lumber Committee, chairman of the Fir Emergency Committee, and adviser to the War Industries Board. None could seriously question that Weyerhaeuser's general manager was doing his fair share.

For a time at least, charges of undue influence seemed akin to the proverbial tempest in a teapot. Nonetheless, Long had to appreciate those who rallied to his support. The *Tacoma Tribune* of February 6, 1918, offered a vigorous defense, asserting that there was "no more highly respected business man in the city of Tacoma than George S. Long." And the city's Commercial Club adopted the following resolution: "The aspersions cast on George S. Long's character were either made maliciously, through misunderstanding or through ignorance of the facts. Mr Long has devoted many years of his life in the service of the community, the state and the nation as a civic leader of clear vision, unfaltering courage and unquestioned honesty."

But if Long thought such charges were now past, he was sadly mistaken. On February 28, the *Tacoma Times* featured a front-page article headlined "Weyerhaeuser Co. Timber Hoarding Called Disloyalty." What followed was a supposed exposé about a "completely equipped logging camp" in the midst of "one of the richest timber belts in the state," served by "a well constructed railroad line," which was "lying unused." The explanation? "George S. Long, memorialized by the Tacoma Commercial Club as a patriotic citizen, is refusing to send one stick of this badly needed shipbuilding timber, which could be transported with ease and speed over that railway line to any shipbuilding plant on the Sound."

The source of these charges was a representative of the Hercules Sandstone Company, "owners of both the camp and the railroad." The editor of the *Times* not only credited the story, but went on to inquire, "What kind of patriotism is it, what idea of national service is it . . . that leads the Weyerhaeuser Company to hold back government war work?"

As for the "unused" logging camp, H. P. Scheel of the Hercules Sandstone Company had found Long to be something of an irritant over the years, and now seemed the perfect time to get even, and just possibly to make some money in the process. Back in 1915, Long had reluctantly permitted construction of a railroad of some twelve miles running into company timber southeast of Tenino. This allowed, through a carefully drawn lease, access to a quarry. Long had made clear to Scheel, "over and over again, so there could be no possible mistake, that our timber along the line of the road would not be for sale, and that in the construction of the road to reach the quarry he must not think for a minute that he could go to any expense in the road extension that would justify it, except for what he could get out of his quarry." But Scheel chose not to listen, and immediately began trying to interest others in the timber now made accessible by his railroad. Scheel should have listened; the timber wasn't his to sell. It was Long's, and Long wasn't selling.

Long tried to explain himself (letter to the *Tacoma Times,* March 1, 1918), and his explanations had little to do with the timber and everything to do with labor: "If there were the men to handle it, there would be three times as much available timber as is needed to supply the shipyards of the Northwest." Without the men to cut the lumber, there simply was no justification for opening up such tracts. "To start cutting a hole in the heart of a timber belt like this would be like taking a bite out of an apple and throwing the rest away. . . . To cut such an opening would be to expose the entire district to dangerous fire hazards." But as usual, people believed what they wanted to believe.

F. E. Weyerhaeuser sympathized with Long and agonized over the slanderous charges against the company. But, typically, he said nothing publicly. One of his nephews, Frederick K. Weyerhaeuser, felt differently. At the moment, F.K. was stationed at Foggia, Italy, serving in the U.S. Air Service as a bomber pilot. Although he learned of the Tacoma controversy about a month after the fact, that didn't diminish his indignation, as his brother Phil soon understood via the mails.

> It's alright to be conservative and not talk much, but not when someone calls you a traitor. I wish you'd tell Dad that. Our people don't realize the value of publicity. We let people rave about the wicked Weyerhaeuser corporation, etc., and don't say anything because we know it's not true, but 9 people out of 10 that read that never do know better and suppose it to be true. If I ever get to have any power in business you can bet I'm not going to let anyone talk that way about me or my Co. without calling for a showdown. If you don't, they think the accusation to be true. Of course, I know the W. Co is doing everything it can to help the Govt. and Mr. Long could not have done more for the Govt. He certainly is a wonder.

F.E. and George Long only wished that they could return to a time of doing business under normal conditions. Recently the two of them had been pleasantly jousting over some pricing disagreements between Boner and Thompson, leading F.E. to suggest the creation of a committee of resolution to adjudicate such disputes in the future. The committee would consist of Long, Louis Case of the Sales Company, and F.E. But Long wasn't looking for another assignment, and in turn recommended that they simply "let the question rest a little while until we can get the bunch together some time and talk it over a little."

There were other differences. F.E. wished that Long felt more enthusiastic about the advertising campaign. At issue was that plan to standardize and refine the grades of the fir lumber produced by the Weyerhaeuser Timber Company. To be able to recommend Weyerhaeuser lumber on such a basis had large appeal to those who thought advertising the key to future success. A guaranteed product was just the thing, or so they argued. Long and his plant managers were less certain, doubting that the effort of special grading would be worth the added expense. The only real difference between two two-by-fours, they maintained, was price. And special grading simply added to the price.

F. E. presented his case skillfully. "I should not be so insistent were it not for the fact that we all consider you the father of our White Pine grades," he said, recalling the day, years previous, when George Long had overseen that program for the Lake States lumbermen. He also warned against putting too much stock in Boner's and Warren's contention that "at the present time there is no money in making additional grades." Even if that were true, F. E. urged that they think in longer terms: "I am sure the final results will be good." Both Long and F. E. knew, however, that their good-natured debate was far from over.

One thing they agreed upon fully: the eight-hour workday was at hand, as well as greatly improved conditions for the laboring class. There was, however, a new and vital ingredient involved—the government. It wasn't a pretty picture: an eight-hour day for ten-hour wages, and time-and-a-half for overtime. But there was reason for optimism. As Long reported to St. Paul, it seemed likely that labor would grow "more contented and not be shifting about so much and really put in more honest effort." Over recent months he estimated that efficiency had dropped by as much as 20 to 30 percent.

Although his responsibilities were supposedly restricted to aircraft spruce, one government agent, Colonel Disque, had clearly become a dominant personality in the Northwest. Long was one of his strongest supporters. As he explained it, the colonel was, of course, "an army man, but he certainly is a good judge of human nature and is trying to solve not only the Government's problem, but the biggest problem of all, so far as it pertains to the industry on the Coast, and that is the labor problem." Allowing that Disque would "make a lot of mistakes," Long nonetheless doubted that anyone else could do as well: he was told to get spruce, and he was clearly going to get spruce.

George Long had resigned himself to being the "good soldier" when it came to dealing with Disque and meeting the spruce requirements. But he did try to lessen undue damage to the forest. He was much more concerned with the indirect damages that could result from selective logging than he was with the price of the lumber. As he observed to Ralph Burnside, president of the Willapa Lumber Company, he would "rather give them a present of $2.00 per thousand . . . if they would not enter our land and cut it the way they will have to under their necessities, but that simply illustrates what we think is the real damage done us and we did not take that into consideration in making them a price."

One of the most difficult tasks was to look beyond the war, at the world that would follow. When Long did that, he invariably thought of getting those spruce loggers out of his woods, of an end to labor shortages, and of Baltimore, perhaps in that order. His Baltimore attorney reported a recent purchase of forty-five acres for $240,000, this adjacent to their sixty-six-acre property. Long was delighted. The Timber Company now possessed a harbor frontage of nearly 1,100 feet, and Long was beginning to think about "the kind of development we will want to make there ultimately." Al Onstad, whom he now referred to as "our mill architect," would soon be dispatched to consider the development on the site, because part of the facility would certainly include a small resawing plant. The key to the operation, however, would be efficient unloading, handling, and storage of lumber. Then there would be the continual problem of the return cargo. What were these lumber ships to carry on the westward leg of their voyages? That was a question that would never go away.

Closer to home, Long listened to the plaints from Snoqualmie Falls. Warren was attempting to run the plant day and night, but didn't have enough workers. Long was sympathetic but there was nothing he could do to help. "It is about the worst time a man could have picked out to start a new plant on a day shift," he acknowledged, "and then to try and make yours a night and day stunt is simply attempting that which is a mighty hard game." But it was a game they had to play. They were caught up in a maelstrom, and the best they could do was provide for present needs while minimizing damage to future operations.

Finally, in November, the maelstrom ended with the signing of the armistice. Like millions the world over, George Long celebrated, relieved that it was ended, that son George, Jr., would soon be home, and that the Weyerhaeuser Timber Company could return to normal.

TWENTY-ONE

Beyond the Armistice

ALTHOUGH AMERICANS YEARNED FOR IT, "NORMALCY," as presidential candidate Warren Gamaliel Harding termed it, was not easily accomplished. It seemed ironic that what had been the hope and prayer of millions suddenly caused anxieties. In part this was because the war ended sooner than most had expected. Charles S. Keith, president of the National Lumber Manufacturers Association, admitted as much in a letter to Long, recalling that "General Pershing had cabled only ten days prior to the signing of the armistice that unless the forwarding of supplies could be accelerated, the offensive would have to stop." No one, not even the President and the Kaiser, seemed to realize how near the war was to being over.

Long was as unsure of the future as any of his colleagues. He knew only that it would be full of surprises. "These are chaotic days," he wrote to Allen, "and what was a live issue yesterday, is frequently supplanted by another one tomorrow." He also thought the war had ended about six months earlier than most anyone had expected, with the result that business was "more or less [up] in the air"; that included the Pacific Northwest lumber industry, which, he foresaw, was in for "its share of trouble." Suddenly, shipbuilding and airplane spruce, so crucial for a time, were history. It was back to business as usual, whatever that was.

Allen was also less than optimistic as he looked ahead. Some were predicting a brisk market for a few months, but others felt that there would be "a short period of intense depression," production declining in the face of high costs. Allen thought that wages were likely to "fall faster than the cost of living and create trouble." The War Industries Board was preaching curtailment so as to avoid price reductions, while the Labor Department was urging everyone "to speed up," thereby maintaining confidence. The overriding question was the government's continued involvement in the economy. Many who would traditionally oppose such meddling now worried that any abrupt withdrawal would have unfortunate consequences. As usual, Long counseled "a whole lot of patience and wisdom, none of which is hanging around on the bushes where you can pick it off in quantities or of a quality that commends itself as being a first-rate article."

The labor question had an immediacy in the Pacific Northwest, and Long advised Warren on the general subject of readjustment. The soldier-laborers in the woods and mills would soon be discharged, and as a result there would certainly be a temporary shortage of workers. Long recommended that Warren discontinue the night shift at Snoqualmie's big mill, but do nothing beyond that for the time being. Although labor and wage policies would undoubtedly change, none could foresee the specifics. The Christmas holidays would slow things down a bit, but in the meantime Long urged Warren to run "as strong as we can," explaining, "We will never get the true economies out of our investment unless we carry out this idea." And that should be their policy until conditions got "so rotten that it will be impossible to market the lumber."

An unusual factor in plans for Snoqualmie was Long's interest in electric logging engines. It was all very experimental. A contract was being negotiated with the General Electric Company and the Willamette Iron & Steel Works, but the manufacturers wanted assurance that the equipment would be purchased if the experiment proved successful. From a cost consideration, there seemed little doubt of the soundness of the plan. The falls provided an abundance of cheap power. Of course, there would be an initial capital cost for the power lines, materials, and construction, but this was assumed to be a one-time proposition. Here they were wrong, not having considered what the price of copper wire might mean to poachers. Long urged Warren to meet with Onstad as soon as possible and go over all the data and viewpoints he could think of, and then decide the matter.

A more immediate worry involved federal tax policies, particularly as concerned timberland assessments. As seemed always to be the case, there were differing views of what those policies were and what the Timber Company's position ought to be. Stated simply, the problem was this: they were required to set a value on timber as of March 1, 1913; sales subsequent to March 1, 1913, would be considered "profit" to the extent that they exceeded the 1913 value. But, of course, in practice it wasn't that easy. Long's contention was that there was timber, and then there was timber—the difference largely based on accessibility, which in turn influenced value and price. In the jargon of the moment, he didn't want the Timber Company's land handled "in the so-called en bloc fashion." Instead, he insisted, "The handy timber must be worth more than the remote."

One postwar development was especially disappointing to Long when he learned on December 16 that "some meeting of loggers" had passed a resolution which in effect summarily proposed an end to government contract work. In addition, they called for an investigation of the operation of the Spruce Division. Long was appalled, as he made clear in a letter to J. J. Donovan, of Bloedel-Donovan Lumber Mills in Bellingham: "Never for a moment did I comprehend or understand or believe that it came from the real loggers of Puget Sound, Grays Harbor or Columbia River, for the resolution could only be implemented as a criticism of the activities of the Spruce Division under the general supervision and direction of General Disque." But clearly it had happened, and now the "honorable thing to do" was to correct it. As to Disque's integrity, that was "beyond all attack." In fact, they should be praising him: "The real facts are that he was called upon to do, and did do a great many things in bringing about order out of chaos that was the lumbermen and loggers duty to have done themselves, and which apparently they either were unable to do or else were perfectly willing for Gen. Disque to do for them."

Long meant every word he said. Indeed, he was soon urging Disque to stay on, continuing to provide leadership for the Loyal Legion. As he stated for Disque's benefit, if anyone could provide the leadership, "you are the man." Disque had to appreciate the thought, but he decided, no doubt wisely, to accept a business offer elsewhere.

When Long returned to his desk in mid-January 1919, following an eastern tour, he found it piled high with matters old and new. Chief among the latter, as time would prove, was a letter from Oregon lumberman R. A. Booth, of the Booth-Kelly Lumber Company, requesting a meeting prior to the Seattle arrival of R. A. Long "and associates." R.A. headed the Long-Bell Lumber Company of Kansas City, and he and his associates were casting about for a Pacific Northwest investment opportunity. That would come to be, although not as easily as R. A. Long likely presumed. One of the inevitable results was that henceforth there would be much confusion regarding the two Mr. Longs.

R. A. Long met with George Long on the afternoon of January 29, and although George was cordial enough, it was apparent that he wasn't particularly eager to make a deal. He quoted his timber prices matter-of-factly: $3 per thousand for the fir, cedar, and spruce; and 75 cents for the hemlock. This drew no comment from R.A., "but he did ask if our terms could be made reasonably easy," and, as George Long reported to F. E. Weyerhaeuser, "I told him I thought they could be." He also indicated that in his "candid opinion," R.A. would "find something somewhere that will appeal to him as being so much lower in price than we will name, that if he does make his purchase, it will be from someone other than ourselves." George did predict that Long-Bell would be successful and eventually "locate on the coast and become large operators in fir."

Far less significant in terms of scale were negotiations with pioneer logger Charles H. Clemons of Montesano. Clemons was one of those characters George Long liked, not because he was a good businessman—he wasn't—but because his failings were those of the mind and not of the heart. Clemons had fallen on hard times, with liabilities of about $300,000. Out of these circumstances, Minot Davis negotiated a solution by which Clemons could continue operating, with the Weyerhaeuser Timber Company assuming majority ownership of his company. Long and Clemons agreed to the details, retroactive to December 31, 1918. "We will go ahead and organize the logging company with a capital stock of $210,000," Long wrote, "and the WTC will subscribe and pay for $180,000 of this stock and you will subscribe and pay for $30,000." In addition, the Timber Company would "convey to the new company all the property subject to $90,000 mortgage which you have conveyed to the WTC." Little more than a month was required before the new Clemons Logging Company was in operation. The marriage may have been one of convenience, but it proved to be wise and happy.

Meanwhile, Long's antidote to the confused present was simple—work. And so he urged both Warren at Snoqualmie and Boner at Everett to make as much lumber as they could. He agreed that "the whole world is going so crazy right now that one has to have all the common sense that belongs to him to really believe that we are going to get back where we belong at a very reasonable time," but Long still held that "before the year is over we will have a big demand for lumber." Also, as he assured Boner, there would be plenty of logs, "even if you should run the mill night and day." The more they made, he maintained, the lower the costs per thousand feet, and therefore the greater the likelihood of profit even under the most competitive situations.

Soon Long, Boner, and Warren were all worried about keeping up with the increasing demand for lumber. One thing they didn't need was complication in manufacturing and handling, but that is just what was in the offing, thanks to Lindsay's advertising campaign. Onstad was on loan to Lindsay, charged with designing machinery that would, it was hoped, brand lumber both efficiently and inexpensively. Grading and marking several hundred thousand feet of lumber each day was no simple task, and Onstad wasn't finding any quick solutions. It was to be a tiresome problem, Lindsay thinking about his mark, and Long thinking about how much that mark would cost.

Onstad was soon predicting that each marking or branding machine would cost in excess of $2,000, and that since the average eight-hour capacity of one of these machines would be about 50,000 feet, Everett alone would require ten of them. As Long reported to Lindsay, if the engineer's initial recommendations "were put in effect, it would mean that we would have to have a plant almost as large as the planing mill is now to brand the lumber the way we handle it out West." Onstad found himself in the uncomfortable position of working for two masters, one who cared very much about the results of his effort, and the other who cared not at all—and wished he could have been working on the details for the wholesale distribution yard in Baltimore.

There was good news from Baltimore. The legal complications had been resolved, and the city had deeded "Lot 4" over to the Timber Company as of December 19, 1918, for $3,100. This brought the total price for the entire property to $74,388.50—as their Baltimore lawyer had earlier observed, a real bargain. Now what was needed was a manager. In fact, Long already had picked his man. Brother Jim not only knew the lumber business but had become fairly well acquainted with the territory as a result of his wartime service in Washington, D.C. Long's decision, however, had no immediate effect. James was still assigned to the office of the director for ammunition purchase, and as yet he had received no word about his date of discharge. In the meantime, like so many other soldiers, Lieutenant-Colonel James Long waited.

There were plenty of other distractions. Among them were the familiar plaints from Boner and Warren regarding their dealings with Thompson Yards and the Weyerhaeuser Sales Company. Much of the difficulty was simply in the nature of things, but Long did believe that some benefit might accrue if the Sales Company moved its headquarters from St. Paul to a western site. He explained his reasoning to Louis Case, that since 95 percent of the lumber they would be selling would be from Idaho and Washington, the logical place for the sales organization's headquarters would be in the West. He admitted, "We fellows on the Coast" preferred Tacoma, Seattle, or Everett, "and I presume the Inland Empire people may think it should be at Spokane." In any case, "I can realize a vast number of advantages to the mills if the Sales Agency were within telephone call of the mills."

Case replied diplomatically. It certainly could be argued that the Sales Company headquarters ought to be "nearby and in close touch with the mills," but one could also argue that it should be "in close proximity to the salesmen who are in the field." It was, in Case's words, a subject that "merits careful thought." The debate would not be won by logic. Initially Long would get his way, Case moving his office to Spokane that winter. But eventually the two Freds, F.E. and nephew F.K., would move the headquarters back to St. Paul. Neither Weyerhaeuser would consider residing permanently in Spokane, or for that matter, any place other than St. Paul.

The end of the war did welcome newcomers to the Pacific Northwest, those who planned to make their homes and lives there. Many of the veterans returning from the war had finished college. If they were interested, how ought these young men be introduced to the lumber business? In a letter to Dr. E. P. Clapp, one of the major shareholders of the Laird and Norton family group, Long expressed his belief that "the larger and broader field of lumbering" was in the fir district west of the Cascades and that ultimately that would be "the big thing of the so-called Weyerhaeuser development." He had already given the matter considerable thought: "I recently made a suggestion to Mr. John Weyerhaeuser with reference to his own boy . . . and that is to give him a little peep into the various activities of the lumber game. The suggestion that I made was that the boy should spend a little time in a good retail yard to really see how lumber is sold and handled, and get the viewpoint of the farmer and the carpenter and of the contractor, and get a little of that knack of meeting people and trading with them, such as a retailer has to do." After that experience, the young man might be placed in a manufacturing plant, then on to the accounting office, "and possibly at some stage of the game let him do a little knocking around on the road, trying to sell lumber." In the long term, these experiences would be of value even if the young man's future involvement was simply as a stockholder.

One who was returning from the war was young Charles H. Ingram, grandson of Orrin. Charlie had written Long a letter of inquiry, and the recipient accurately sized up the applicant when he observed, "I rather imagine from your past experience that you are not seeking for clerical work but want something in the way of good active outdoor experience." He promised Charlie that he would reply, "just as soon as I make the rounds of our plants." Few applications were to prove more important.

F. E. Weyerhaeuser, along with Louis Case and George Thompson, spent several days in Tacoma in mid-May discussing, most significantly, the retail-yard network question. Specifically, should there be additional purchases? Long wanted this to be considered within the context of other financial requirements. His reasons soon became clear: "The net conclusion that we all reached was that the most important immediate activity of the Weyerhaeuser Timber Company should be to get in shape to manufacture more timber into lumber and put it on the market where it could be sold." As to the means of marketing, should they continue expanding their own retail network? This was an expensive exercise, as Thompson Yards had demonstrated all too clearly. The total investment, including $3 million of capital stock, now approached $5 million. Was it worth it? Long had his doubts.

Thus a critical decision was reached: There would be no further investments in retail yards for the time being, and when surplus funds became available, these would be committed to "the activity of making more lumber." Long was pleased. He might have been even happier had there been a way to sever the Timber Company's connection with Thompson Yards entirely, but they were now too deeply involved. The eggs could not be unscrambled.

In any case, Long was so encouraged that he began thinking about nearly forgotten subjects, such as what to do in Klamath Falls. Coincidentally, he heard from William P. Hopkins, trustee of that timberland morsel which had been coveted for years. He responded with atypical enthusiasm, noting that their properties had lain fallow for too long, and "the epoch of actual development is one that should not be postponed, if we handle our property for the best results." Long and Hopkins already understood that by combining their interests, they could have a "magnificent working unit." To this Long added, "I likewise feel, without any expression of optimism or egotism, that our knowledge of manufacture and marketing, and our facilities for marketing, ought to place your share of the interest in an amalgamated company . . . on a level where you would get the very best possible obtainable results."

From all indications, Long and Hopkins were of one mind as to the merging of their properties. They also agreed, for the time being, that it was probably just as well to keep quiet. Thus, when Walter Ryan was sent to Klamath to do some preliminary railroad surveying, he was told only that "we are all in there talking about land exchanges."

When the last shareholder headed homeward after the

annual meeting, Long began to plan for a Klamath Falls tour. He forewarned Jackson Kimball that sometime in mid-July, he and George, Jr., would pay a visit to Klamath County, and added, "I won't let any small matter deter me from making this trip, and I rather imagine there are others of the Long family that will be very keen indeed to taste again some of the hospitality of that wonderful country."

The Klamath pine would look especially good to young George, after spending five months salvaging burned timber in southern France. Father and son made their plans, the elder indicating that he wanted to share some hours with Kimball and Ryan "in the woods," and later, in the company of Hopkins, to "look over the situation that we all have in mind, as has heretofore been explained to you, about the possible exchanges of timber, etc."

Long and Hopkins negotiated easily, but such was not the case with the continuing discussions between Long and representatives of the Long-Bell interests. This was to be an intriguing contest. The Tacoma Long was in no hurry and, at this point anyway, neither was the Kansas City Long. "Like your good self," R. A. Booth wrote to George, "Mr. R. A. Long means to be extremely careful in his dealings and we are probably making good progress by going slow in the beginning."

So Long found himself playing a waiting game—one that only he was in a position to win. R. A. Long was a challenge, of that there was no doubt. He was pursuing his investigation of western Washington timberlands, but by now it was clear that Long the buyer was more eager than Long the seller. In a meeting on August 11, they finally agreed to send separate cruising teams into the Cowlitz County tract. The cruisers would probably take about three months and, as George Long now saw it, "this looks rather favorable to a probable sale."

Long spent the last two weeks of August traveling, first to Baltimore where he reviewed plans for the port development with brother James and Al Onstad. James Long, on duty in Baltimore for a couple of months now, had attended the annual meeting to respond to questions, but more important, to get acquainted with his new colleagues and some of the West Coast operations with which his facility would be associated. George Long was anxious for progress in Baltimore. His most frustrating problem had been freeing Onstad from his responsibility with those damn branding machines. A decision regarding the machines had, however, apparently been reached: they were going to test them out at the Rutledge Timber Company in Coeur d'Alene, Idaho. Now Long insisted that Onstad concentrate on Baltimore. Lindsay and his cohorts would have to wait. Within days of Long's departure from Baltimore for St. Paul, James contracted for the construction of Baltimore's dock and bulkhead.

The meetings in St. Paul proved rather more complicated than expected. There was discussion, of course, of the possible Long-Bell trade, "but a much larger portion of the time [was] devoted to Thompson Yards development on the Midway and our own developments at Baltimore, and then certain questions pertaining to the Weyerhaeuser Sales Agency." The Sales Company received the most serious attention, resulting in a decision to reorganize. This involved the withdrawal of the Minnesota mills, those in Cloquet and Little Falls and the Northland Pine Company operations as well as the Warren, Arkansas, plant, since they were "practically cut out." Accordingly, a new agency would be organized with headquarters in Spokane, as George Long had earlier recommended, "the directors of which should be the presidents of the western companies, or their General Managers . . . and the Sales Agency thus organized would have complete and full charge of the sale of all lumber of all western mills at market prices." Coincidentally, those who previously had been serving as sales representatives of their respective companies would hereafter be in the employ of the Sales Company.

These were buoyant days, encouraging operators to think in expansive terms. At the same time, markets could fall as surely as they could rise, and today was not forever. Thus Long worried a good deal about sales strategy, and so too did Boner at Everett and Warren at Snoqualmie. As he wrote to the former, "One of the hardest lessons that I ever had to learn was to sell lumber very freely on what appeared to be a declining market, and I am just as sure as anything that we are facing a declining market." Long therefore advised Boner to load up with all the business that he could get, even at the risk of damaging their reputation for providing prompt service.

That may have been sound advice, but at the moment Boner had a more basic problem: he had the orders; he had the lumber to fill the orders; but he didn't have the cars to carry the lumber from mill to market. They were making a million feet a day, and possibly shipping four or five hundred thousand. One potential solution was Baltimore, Boner urging that they "hasten with all possible speed" the development of that project.

Despite Long's best efforts, Baltimore was too big and too complicated to be hurried. As he explained to Boner, there was much dredging and filling yet to be done, and "I cannot conceive how it would be possible for us to hope to ship much lumber around there before the first of next March, but I have today written our manager at Baltimore to make up what, in his judgment, will be the initial order that they ought to place with the mill."

Louis Case paid a mid-September visit to James Long in Baltimore, and they agreed to early arrangements between the yard and the Sales Company. Because of the frequent fluctuations in prices that the Everett mill made based on market conditions on the coast, it seemed advisable that the Baltimore branch deal directly with Everett, sending copies of orders to the St. Paul office. The final decision as to whether Baltimore would avail themselves of the services of the Sales Company could be delayed until experience indicated the best course to follow.

There was, however, the reorganization of the Sales Company still to effect. This took place in October, when Long and his western colleagues turned to the Sales Company question following a seminar on the federal income tax conducted in Spokane by Major David Mason. Since all were together, Long decided that they could "kill two birds with one stone," Sales Company business being the second bird. The new articles of incorporation were agreed to with very few changes. The associated companies would support the sales agency on a pro rata formula, and the trustees would be representatives of the participating plants.

The usual day-to-day activities were suddenly and tragically upstaged by an Armistice Day event in Centralia, Washington. A parade was held by the American Legion, and the marchers had as a destination the IWW headquarters. In the excitement of the moment, shots were fired and three paraders fell dead. George Long sent a note of condolence to the father of one of the victims, killed, in his words, "at the hands of the enemies of mankind." Although the IWW had made little progress in its western Washington campaigns, labor unrest continued, and for good reason. The Centralia tragedy served only to dramatize the tensions that remained a part of life in the Pacific Northwest and elsewhere.

It may seem ironic that Long was, at this time, making decisions about how to handle the company's fast-accumulating profits. Now, in the fall of 1919, he sought the counsel of the Weyerhaeuser brothers on the question of dividends. F.E. thought that a 6 percent figure was about right, even though the Timber Company could easily have managed double that amount for the year. But as F.E. saw it, a high dividend would set a bad precedent and would "make them unhappy in the future." More important, "We are needing so much money in the development of our properties that I dislike to see any liquidations made when we can put our money safely and profitably into our own companies." That is the way that father Frederick would have reacted.

Basic to the question was what the Timber Company ought to become: "One where we sell large blocks of timber, or whether we go slow in that direction and proceed with some plans about further development of manufacturing plants." In this instance, F.E. expected that Thompson Yards would need a large amount of money in its development of the Midway property, and "personally I would much rather see his paper taken up by the Weyerhaeuser Timber Company than make an extraordinary dividend." F.E. closed, allowing, "Excepting our family, all of the stockholders are looking for dividends." George Long, of course, already knew that.

Now, surprisingly, the St. Paul deliberations resulted in something of a compromise. Long was given a free hand to proceed in his talks with Long-Bell. In his subsequent meeting with Booth, still serving as an agent for Long-Bell, Long held firm. Booth insisted that R. A. Long required additional timber, and George agreed that the "area might be increased in the immediate vicinity." But Booth seemed "bitterly disappointed" that Long was unwilling to offer the large tract earlier mentioned in the Grays Harbor district, "and to which Mr. [R. A.]

Long's attention had been called at the time of our first interview with him in the winter of this year." Booth could hardly have been accustomed to dealing with one so adamant. Nothing changed: The deal would eventually go through, but it would be on George Long's terms.

He was satisfied with the pace of the Long-Bell negotiations. There was no need to hurry. And it was essential that both parties be certain of the propriety of the arrangement: The Weyerhaeuser Timber Company wanted Long-Bell to succeed, for a number of reasons. Most obviously, payment would depend on that success. And in a larger sense, Long-Bell was about to do what George Long eventually hoped to: build a manufacturing plant equal to the forest resource. Thus he could observe and learn while another spent the money and made the mistakes.

Long was less patient concerning the progress at Baltimore. One of the most irritating aspects of that situation had nothing to do with Baltimore. Onstad was once again elsewhere, this time in distant Coeur d'Alene, trying to get the branding machines working. Long might have found this amusing, had Onstad not been wasting valuable time. Lindsay reacted almost sadly to the problems experienced at Coeur d'Alene: " 'Tis an ill wind that blows nobody good."

Lindsay was clearly getting on Long's nerves. If only those in St. Paul could find ways of expressing their interest that didn't waste time or money. But if Lindsay was an annoyance, the federal tax questionnaire was a real aggravation. As Long informed Allen, "To answer the questionnaire intelligently is going to take a colossal amount of work in the preparation of maps and schedules and information as to prices." In short, he had realized the impossibility of meeting the December 26 deadline and had already requested an extension. Still, he hoped to have the material "in pretty good shape very close to the time set by the government."

Lumber sales were always a worry, and such was certainly the case in late November when Long ordered Sales Company personnal to stop selling. They probably should have done it two weeks earlier. Prices were skyrocketing, and deliveries were impossible as a result of the continued shortage of cars. As Case saw the situation, "The erratic prices on lumber can be accounted for only by the fact that the buyers see a tremendous demand ahead and are trying to get under cover and are willing to place orders with anybody that they think can furnish the stock and at any price they ask." Case also warned Boner that getting salesmen to stop selling was easier said than done. But the word had gone forth.

Long thought he knew the reason for the "abnormal demands" that encouraged sellers to assume there were no limits. First, the country was building to catch up on what had not been built during two years of war. Second, labor everywhere was producing less than at any time in American history. And third, the railways were gradually slowing down in their efficiency and were not handling the freight. Sooner or later, the problems would be resolved, but, he predicted, "It is going to be a slow game." Both the Everett and Snoqualmie yards were full—120 million board feet of lumber piled at Everett and half that amount at Snoqualmie—with the result that unless they got relief soon by shipment, they would have to shut down the mills, because they would not have room to pile the lumber.

At the first meeting of the newly organized Weyerhaeuser Sales Company, held in Spokane, Long was nominated for president, but he respectfully declined, citing his workload. He also simply did not want to assume that responsibility. Instead, Tom Humbird became the president, Bill Boner, vice-president, Allison Laird, secretary, and Huntington Taylor, treasurer. Case prepared to move his office from St. Paul to Spokane early in the new year, with all of the general salesmen of the various companies now to be in the Sales Company employ.

Humbird took over immediately, and his initial concern was that Case not continue to make all the decisions himself. Humbird thought the men on the road were treated like "orphans," and that there was "an utter lack of the spirit of pride in the organization." He concluded that a principal cause was Case's reluctance to delegate authority. He urged Case to draw the sales managers into policy making: "This not only involves the initiative on the part of all these gentlemen, but they having helped to decide the best method in the case, are not in position to find fault with results, and they are made to feel that they are an important part of your organization." And he added, "I am strongly of the opinion that any plan that does not place responsibility on these sales managers, involving their initiative and in this way interesting them in the work, will prove a failure." Long was already relieved that Humbird was the one responsible.

For the moment, Long was deskbound, wrestling with the tax questionnaire. As he informed lawyer Stiles Burr, "I think by making a herculean effort I might be able to get the Weyerhaeuser Timber Company questionnaire, with all of its details, prepared by the middle of January," after which he would travel east to meet with Major Mason and resolve any outstanding questions. Another very important matter would be coming to a head early in January, he noted: the negotiations with Long-Bell. Given those negotiations, federal tax determinations, the Baltimore development, production decisions, labor problems, and plans for Klamath Falls, the start of the new decade promised to be full of challenges.

TWENTY-TWO

The Twenties Begin with a Roar

GEORGE LONG'S PLANS FOR A WASHINGTON, D.C., VISIT
were delayed, largely because of matters relating to the new
federal tax. How was it possible to describe in detail every single
transaction? Long's dilemma was the price of doing business for
two decades in a big way. With a sigh, he now began to think in
terms of a March departure.

If Long was frustrated with his tax problems, he had to
be reasonably satisfied with the progress of the Long-Bell purchase. The cruising of nearly 24,000 acres was completed in
mid-January and evidenced a total of about 2.5 billion feet.
Long described this as being "a very magnificent stand of timber [and] indicates, we all think, about as high an estimate as
you could squeeze out of it." In early February he met for
several days with representatives of Long-Bell at the Blackstone
Hotel in Chicago. And when they parted company, "the only
thing left is to have signatures attached."

In Tacoma, the tax questionnaire awaited, now due on
March 1. As Long lamented to Tom Humbird, he was "balled
up with a terrific amount of work. . . . Everybody around this
office will have to put in overtime to get our document in shape
to send forward." Nevertheless, Long carved out time to attend
a mid-February meeting of the advertising committee in Spokane. Thus far no one seemed happy with the Sales Company
arrangements, but that may have been inevitable. The market
was simply out of control. Long subsequently reflected on the
situation in a letter to Boner: "I do not know anyone quite wise
enough to know just what to do these days, but one thing is
sure as Fate itself, that things are drifting in a direction where

someday we are going to have a horrible big smash." He added, "I very sincerely wish that the whole country was on a slower pace, but they are alike everywhere and in every line of business, and we are all caught in the current and seem to like it."

Long was not the only one worried about inflation, even specifically inflation in the lumber industry. Increasingly, expressions of concern were being heard in Washington, D.C. Members of Congress inquired of the Forest Service why prices were rising: did it have to do with monopolistic ownership or forest depletion? These were the two biggest bugbears to be confronted, on the surface a logical explanation of what seemed to be cause and effect.

Word of the developments in Washington, D.C., came to Long via E. T. Allen. In reply, Long reported on the recent meetings and "the attitude of the Weyerhaeuser interests on calling a halt on the skyrocketing of lumber prices," and he added, "Personally I have been preaching it myself for more than three months past."

An official announcement dated February 24, 1920, detailed the Sales Company position, this being "the result of a conclusion of all of the managers of the entire so-called Weyerhaeuser interests." The current market conditions were noted, particularly as concerned the recent rapid and continuing inflation of lumber prices, culminating "in what is termed an 'Auction Market,' which is always irregular and uncertain, and most demoralizing to buyers and consumers, with its constant threat of price advances not to be measured or foreseen." The key ingredient of the announcement was a commitment to price stability, that "at least until June first, which covers the whole period of Spring buying, [the Weyerhaeuser mills] will make no advance in prices over the scale in effect early in January. . . ." While many welcomed the news, others wondered about the influence such an action implied, even though tucked away in the announcement was a reminder that the combined output of their mills was not great enough to have a decisive influence on the market. This was intended to be but an example of responsibility to the industry, a small first step.

One incident in particular raised the specter of trouble ahead. The Forest Service had placed a parcel of Montana timberland for sale in the usual manner, calling for bids. But when the Bonners Ferry Lumber Company turned out to be the highest bidder, the Forest Service "deemed it best not to confirm the sale," because, according to Allen, "it was a possible monopolistic situation." Long had little interest in the details of that aborted transaction, but he was concerned about any accusation of Weyerhaeuser monopoly.

It was easy for outsiders to assume that the Weyerhaeuser Sales Company was a piece in a monopolistic scheme. True, it would serve the affiliated mills, but the intent was to do away with "lost motion, and a multiplicity of expense." The purpose of the Sales Company, as Long pointed out, was simply to provide better service to the public. Lower expenses made for lower costs, and buyers preferred to deal with sellers who offered broad services at low prices. Long didn't wish to pretend that the Weyerhaeuser interest was smaller or weaker than it was, but he did want to remind one and all that the production of the affiliated mills was not in excess of 2 or 3 percent of the lumber purchased in the United States, and even if they were to double their production, which was hardly probable, they would not produce 5 percent of the lumber consumed in the United States. Monopolies, Long repeatedly pointed out, were possible in a good many American industries, but not in lumber.

Nevertheless, some Forest Service personnel seemed convinced that the Weyerhaeuser interests were inclined to monopolize white pine. Long was reassured by Allen's position, serving as a bridge between his Tacoma desk and Forest Service leadership. As Long described it: "We have a medium in Mr. Allen of very potent strength. He believes everything I tell him and I only tell him the truth, and the Forest Service believes Allen." It was indeed a "fortunate alliance."

Amid the frustrations and uncertainties, two important steps were completed. Long mailed in the General Forest Industries Questionnaire on February 25. Accompanying the questionnaire was an expression of interest in meeting with Major Mason during Long's next trip to Baltimore, at which time he would be happy to provide "any further or added information that may be required or desired in connection with our return."

The second major accomplishment was really a lawyer's exercise: the final details of the Long-Bell purchase. George Long received the signed contract on February 24, with a check for $503,384.17. Everything was in order. They only hoped that they had played their cards so quietly that officials in Cowlitz County would not consider reassessing timberlands.

With federal taxes and the Long-Bell purchase behind him, Long turned his attention to the publicity campaign. He saw much of this as being unnecessary or, worse, unnecessarily expensive. While there was some pleasure in seeing boards smartly stamped with the company logo and grade, stamping required time, machines, personnel, and, therefore, money. Long was convinced that buyers of lumber made decisions based on price, not gimmickry. As he put it, they should be selling lumber "instead of talking about it."

Long addressed his concerns to a Pasadena-bound F.E. In the wake of the Sales Company determination to stabilize prices, the advertising proponents wanted to go full speed ahead, viewing the price announcement as "a big opportunity to start our advertising scheme." Long considered this to be an end run around the stated policy of proceeding only "after it had once again been submitted to the stockholders of the various corporations for their approval." If now was the time to begin, Long urged that they do something positive in the way of education, informing the public "as to who the so-called Weyerhaeuser interests are, and what their facilities are." He probably wished that that would be the end of it, although he knew otherwise.

It wasn't a happy F.E. who responded. He thought Long and some others were dragging their feet. "Referring to our general advertising," he later observed, "I am disappointed that some of our people are still very much in opposition to it, and at the same time, as I believe, have never given the subject very thorough consideration." Although he denied having "urged the adoption of our General Advertising Campaign," he did acknowledge that he had always supported it and was "now more than ever convinced" of the correctness of that commitment.

At Long's behest, Humbird reminded the St. Paul crowd of the executive committee action promising shareholders one more chance to vote on the advertising campaign. And if that proved an insufficient brake, Long "had another one in the background which I was going to spring with all the force that I could bring to it, namely, that it looked to me like rather a cheap performance to go ahead and announce our program as to prices, and then to follow it up with a whole lot of advertising of the kind which they have already got mapped up, and it would lay us open to the very natural criticism that the only object we had in view in launching our cheap prices was to get the benefit of a lot of cheap advertising."

The concern for public relations came more naturally to Long than it did to those in St. Paul, no doubt due to experience. In recent weeks, it was public relations with the Forest Service that had been foremost on his mind. Worried that the spirit of cooperation had suffered unnecessarily because of relatively minor and passing differences, Long received with delight the news that his lawyer friend A. W. "Gus" Clapp had accepted appointment as an ex officio member of the committee of the National Lumber Manufacturers Association "to deal with or cooperate with the Forest Service." As Clapp saw the problem, "Lumbermen have allowed themselves to ignore the Forest Service; to treat them as theorists and to reject for that reason any idea of co-operating with them; and I am sure that this is not the right policy." Long agreed completely.

Less than a week earlier, Chief Forester Henry S. Graves had written to Long on a variety of subjects, including state purchase of cutover lands. Long had, of course, been advocating such a policy for years, and now Graves promised his "hearty support . . . if there are any ways in which I can assist this movement in which you are taking the leadership." Like Long, the Chief Forester looked forward to the day when taxation and fire control could be managed in such a way "to make it possible for the private land owner to take a definite part in timber production." But for the moment, "the public must clearly lead the way."

The support was encouraging. Graves, however, was on his way out as Chief Forester, and attention naturally turned to his probable successor. There was some brief fear that Gifford Pinchot might be named, but apparently that wasn't seriously considered. Colonel William B. Greeley, who had commanded the forestry regiment in France, proved to be the designate, and though Greeley was hardly an unknown quantity, Long initially offered a reserved expression of support. "I think Mr. Greeley is a very able fellow," he observed in a letter to Clapp, "but a very determined man when his ideas are once formed." And he added, "I do not think he will be vacillating and we at least will always know where he stands, and that is worth something." In fact, it was worth a good deal, and as the years passed, Long and Greeley grew closer together, both professionally and as friends.

In late spring, according to plan, Long went east. The most important item on his agenda was an appearance at an Internal Revenue Department hearing conducted by Major Mason, which went as well as could have been expected. All of the valuations in Washington were accepted. There were some differences in both the Oregon fir and pine districts, but in general Long had no complaints: "We are all inclined to think that we got very reasonable treatment and our attorneys especially were very much elated."

Nevertheless, Long was relieved to leave tax matters behind and head to Baltimore and a meeting with brother Jim. Progress on dredging and construction had gone slowly, primarily because of abnormally severe winter weather. Still, the pier was nearly completed and work was proceeding on a railroad spur. Lumber had begun to arrive—by rail—and machinery had been ordered, including a crane, "but everything is coming slowly, and the chances are that we will not get that property in shape to receive lumber by vessel much earlier than October or November, in fact we may be fortunate to accomplish it by that time."

Accompanied by F. H. Thatcher, Long continued on to New York City, where they joined William Carson and Charlie Weyerhaeuser to confer with representatives of the United States Steel Corporation "about the possibilities of their hauling lumber for us from the Pacific Coast." For U.S. Steel this would be a back haul, their own cargoes moving almost entirely east to west. Like it or not, Long was beginning to learn about a completely new subject, and as he learned he began to speculate on matters of tonnage, "and the way I have it figured out is that freight rates by rail are apt to go up and freight rates by water apt to go down before a great while, so that ultimately the moving of lumber by water to Baltimore will have relatively even a better prospect for a successful venture than it has now." However promising the future, it was far too early to consider an "experiment of building ships and I have not yet heard of any for sale that were offered at a price that seemed conservative enough for our wants." Once again, Long was willing to wait, to sell or to buy, until the right moment. Patience meant knowing when to act.

The only major disappointment on this junket took place in St. Paul, where Long tried to get a group together to discuss the financial situation of the Thompson Yards. Unfortunately, he "found everybody away from home." Long wasn't happy. For F. E., he noted the present cost of lumber, "as well as all kinds of material that [Thompson] buys, and with his policy of stocking up freely, it is taking a colossal amount of money to handle his business." The Timber Company had provided Thompson with nearly $1.2 million over the past two months, and Long hoped that he would "be getting over the peak pretty soon and begin to have money to distribute, or at least to pay some of his debts."

Accordingly, Long invited Thompson to attend the annual meeting of the Timber Company, as most of the stockholders in Thompson Yards, Inc., would be in attendance. This would provide "the first real good opportunity" to discuss the details of the situation. In preparation, Long instructed Thompson to draw up "a condensed statement which will reflect your outstanding liabilities in shape of notes to banks, individuals and open accounts to individuals, including not only the accounts which you owe for lumber and miscellaneous purchases, but the accounts which represent borrowed money, in other words, something of that kind that will be a statement of your actual liabilities as of June 1st." He also wanted Thompson to detail how much material was on hand at the various yards and to offer a prediction as to future outlays. In short, he wanted to know exactly where matters stood. And he wanted others to know as well.

In some ways it must have seemed that the concern with Thompson Yards should not rest with George Long. But fair or not, Thompson was a problem for the Timber Company, and therefore a responsibility. Thompson Yards was intended to be a major seller of Weyerhaeuser lumber, and on that basis it had incurred a very large indebtedness. Long wasn't going to see that money lost. It belonged to his company.

On the eve of the annual meeting, a special confab was

held to discuss the advertising campaign. Most of the principal stockholders were in attendance, as well as members of the Weyerhaeuser Forest Products Publicity Executive Committee. In the course of the deliberations, George Long achieved a partial victory: the advertising campaign was authorized to proceed, and lumber would be trademarked, but no species and grade marking of lumber would be instituted. To Long, this constituted a triumph of logic and reason, and for the immediate future he would keep other misgivings largely to himself.

The annual meeting itself went well, with the largest attendance ever. Long couldn't help but note the number of young people, with "nearly all, if not all, of the old crowd having disappeared." Such gatherings, of course, are made all the easier by good times, and good times there had been, "although our profits are not very large." From a price consideration, the figures were impressive. Everett lumber shipments for the month of May over the past several years averaged as follows: $11.49 per thousand in 1915; $13.44 in 1916; $14.20 in 1917; $22.43 in 1918; $22.68 in 1919; and $45.93 in 1920. But expenses had also increased, "at an astonishing rate," in Long's words, though admittedly at a far slower rate than prices.

Less pleasant was the review of problems confronting Thompson Yards. If no quick answers were forthcoming, one thing was clear: the Timber Company was so deeply involved that it couldn't afford to let Thompson Yards fail. Prior to the meeting Long had received a pessimistic report from George Thompson, and he now found himself in the unfamiliar position of having to encourage Thompson, a man he had previously been trying to discourage. "Everything has been too giddy," he wrote to Thompson, "and wise men in the country thought that it undoubtedly would topple over, if there was not a stop in the upward flight of everything and a real inventory taken of our actual situation." This was an expected period of adjustment. There was only one thing to do and that was "to keep on feeling optimistic and cheerful and bucking the game with all your old time vigor."

There was one major setback: Long failed to elicit support for any commitment to an early Klamath Falls development. He shared this disappointment with Hopkins: "I fear our crowd being from the East were a little more pessimistic than we are out in the West, for they seemed to entertain the notion that it was rather an unfortunate time to make such an investment." Long worried that Hopkins might not wait on a mere promise, but still he promised: "I feel it will not be long until I can bring about a purchase."

Meanwhile, the advertising issue was white hot. Lindsay, having assumed that his campaign would be automatically approved in its entirety, was clearly surprised at the qualified approval. He saw the thing so simply: they should organize a single advertising effort, "shooting at one central bull's eye," instead of spraying shots at random. How could anyone oppose that? "I now realize," he acknowledged, "that I never fully sensed the deep-seated feeling of the Laird and Norton interests against this Publicity Campaign, my interpretation of their general attitude being toward liquidation rather than expansion, in other words a general tendency to retrench rather than to develop." What made it all the more painful was what Lindsay understood so well, that enthusiasm was "everything, and with me I so well know that this comes largely as a result of close and sympathetic understanding, cooperation, and encouragement in our publicity program, I having been so constantly put on the defensive, and which has made it hard indeed."

Long knew he was on the horns of a dilemma. There was going to be an advertising campaign, but what form would it take? Long decided to lend his influence to the cause, trusting he could also influence its conduct. He attended a meeting of the Forest Products Executive Committee in St. Paul on July 13 and offered the key resolution, that as of June 1, 1920, each participating company would commit to the central advertising program on a basis of 45 cents per thousand of its monthly cut, and that from such credits Weyerhaeuser Forest Products would receive remittance as needed, "it being understood that the active campaign of advertising will be inaugurated on or about October 1st, 1920."

The resolution passed unanimously. Long, however, still couldn't convince himself that in the end the publicity campaign would be worth the expense and the inherent dangers. Costs were a constant worry, but Long also worried more about the content. Would he get advance copies, he inquired of George Dyer, Weyerhaeuser Forest Products' new advertising consultant, "especially during the early epoch when the foundations of what we propose to do are to be set forth. . . . The thought for all this," he explained, "is that we are making public sentiment as well as simply advertising lumber, and we want to build it up on a substantial basis and in a way where it cannot be picked to pieces by any erroneous statements that might possibly creep into our advertising." The point was simple enough: Dyer, whose company introduced the public to Gillette razors and blades, may have known advertising but he didn't know Weyerhaeuser.

Long had allowed himself to be caught up in an advertising campaign of which he didn't fully approve, largely because many Weyerhaeuser shareholders were on the other side. The tables were reversed, however, when it came to Klamath Falls. If Long was to have his way, he had to get the support of the Laird-Norton shareholders. That wouldn't be easy.

While in St. Paul, Long had met with William Carson, Horace Irvine, Harold J. Richardson, Charlie and Rudolph Weyerhaeuser, "and they all seemed to concur in the idea that it would be worth while to try and buy the Hopkins Tract." Long now informed F. S. Bell, chief spokesman for the Laird-Norton group, that William Hopkins had agreed to extend the option another sixty days, "by which time I want to make further and complete and satisfactory investigations and feel reasonably sure that we will be able to buy it on the basis of what we find to be on the land." Long sent Thatcher a telegram with the same information, acknowledging that "our people did not authorize me to purchase nor to negotiate but I hoped they would change their mind." He closed sounding flippant, but he wasn't: "Wish you and Fred [F. E. Weyerhaeuser] could join in a telegram to me . . . to buy it!"

Thatcher did as requested and, as Long knew he would, F. E. argued in favor of the purchase. Then by letter Thatcher detailed his ambivalent feelings on the subject: "I could see but one thing to do in reply to the request of your telegram, and that was to comply. The only other thing I could have done was to have answered you, the same [way] you answered Mr. R. A. Long in our Chicago negotiations when he had logically brought you to his position and you had admitted it, and when he asked you why you did not 'come across,' your reply was 'I don't want to.' My feelings on getting your telegram were the same as your feelings when Mr. Long appealed to you. But I did not have it in my heart to so say in reply to you."

Thatcher "guessed" that the group favored the purchase, but development was quite another matter. In Thatcher's mind, there were too many other expenditures in the offing— Baltimore, for example, as well as a fir manufacturing plant at Willapa Harbor. Where would it all end?

The option on the Hopkins tract purchase was to have expired on September 15, but Long got another week's extension while Minot Davis and a crew labored to confirm the details. On September 17, Long telegraphed shareholders that he would proceed with the purchase unless the "majority of trustees advise to the contrary by wire." And he closed assuring, "Our finances are such that we can handle this deal." The enthusiasm of the responses varied, but Long got the necessary approval. On September 30, he wired that the Timber Company made the purchase "on terms outlined in agreement of July twenty-eighth." In round numbers, this involved 40,000 acres, approximately a billion feet of timber, for a price of $2.5 million. Thus the lengthy battle came to a happy conclusion.

Business continued, in Long's words, "on a slow bell," so slow that he was talking about shutting down operations on December 1. But despite the present depression, there was lots of cash in the till, with more on the way. For example, $750,000 was due from Long–Bell in November, the initial payment for the second transaction. The trial balance on October 31 read $600,000, not counting $2.8 million of Treasury certificates. Bell inquired what dividend might be expected, noting that many were assuming the 10 percent paid in 1919 would be repeated. Even Long, ever the hoarder, was hard pressed to argue otherwise. "There is," he admitted, "no physical reason why the company could not, if we saw fit, make another dividend at the present time, but . . ." And he had several reasons why not. Thus Bell read the litany, from new investments to emergencies.

Bell agreed that they could differ over the size of the dividend, but he urged that a dividend be paid. As for Long's desire "to hang on to the cash," Bell countered, "too much money on hand is almost as bad as too little." In addition, he reminded Long "that a collateral element in the policy was to give the stockholders, little as well as big, some assurance that their income from the company would be somewhat approximate to its earning power, and so perhaps give the stock a broader market for those who might find it necessary to sell." It was a familiar refrain.

Several supported Bell's contention that they could safely declare a 10 percent dividend, but, as usual, Long received support for "the old conservative policy" from the Weyerhaeuser brothers. John cited "the heavy indebtedness of the Thompson Yards, and realizing they may soon be demanded the money, I do not know where the money will be coming from unless the Weyerhaeuser Timber Company can finance it." F.E. responded directly to Bell: "I thought the man who was responsible for the financing of our subsidiary companies, such as the Thompson Yards, should be consulted before any large distributions are made." And that man, of course, was George Long. Given the differing views, Long felt comfortable following his own inclinations. The dividend would be 6 percent, "anticipating that a formal action of the trustees will follow some time subsequent to ratify this action."

As 1921 commenced, American political activities were somewhat divided between Washington, D.C., and that most unlikely small town, Marion, Ohio. There President-elect Harding held forth, smiling and jovial, meeting this delegation and that, a sometime listener but always the good host. One of those delegations represented the National Lumber Manufacturers Association, conferring with the president-elect for some forty-five minutes on January 8. As Secretary Wilson Compton reported, their purpose was to inform Harding of the lumber industry's support for the legislation before Congress.

At the moment, however, Long was more concerned with the Federal Trade Commission's report on the lumber industry. He worried that the recent past had put lumbermen along with everyone else in a vulnerable position, especially regarding prices. "The theory seemed to be," he observed in a letter to E. T. Allen, "that whatever price the other fellow was willing to pay was the right price to ask," and the only ones not participating were "those who were so situated that they could not float with the tide." In this, lumbermen were no better nor worse "than the average in any other line of commercial activities," and the trade commission "could say the same thing about the shoe business, the wool business, the steel business, the sugar business and possibly of all food products, and I do not know but what about everything else." In short, events had proved once again that " 'War is hell,' and we were all in the swirl of the cesspool not only during the fighting epoch, but in the two years following."

In fact, Long was feeling a bit defensive, believing the Weyerhaeuser record was better than the rest. Thus he reminded attorney Stiles Burr that the "Weyerhaeuser crowd did show some marked signs of sanity" when the rest of "the whole world was crazy in the early part of 1920. I do not presume that we will get much credit for what we did, and yet it would be interesting if this little performance of ours could get into the record in some way which the Trades Commission is passing on to the public."

The slights against Weyerhaeuser may have been imagined, but the hard times were real. And as always, Long had to be the team leader, sounding optimistic while preaching the virtues of parsimony. He discussed the problem with J. E. Rhodes, now general manager for the Southern Pine Association. The nation, he predicted, was "in for a reconstruction of values and unfortunately we have rather a high peak to work from," with lots of accompanying uncertainies. Still, Long thought the United States was "weathering this ordeal in rather a manly way and we are making real progress in settling down to a basis that will have a solid foundation under it." What to do about wages was, however, a puzzlement. On the one hand, Long didn't wish to turn the clock back and he knew that any reductions would cause hardship, but finding themselves operating on "a close margin, we are greatly handicapped in marketing our wood by paying for eight hours labor what other districts pay for ten."

The question of wage and salary reductions was by no means cut and dried, and the situation seemed doubly difficult at Snoqualmie Falls, the company town. Long sent a letter of inquiry to Rod Titcomb, son-in-law of John P. Weyerhaeuser and recently appointed assistant general manager there. Long began by describing a recent discussion with "a gentleman who has some very pronounced ideas about wages and the cost of living," who also ran a sawmill and a store in Snoqualmie, and he shared a list of wages "very much less than any schedule that I could approve," along with a list of the retail prices he charged in his store. Long was curious as to the relationship of wages to expenses, thinking that would most accurately reflect the circumstances of the working man.

At Long's directive, Titcomb did some research, and the figures he uncovered corroborated Long's contention: Weyerhaeuser had tried to be fair in keeping prices as low as possible, "in some cases hardly paying for more than the handling." For example, the retail price for potatoes was 12 cents per pound on July 1, 1920, and had fallen to 2 cents as of February 1, 1921, with wholesale prices dropping from 7.5 cents to 1 cent; similarly, the price of rice had fallen from 20 cents to 6 cents a pound, and a pair of overalls, $3.50 to $1.80. And so it had gone. Titcomb further claimed that the community appreciated the fair dealing. "Only yesterday," he noted, "one of our old men told me he thought we were doing a wonderful thing, that this was the first Company he had ever worked for that gave credit to the men as generally as we have during the shut-down."

Long seemed genuinely relieved: "I feel that we not only can afford to be very liberal in our idea of profit taking, when we sell food, clothing and other supplies to our employees, but I think at this time particularly we should be very watchful to see that our employees especially get the benefit of every reduction that can be made." Clearly, that would be the policy. Although Weyerhaeuser was not to blame for the hard times, "any way that we can consistently and safely reduce some of the burden on our men, I am very well pleased to have this done, as apparently you are doing it in a very sensible and practical way."

In late January, Long went to St. Paul to attend an important meeting of the By-Products Committee. The members agreed to undertake the manufacture of wallboard from sawmill waste by means of a new organization, the Wood Products Company, to be located at Cloquet, Minnesota. Cloquet seemed the logical choice, "where they have more or less abundance of power, plenty of waste and likewise where the Northwestern Paper company (a subsidiary of the Weyerhaeuser Group) have a number of very expert men familiar with the details of making paper and various other kinds of by-products from wood." It was presumed, as Long explained to others, that this was to be "somewhat in the nature of an experiment," and if successful, "then we would probably install similar plants at any or all locations where the raw material and the power and the market were available."

Attitudes regarding "experiments" are generally ambivalent, and so it was in this instance. By-Products Committee Chairman Carson warned against undue optimism: "I wish to say that we all feel that we are starting out on a new enterprise and do not look forward to immediate success," adding, "We realize that we have many problems to solve and that we must look forward to a number of disappointments." Few assessments carried greater truth.

Carson didn't need to preach patience to George Long. For those who watch trees grow, time assumes a different quality. Still, even for the likes of Long, the moment could be important, and such was the case as he observed political developments. The recently published Federal Trade Commission report implied that timber holdings were controlled by a monopoly, and that the industry "as now constituted was obstructing any constructive legislation that would guarantee a perpetuation of our timber supply." Judge L. C. Boyle counseled, "We must defend ourselves against the charge that we are obstructing constructive legislation in order to work out selfish purposes."

Long's response was lacking in enthusiasm, to say the least. Not only had timber and lumber interests never controlled their affairs to the public's detriment, he knew it wasn't possible, laws or no laws. Indeed, he recalled when nearly anything was allowed and when operators talked a good deal about collusion, and the talk never came to anything: "At no time within the writer's knowledge, even during the old days when it was not illegal, have I known the lumber industry to get behind a propaganda for higher prices or an agreement pertaining to higher prices that had any more strength in it than a rope of sand." As Long saw it, "The industrial life of our nation shows epochs of very low depression, many irregular levels as to values, with occasional peaks when values are absurdly high, but when analyzed it is observed that this pertains to nearly all classes of commodities and therefore becomes either a national characteristic or national drift that is part of our life. Every time the peak gets too high, it breaks in a perfectly natural way and as is manifest in lumber prices today, everybody goes gunning for himself and prices are all shot to pieces and all ideas of a trust or an agreement vanish in thin air, and I think the public knows it." Further, the public continued to demand products of the forest. Loggers, as Long so often observed, didn't cut down trees for the exercise. There should, of course, be greater utilization of the resource, but that wouldn't happen until it was cost effective. All in all, he was convinced that the market would prevail. It always had.

TWENTY-THREE

Forestry and Business

AS HARDING TOOK THE OATH OF OFFICE ON MARCH 4, 1921, George Long wondered whether the cure for his business problems rested in the other Washington. Little good news seemed to have come from that locale lately; and even less from eastern and midwestern markets. In fact, the question on the minds of most Pacific Northwest loggers and lumbermen in early 1921 was whether to operate, shut down, or stay shut down. Long wasn't sure. But, as he informed Warren at Snoqualmie Falls on April 1, his instincts told him to operate, "whether or not we are losing any money or making a little."

Given the business conundrum, Long welcomed the excuse for a visit to Klamath Falls, even if that excuse was the intervention of a pest, the pine bark beetle. In taking leave of his Tacoma desk, he interrupted his "kindergarten" classes for "the Big Four"—son George, Jr., Edmund Hayes, another grandson of Orrin Ingram, and brothers Fred and Phil Weyerhaeuser. George, Jr., and Fred tagged along on the Oregon tour. While they were gone, Ed Hayes was sent to Snoqualmie, Long asking Warren to introduce him to sawmilling. This was not busy work, as Long explained to young Hayes: "The timber at Snoqualmie is of a variety that requires the logs to be cut with more special reference to getting out of them whatever they will make in the way of good lumber." He likened Snoqualmie to "the old white pine saw mill operation in Wisconsin, where the mills . . . sawed their log into that which it was best adapted for, regardless of whether they had an order for the lumber or not."

Sadly, Warren would be of little assistance to Hayes; he died of a heart attack in early April. Long felt the loss in many

262

ways. He described Warren as "a fine gentleman, thoughtful and considerate . . . a friend, but also a stalwart prop in our organization." Rod Titcomb had been handpicked as Warren's successor, but in fact Titcomb was unprepared. Long would have to pay closer attention to the Snoqualmie operation in the years ahead.

Almost coincident with the passing of Warren, F. H. Thatcher died after a series of strokes. Although he and Thatcher had differed on occasion, Long was sincere in expressing sympathy, recalling Thatcher's "geniality, his kindliness, his interest in others and his hearty willingness to always lend a helping hand in any issue or question calling for either assistance or advice."

Meanwhile, Long was left with the unpleasant business of doing business in 1921. The brightest spot on the horizon was Baltimore, and even that troubled. For example, the initial shipment of lumber was delayed until April 29, with delivery likely to be about May 20. This was "a little late to get started," he admitted to his brother. In any case, Long and Boner promised that when that shipment finally arrived they would be dockside "so that we can test out the working of our equipment."

Weyerhaeuser's plans for Baltimore were ambitious. They intended to have a complete stock of materials on hand, providing customers with whatever was requested. Specialization might come later, "when we get the drift of the trade," and most likely they would carry those items which were "difficult for the yellow pine people to furnish." This was typical of George Long's approach. "My own belief is that we do not want to be too small in our ideas of quantity, [and we should] send . . . twenty-five or thirty million feet of lumber, as well assorted as we can guess it, and be in a position to really go after the trade in a large way and at once, if we should develop any autumn trade." Long was now admitting that any spring trade was impossible, although he retained some slight hope for "what might be called summer trade, if there is such an animal."

Though thinking big and thinking ahead came naturally to Long, most worries were of the day-to-day variety. Long kept the Clarke County Timber Company logging shut down, and with reluctance allowed a start-up at Cherry Valley's operations because they had some "fifteen to eighteen million feet of logs in the woods, which ought to be removed before the possibility of a forest fire." At Everett, Boner ran old Mill A in February and March, and then closed it down and started up Mill B, so as to give its crew a little work. Everything was at a snail's pace, the industry operating at about 30 percent of capacity. "We are all just a trifle hopeful," Long informed Bell, "that things have reached the bottom and there will be a gradual improvement, although we do not look for much that can be called improvement during the current year."

Long and Boner traveled to Baltimore in mid-May, where they inspected the new yard, still awaiting its first large shipment of lumber. Clearly there were going to be difficulties in off-loading and moving the lumber from dockside to the 174 storage bunks. Baltimore would bring another problem to the fore, one that would never be resolved, at least in Long's lifetime—overcoming the prejudice of eastern buyers against hemlock. Long knew that hemlock could be a useful wood, but it had a terrible image. For years, the market discriminated against the eastern variety (*Tsuga canadensis*), and for good reason. Western hemlock (*Tsuga heterophylla*), however, was "in every way far superior." How could they make that difference understood? It wasn't enough to distinguish "East" from "West." As a semi-serious student of the American frontier, Long knew that those were elusive terms. "Some of the New Englanders think that Western Pennsylvania is out West, and they are very positive that Michigan and Wisconsin are in the wild and wooly West." Long preferred the term Pacific Coast hemlock, "or if I wanted to be selfish, I would have said Washington Hemlock." But what buyers continued to hear was hemlock, and that was enough.

Long returned barely in time to prepare for the annual meeting. The meeting had the usual mix of good news and bad, with the bad predominating. For the first seven months of 1920 they had had plenty of orders, but no cars for shipping, and as a result business suffered by some 50 percent. Then the market disappeared. The story of lumber was little different from the other staple commodities, especially agricultural products, but the slump in both price and demand came so fast that it probably assumed a more serious aspect at the end of the year than even the most pessimistic prophesied. The figures spoke for themselves. In April 1920, the average price received at Everett's Mill B was $49.51. Exactly one year later, that price had fallen by more than half, to $20.85 per thousand. Fortunately, in the case of the Weyerhaeuser Timber Company, lumber sales were only a fraction of the whole. Timber sales for the year had been "abnormally large," this the result of the two transactions with Long-Bell, totaling nearly 57,000 acres, 5,331,542,000 feet for the sum of $14,119,965.75. In short, the wolf had been kept at bay.

It was a rather awkward time to be thinking about jumping into the eastern marketplaces, but if they were to do business out of Baltimore in the fall, now was the time to make arrangements. Hopes for early profits were slim. Long wrote to his brother on August 5 in typical run-on fashion: "The real facts are that we are going to spend a million dollars there before we turn many pennies back, even at a cost price, and may spend a whole lot of money there before we make any money; but this is nobody's fault and is all right for the Weyerhaeuser Timber Company, but it isn't a very attractive atmosphere for the average stockholder to look at, and while I have had no complaints whatever from any source, I know that the Baltimore venture is considered possibly in advance of the times, more or less experimental and quite expensive, so that I may be impressed a little bit more than necessary with the idea that we ought not to go to our limit on anything in the way of expenses that could be avoided."

Long's assessment was directed less to the overall picture in Baltimore than to his brother personally. Colonel James Long clearly enjoyed the trappings of the good life, and his brother knew of this weakness. The dollars involved were relatively few, but George wished to avoid conveying the impression that Baltimore was unnecessarily extravagant. His advice went unheeded: James continued to be the dandy and was soon chauffeured here and there by a black driver.

If James enjoyed the fruits of his labor excessively, brother George was the opposite. Indeed, he became less and less able to balance work and family, the latter the loser. Friend Stiles Burr chided him for having failed to take "that long promised vacation," observing, "It will be a sin against yourself and your family, and also against the many who depend on you so utterly, if you don't take a holiday and rest before you grow too tired to enjoy and be benefited by it." Stiles closed by "paraphrasing Lincoln, a man can't do five men's work all the time."

It wasn't easy, of course, to think about vacations when business was so bad. In the late summer of 1921, Long continued efforts to bolster spirits, including his own, remarking to William Carson that he was operating "somewhat on the general theory that it is always a little darker just before daylight." If that were true, autumn should be better. The Baltimore yard had now accumulated between 6 and 7 million feet of lumber and had no choice but to enter the market. In Long's words, "We certainly will have the fun of trying to do business under the most unfavorable of circumstances that have existed for many years."

Marketing was a constant concern. So too was public relations, commonly involving a response to charges about how logging was conducted. Aesthetically and spiritually there was no defense. This was the generation that read Joyce Kilmer's poem "Trees" with shaking heads, if not watering eyes. But as impossible as the task of justifying the destruction of trees may have been, George Long never ceased trying. In the summer of 1921, two "lone women" from Massachusetts were touring the country, and August found them in the Pacific Northwest. On their way down from Mount Rainier, they passed through "a magnificent forest." One tree in particular caught their attention, measuring 44 feet in circumference. Upon reaching Portland, Cora Brown addressed a letter to the "Warehouse Timber Company." "I am writing this letter to show you our great appreciation of the privilege of seeing such a tree growing in its natural, unspoiled state, rather than cut down." And she closed expressing the hope that "you will continue to preserve them and see to it that they may never be destroyed by man."

When the vacationing ladies reached their Concord home, they found a letter awaiting them from George S. Long. He mentioned his pleasure at having read their reaction, which showed "that you are a true lover of Nature." Long also noted that there were countless "wonderful beauties" in the western woods, many of which were accessible only to "hardy woodsmen," and that it was "always inspiring to come in contact with such spots, even to a lumberman." Then George Long became the teacher: "If we build houses, we have to cut down trees and while the mere act of destroying a forest seems ruthless and wasteful, yet when one realizes that the result is houses and barns and everything for which wood is used, then we begin to realize that the lumberman should not be criticised (as he is some times) any more than the coal miner should be criticised for removing coal, or the tilling of soil for the cutting down and use of the crops, or the use of the innocent lamb for our breakfast lamb chops, nor the docile cattle for what it affords in the way of the roast beef. . . . I find so often that in the regions where forests do not have to be used for the wants of the country, that there is a habit of mind developed that sometimes is rather critical when they see our magnificent woods of the Far West utilized."

Certainly the stockholders had little reason to criticize the Weyerhaeuser Timber Company. Despite the depressed business conditions, the cash flow was healthy, and on September 1 the company distributed another 6 percent dividend. "We were prompted to do this," Long explained, because "during the past week we have received from the Long-Bell Lumber Company a remittance of $650,000.00, being the payment of the first deferred amount under our so-called second contract."

Like most lumbermen, Long felt more like celebrating the end of 1921 than welcoming 1922. From nearly every standpoint it had been a most disappointing year. Early in January, Long was called to Washington, D.C., to appear before the House Committee on Agriculture. This was his first such experience, and he was unimpressed. For one thing, a new battle was raging over the proposed transfer of the Forest Service from the Department of Agriculture to the Department of the Interior. On this issue, Gifford Pinchot and Chief Forester Greeley agreed. In Pinchot's words, "The National Forests belong naturally in the Department which has to do with growing of all crops, including tree crops, from the soil," adding that "forestry is a part of agriculture, and is so recognized the world around." The issue would hang in the balance for more than a year.

Long visited Baltimore and then headed homeward, as usual, via St. Paul, spending nearly all of January away from his Tacoma desk. On his return, he found Peter Connacher awaiting instructions on when to commence logging. Long had an answer, but he wasn't sure it was the correct one: "The right thing to do is to get ready to hit the game hard, and if it doesn't come our way we can stop."

There were changes in the wind, some of which involved employer-employee relations. As yet, any legal responsibilities or benefits for injuries and illnesses had been minimal, but greater attention was being paid in some quarters to working conditions and workers' welfare. Long encouraged the initiatives, though they seemed awkward by later standards. Long-time employee O. H. King had been appointed "welfare man" in the Everett plant, responsible for the "Trouble Department," and he went forth with the dedication of a missionary, responding to individual cases, friend to friend. The work was now and then publicized, evoking an occasional response from shareholders. St. Paul attorney Harold Richardson complimented Long on these efforts, assuring him that he for one would "follow the results from year to year with interest." F. E. Weyerhaeuser also approved, matter-of-factly, in a letter devoted primarily to complaints he had heard about the Sales Company.

F.E. was feeling defensive, and Long tried diplomatically to restate the basic question: "There has always been the somewhat debatable ground of whether a dozen corporations could dispose of their products to the same advantage if handled by one agency as against each individual corporation handling its own problems of sales, and I think we have in our own ranks men who entertain different opinions on this subject and possibly Mr. Carson's letter is to open up this subject for another very full discussion."

In F.E.'s view, of course, this subject had been discussed ad nauseam, and here it was, apparently being raised anew. F.E. was convinced that they had designed the best system possible, but no sales system could succeed when there were no buyers. Furthermore, a sales system could work only if everyone believed in it. Some refused to be convinced, and F.E. suspected that George Long was among them. Actually, George wasn't a detractor; he just wasn't a true believer.

But he could and did believe in some things—among them the worth of western wood. Success in the eastern market depended upon an acceptance of the western product. There were many facets to the East-West differences, not the least of which was attitude. E. T. Allen's recent experience seemed to prove the point. Novelist Joseph Hergesheimer was on assignment for the *Saturday Evening Post,* writing a series of articles on the general subject of resources, "particularly western ones." Allen described Hergesheimer as having "both a keen intellect and a robust conceit." Thus armed, he was able to arrive at answers, right or wrong, with amazing speed and complete confidence. As the novelist viewed the forest resource, he immediately concluded that "we are in a bad way," and the only solution required "a total reform of government."

Hergesheimer represented what Allen, Long, and others most despised—an ignorant easterner who thought he knew all the answers. And despite Allen's best efforts to fill in a few gaps in Hergesheimer's forestry education, he left Portland as arrogant as he had arrived—and, apparently, none the wiser.

With or without the "help" of individuals like Hergesheimer, the forestry movement was advancing. If some wondered about the particulars of policies to be adopted, more worried that things were progressing too slowly, regardless of the direction—in short, that the forest wouldn't outlast the deliberations. As for Long's own efforts in behalf of forestry, or "reforestation," as he preferred, they had lately been considerable. In mid-March he talked to the University Club of Tacoma "and met with a very hearty response." Then in early April he addressed the Tacoma Commercial Club and Chamber of Commerce. At the conclusion of that discussion there was a motion to appoint a committee to encourage the state to "adopt a practical forestry policy."

The lumber industry could afford to be generous. Business was again booming, not only for Weyerhaeuser, but for everyone. Long, well remembering the bad times, declined to join in any loud celebration. Still, even he had to admit that the reports were impressive. The best news came from Baltimore, where April shipments exceeded 5 million feet. By June, Boner was willing to declare the Baltimore venture a complete success, predicting "a fixed trade there of from sixty to a hundred million feet a year." Moreover, he figured they had only just begun to tap the territory. Wasn't it time, he inquired of Long, to think about building a sawmill "at poor, old South Bend," and then establish a yard in Boston, "or some nearby point?" Perhaps it was, but Long's plans were in place and not likely to be altered.

At the Weyerhaeuser Timber Company annual meeting on May 25, the sad figures for 1921 were far more easily accepted amid the good news of 1922. Long reminded his audience of the closing paragraph of the 1920 annual report: "The lumber industry on the Pacific Coast for the year 1921 is probably facing a reduction in output from 50 to 60 per cent of normal and we apparently have in store a year that will be devoid of any profits whatsoever." And, he further noted, "The record for the year confirms this prediction."

But things had turned around. In fact, business was so good that Long was soon advising his brother not to push Baltimore sales. At the same time, he thought they had done well thus far, that their prices had attracted the buyer, "and when this is coupled up with lumber that pleases him and with service which charms him, we have made an impression which is an asset that will be reflected in the future." Closer to home, the mills had almost more business than they could handle. At Everett, for example, "it is hard to get men enough to run our mills night and day and a further situation which is even more alarming is the probable shortage of logs." Complicating that shortage were forest fires; two of the camps which supplied Mill B had burned in mid-May, and, according to Long, "Mr. Boner will have to scratch around pretty lively to get logs enough to run."

Once again, it was amazing how quickly the economy could turn from bad to worse and then rebound so dramatically. The depression in the postwar lumber markets was primarily the result of a calamitous drop in agricultural prices. As seems to be forever the case, the inflated prices of the war period were accepted as normal; too many believed that a corner had at last been turned, when they had merely experienced an artificial bend in the road. And with the wartime demands and the accompanying eastern credit, bankers had lent and farmers had borrowed to a fare-thee-well. Wheat prices, for example, which averaged less than a dollar a bushel in 1914, rose to $2.30 in 1917. Few farmers could avoid the temptation to buy additional acreage, animals, and machinery on credit, with little thought of the dangers involved. When the prices plummeted, interest payments came due and passed, unpaid. Farms were foreclosed and banks, especially smaller state banks, failed at an alarming rate. In 1922 alone, those failures totaled more than five hundred.

That same year, however, as Boner noted, there was cause for optimism. The factors were many, and while the lumber market did not exactly mirror the general economy, it nonetheless reflected the changes. Presidents Harding and Coolidge, and perhaps most importantly Secretary of the Treasury Andrew Mellon, presided over the new order and preached its gospel in various forms, always emphasizing that the business of America was business. The facts seemed to support the belief. In 1914, the United States was a debtor nation, to the amount of $3.5 billion. At the war's conclusion, it was a creditor, owed some $13 billion. And this was just one ingredient in the recipe of prosperity. Fueled by tax reductions on the well-to-do and on corporate profits, Americans became their own insatiable consumers, purchasing with abandon all the latest, from radios to automobiles. Domestic production would double during the decade of the twenties. We know, of course, how suddenly the boom would end, but it was great fun while it lasted.

Meanwhile, the summer of 1922 continued hot and dry. Long described conditions and his ambivalent feelings, that it had been beautiful since mid–May, "causing us considerable anxiety on account of the forest fires that have developed, although we had such a long, tedious, unpleasant winter that we are happy to find real summer weather so continuous." Scattered showers fell in late June, but no general rains relieved the fire hazard. Long worried and advised those in the woods to be alert and do whatever was required to lessen the danger. He instructed Cherry Valley's logging manager "to keep quite a strong force available during the July 4th shut down, not only to watch the fires but really put a force on to clean up as many as you can of your little slumbering fires, for if we get another hot day with an east wind we will have another visitation and probably a worse one than the first."

The best efforts by managers and men would mean very little unless nature cooperated. Without rain, fires would rage. Long reported in detail to the Timber Company trustees on the losses sustained thus far. The problems had grown serious in

May, and June had been the driest on record in western Washington. Fires "seemed to spring up in every logging camp, and in most cases were beyond control, running rapidly through the slashings, doing much damage in the burning of equipment, bridges and felled timber." The Snoqualmie Falls Lumber Company experienced the first serious fire in its history, destroying little standing timber, but burning over nearly all of its cutover lands and consuming some 3 million feet of logs. Fire also burned through Cherry Valley's holdings, and in the Clarke County Timber Company operation there were losses of some $1,500 worth of equipment and 1 million feet of logs. And in mid-June, fires destroyed the mills of the Mineral Lake Lumber Company, burning up 1 million feet of lumber. Mineral Lake's total loss was estimated at $170,000, insurance covering about $130,000 of the amount. In short, "The situation is one of grave alarm," Long expressed to major shareholder Horace Irvine, "in that there are scattering fires all over the western part of the state, the weather still dry and hot, and if we are again visited with a hot east wind and it would prevail for a day or two, the possibilities of grave losses are very imminent." It had been twenty years since Yacolt, and too little had changed.

During the second week of August, general rains fell throughout western Washington, and the drought ended. Loggers returned to the woods and cruisers who had been on fire patrol finally got a bit of rest. F. E. Weyerhaeuser was among those who welcomed the good news that the fire danger was past and that the Timber Company had suffered only minor losses. F.E. was gradually assuming a more active business involvement in the summer of 1922, after preoccupation with the lingering terminal illness of his seventeen-year-old daughter Virginia. F.E. had been missed, for it was he to whom George Long looked for support. In a July 24 letter, Long made general inquiry as to future policies, reasonably confident that F.E. would say, in effect, keep up the good work. And that is precisely what he said.

Claiming lack of sufficient knowledge or up-to-date information—"My friends say to me that I am 'three years ago' "—F.E. nonetheless expressed an idea or two: "I have not changed very much in the opinions I have held for a long time—that we should move conservatively but steadily toward a much larger development in the manufacture of lumber from our own timber; that we should buy timber occasionally when we can get it at a reasonable price that will block up our own holdings; that generally speaking, we should confine our sales of timber to actual operators who have need for some of our timber that we can spare without jeopardizing the future development of our own holdings—in other words, that we should serve the loggers in holding timber for them very much as you have done in the past, and that we should gradually raise our timber prices so that we are well paid for holding the timber."

George Long could not have stated it any better, although he would have emphasized taking advantage of any opportunity to sell freely "in localities where it will not disturb our own future program for logging and milling," thereby acquiring capital "to take care of our rapid development if we ever want to go to it." Such plans were inevitably linked to the dividend policy. Again F.E. and George were in agreement: they should keep annual dividends to something less than 12 percent.

Largely because the business climate had improved, Long was feeling optimistic. In that spirit he occasionally looked ahead and thought the unthinkable—that the forest could be renewed, although he remained convinced that reforestation would be accomplished only through public auspices. In the meantime, he addressed himself directly to the most volatile issue currently before the public: the fear of a timber famine. The figures Chief Forester Greeley had been quoting—that at the time of the white man's arrival, there was some 5.2 trillion feet of standing timber and now there remained only about 1.6 trillion—abetted fears. The problem, as Long saw it, was that Greeley's appeal to figures on "virgin" timber grossly underestimated the amount of timber available "for lumber purposes." In other words, what so few seemed able or willing to understand was that the forest was a living thing, and was never exclusively a community of beautiful old-growth monarchs. It was all well and good "to sound an alarm," but if they wanted useful answers to legitimate concerns, they had best use the most accurate information available.

To confront the problem effectively, the state must address a number of points:

First, of course, that the State should co-operate very heartily and earnestly with the federal authorities in this movement.

Second, that the State itself, by liberal appropriations should provide funds to protect the standing timber as well as the young growth timber from forest fire destruction.

Third, that the State itself should acquire by purchase or by gift or by condemnation, logged off lands suited for timber growth and not for agriculture and adopt the policy of forest growth.

Fourth, that as early as the laws of the State of Washington can be revised so as to make it legally possible, that special legislation be passed making it financially possible for individuals to set aside lands suitable for forest growth and under a tax burden that will enable them to keep these lands intact until the timber is of an age and size suitable for use as lumber and I question a little whether we should attempt to define what this legislation should be at the present time, because it is through such details as this that one will stir up a hornet's nest, but that we submit to the State the favorable consideration of this program.

Then again I think the state should make a survey of its logged off area and even its timbered areas to give a comprehensive report of what proportion of it is better suited for forest growth than for any other purpose.

Long shifted the responsibility of presenting these points to the State Forestry Conference onto E. T. Allen, and he and Boner headed east, first to St. Paul for several important meetings, and then to Baltimore. Joined by F. E., the group arrived in St. Paul on a Tuesday morning, October 3. The Thompson Yards situation was high on Long's agenda, and he had wired ahead to Horace Irvine urging that all who were available attend the meetings. But important meetings don't necessarily result in important decisions. And so it was with the Thompson Yards discussion. It was difficult to face the real problem: George Thompson himself.

One item, however, approached resolution. Boner had been frustrated in obtaining ships for his Baltimore trade, and Long was on the verge of approving a big step, the purchase of their own vessels. He encouraged Boner to pursue the matter, and should they find ships costing between $20 and $30 per ton, "in good shape and built by yards where we can have reason to think the workmanship was dependable, then I would be in favor of buying one or two of such boats and put them in the lumber trade for our own use." He further thought it would be "a very simple matter to organize a small corporation to take over the venture and allow the stock in it to those we have invited to join us." He was wrong about that.

TWENTY-FOUR

1923, a Very Good Year

AS 1923 DAWNED, STILL UNANSWERED WAS THE QUESTION of how Weyerhaeuser was going to handle its Atlantic Coast requirements. Was it best to align with experienced shippers, perhaps taking a partial interest in a new organization? Or should they go it alone, purchasing their own ships? Long assumed that those in St. Paul would favor some sort of limited entry into the field.

As he saw it, the problem with a shipping partnership was that the partner in charge would inevitably operate to his own advantage first. Furthermore, Long was sure that the intercoastal business was going to be "one of very sharp competition" and that a situation would develop quite soon that the man who controlled the westbound tonnage would control the business. Long seemed prepared to purchase as many as four boats, provided the price was around $30 per ton, or "less than one-half what they can be built for at the present time in America and about 60% of what they can be built for in Europe, so it looks like a safe investment insofar as getting the bulk of our investment back, if we get tired of the arrangement." This was the policy implemented.

Another possible investment opportunity was improvement of the waterfront property in the vicinity of St. Johns near Portland. S. M. Morris, vice-president for Long-Bell, asked Long about plans for building a sawmill, "should you have such plans." Morris left no doubt as to the reason for his interest: "Should you be planning a mill in this vicinity we would like very much to have you as our neighbor on the west side of the Cowlitz River on the property we now own." Thus was opened what would become a lengthy and noteworthy negotiation, as well as another test of wills.

Initially, Long played his cards in typical fashion. R. A. Long had requested by telegram that they discuss possible mill-site properties along the Cowlitz River, on land now owned by Long-Bell. George Long responded as follows: "Dame Rumor has been rather active, and you know how vigorous a damsel she is, for as a matter of fact we have not seriously considered building a mill anywhere on the lower Cowlitz River to take care of such timber as we might develop lying on the eastern side of the Northern Pacific Railway in Lewis and Cowlitz counties, but what we really have had in mind was to possibly do some logging there in the next few years and ultimately have a mill somewhere on the Columbia River. We have a very excellent mill site near Portland but believe one near the source of supply of logs would be equally as good and therefore are willing to consider seriously the overtures which come from your company." That mill site near Portland was the "bird in the hand" that George Long held up for all to see, but he never intended it for anything other than a bargaining chip.

Some miles to the south, the Timber Company's investment in the Klamath Falls area had been held in abeyance while awaiting a rail connection, or, preferably, connections. There were some who had clearly grown weary of the wait. F. S. Bell left no doubt where he stood: "I am strongly of the opinion that it would be good business to sell the Klamath timber at a right price and devote ourselves to marketing fir." He admitted that he had no idea what a right price might be. Long, of course, felt otherwise: "Personally, I have always entertained the hope that we would build a mill down there and manufacture the timber, but in this I was rather influenced by sentiment; recognizing that we had such a magnificent property and that it would get us back into the old game of manufacturing pine." In this he had the support of, as he called them, "the Weyerhaeuser boys."

Bell wasn't prepared for a fight to the finish, but he did want to discuss differences, adding that he and the late F. H. Thatcher had agreed "pretty well on general policies, but often differed radically on details." Bell said that it was Thatcher who had vociferously stated the case for early liquidation: "He took his most extreme position in argument with F. E. Weyerhaeuser, whom he told, I believe, that he thought the whole holding of the Weyerhaeuser Timber Company ought to be closed out within the next ten years."

Unlike Thatcher, Bell supported manufacturing efforts, but within limits. He thought they had far more timber than they would ever need and advised that if buyers arrived on the scene, Long should "seize the opportunity, getting of course all we can for the timber and certainly reserving for ourselves ample blocks of timber for manufacturing operations of many years' life." As for Klamath, "Mr. Thatcher and I were absolutely agreed that it was the sort of thing which we should sell rather than operate." His main objection to the tract involved the species (pine), since he viewed Weyerhaeuser's "best ability in the fir situation." A secondary concern was the location of the timber, "far from the sea and we hoped that water transportation would make most of the fir holding very close to the Atlantic market."

Finally, Bell suggested that while selling Klamath would probably be supported "by many elements in the stockholding," a controversy would be unfortunate because "it would possibly harden their [Weyerhaeusers'] indisposition to let anything in the way of timber go." The Weyerhaeusers had "grown up on timber and naturally have an intense desire to increase their holdings." Bell feared that this had become habit, and advised Long that "as opportunity offers you sound out your leading directors, letting them see that the question has two sides in your own mind and that you are not quite ready to commit yourself either way." In truth, however, Long was already committed.

Meanwhile, plans for branching out into shipping were slow to be implemented. The first of the potential vessel purchases, the *Argus,* wasn't even available for inspection, and there were many other questions awaiting answer. George advised his brother against closing any purchase "until we decide definitely on what our future plans will be. . . . This, of course, upsets our entire shipping program and we feel a little bit at sea out here." He was still "at sea" as he headed to New Orleans on March 13, to attend the annual meeting of the National Lumber Manufacturers Association.

Long would have preferred to stay at home, but, given the pending federal legislation, duty called. As he explained to Jim Rhodes, secretary of the Southern Pine Association, it was crucial that they "map out a program of coordination" between the various lumber sections of the National Association, "to lend their efforts towards working harmoniously in an effort to assist the senate forest committee to get the right kind of information pertaining to forest and lumber interests."

Long would later describe his contribution in the usual matter-of-fact terms. When the editor for *The Timberman* requested a copy of the remarks he had made at the meeting, Long indicated that he had "simply sat with the Forestry Committee and formulated a report which I presented from the floor with very few comments, which were not important and which were far away from being anything worthy of republication." The fact was, however, that his mere presence was as important as his presentation.

Not everything awaiting George Long's return to Tacoma pleased him. A continuing irritant involved the work of Weyerhaeuser Forest Products, the advertising arm of the affiliated companies under the leadership of Carl Hamilton. What immediately concerned him was the lack of progress in the house-building program: "I know you will vote me a nuisance and a bore when I again say that I am beginning to believe that all the houses will be built before you get your house program started." And he added, "This year and last year are the biggest years we are going to have for a long while in home building." It had been nearly a year since the first booklet was published, and that "was confined largely to the nice homes, which constitutes about 5% of all the homes that are being built in the United States." In Long's opinion, they were about to miss the boat—if they hadn't already done so.

There was more promising news on the shipping front. His brother's research concluded that a large ship of 12,000-ton capacity would best serve their needs. One, the *Pomona,* was soon to arrive in Seattle from Yokohama, and Boner made plans "to look her over and get further impressions of this type." Boner boarded the *Pomona* and was duly impressed, so Long decided to see for himself.

The *Pomona* was being loaded with flour at the Fisher Flouring Mills Company in Seattle, and, as Long subsequently reported to the Weyerhaeuser directors, it "was a magnificent ship in every way, so far as a layman could decide, and while we had one or two experts tell us about her, they all voiced the same sentiment as did the Captain and the Mate and such of the crew as we could confer with." Long subsequently made a deposit of 2.5 percent of the purchase price, akin to "earnest money" but refundable should formal inspection indicate serious deficiencies. And he did the same with the *Hanley,* a sister ship. Both vessels had been built after the war by the Union Iron Works of San Francisco, a subsidiary of Bethlehem Steel Corporation. Each had a dead weight tonnage of 11,724 and a price tag—computed at the same $30 per ton—of $351,720. It was estimated that they would have a carrying capacity of some 6 billion feet of lumber.

Realizing that the ship commitment would come as a surprise, Long hastened to assure his directors that he had acted wisely. Most of them had assumed that there would be some interim contract with an established firm, but, as Long explained in a May 11 report, it was "necessary for us to have a dependable movement of lumber at regular intervals at Baltimore, and we have not felt at all comfortable in facing the possibility of having to charter ships for every voyage that should be made." Still, they would not be going it alone. Arrangements were in place with Holder, Weir & Boyd, who had "a very excellent organization in the East to solicit and get westbound freight."

It was experimental, but, considering the purchase price, there seemed small likelihood of loss. Long even saw "a slight chance of our making a little money," beyond the fleet's primary function of delivering Weyerhaeuser lumber to Weyerhaeuser Atlantic terminals. "It may or may not lead to further investments in ships," Long warned, "depending of course largely on our necessities and our success." Thus did Weyerhaeuser enter a new phase of operations, one that carried with it unimagined challenges.

Then, out of the blue, came word that President Harding might stop off in Seattle in mid-June on his way to Alaska: "May we count upon your willingness to entertain him and small party at Snoqualmie?" Long's answer was predictable: "Would consider it rare privilege to carry out suggestions contained in your telegram." The president's plans, of course, were subject to change. His ill-fated trip to Alaska would take place, but there was no Snoqualmie interlude.

At about the same time, Rod Titcomb, manager of the Snoqualmie mill, learned of a possible June 1 visit "of eighty or ninety Senators and Congressmen," and he encouraged them to tour Snoqualmie Falls. As he reported to Long, not only would this be "a wonderful opportunity" to show off their facility, but they could also see "some of our over-ripe timber, and we could possibly impress upon them some of our ideas of reforestation." George Long had enough to do without playing host to visiting dignitaries. Still, one didn't say no to congressmen and senators, and certainly not to presidents.

All of this was complicated by preparations for the annual meeting of the Timber Company. Writing to the directors regarding the event, Long mentioned overtures recently received from the Long-Bell Lumber Company concerning "15,000 acres of low valley land, abutting the Columbia River," where they would have deep water and also enjoy rail connections with the Northern Pacific, the Great Northern, and the Union Pacific, "and there are some rumors" that the Chicago, Milwaukee & St. Paul might soon join the crowd. The site was, in Long's words, "a most excellent one in every way." Accordingly, he hoped that as many of the directors as possible would "go down there in a body to meet Mr. [R. A.] Long and his associates."

The annual report for 1922 was, of course, favorable in most respects. Long detailed the "rather unexpected return to active and profitable business," given that trade, "especially as it pertains to Europe, continued to be below normal." But there were "some marked improvements in other directions which seemed to fill the gap," Long noting increased Douglas-fir shipments to Japan, California, and, of course, the Atlantic Coast. The good times were reflected in the Timber Company's figures. In 1922, Everett produced 226 million feet of lumber and sold 230 million feet; and Snoqualmie had produced 103 million feet and sold 108 million feet. As important, Everett's prices had increased from $20.07 per thousand in 1921 to $22.03 per thousand when sold at Everett, "and over $40.00 per thousand for the lumber sold at Baltimore."

It was Baltimore that provoked the greatest interest.

There, net profits for 1922 amounted to nearly $320,000, and Long acknowledged, "We probably could not have had a more favorable year to experiment with the eastern trade than we had in 1922," for by then "the freight rates had gotten down to a fairly satisfactory figure and the demand was good." It seemed possible that they could sell upward of 100 million feet out of Baltimore alone in 1923.

Although the Longview mill site occupied the thoughts of most shareholders, Long was nearly as interested in pushing matters forward in distant Klamath Falls, including the purchase of a Long-Bell tract. Long was ready to move forward, requesting a legal description, "so that we could plan our activities during the summer to make the proper examination of the actual land involved."

Klamath had been receiving its share of visitors. Chief Forester Greeley toured the area in late May and was shown the pine beetle project in which Jack Kimball had been involved. Kimball was justifiably proud of the results: "The particularly gratifying fact is that wherever we have conducted our artificial control, the woods show almost no new infestation." Greeley was followed a couple of weeks later by a party including Horace Irvine and William Carson. Kimball thought his guests "were very much pleased indeed with our mill site and the timber which they saw," and added, with pride, "I very much hope that they went away with a realization of the wonderful property belonging to this Company." George Long hoped so too.

It seemed the season for touring. Long learned of plans for a visit by members of the Senate Select Committee on Reforestation. How best to conduct such an important tour? They needn't visit a Weyerhaeuser facility, and indeed, Long worried lest there be "too much Weyerhaeuser," which might give the impression that "the Weyerhaeusers were trying to hog all of their attention." Still, he had to admit that "one of the most agreeable and spectacular trips is of course the trip to Snoqualmie Falls, which enables one to see the new plant at that place and easily enables them to get to the woods and see the logging operations."

Chief Forester Greeley had left no doubt as to the importance of the senators' visit. Long informed Allen that Greeley thought "the lumber associations as such had as yet not made just the right kind of an impression," and he therefore hoped that "we will be able to correct this impression in the Northwest and I think we will be able to do so." He sent a similar message to others, urging an all-out effort which would have the senators appreciating "what we have attempted to do in the Northwest and if we can get it over to them in the right kind of a way I believe it will be a valuable thing for us in every way."

Meanwhile, the Thompson Yards situation was heating up again. During a St. Paul meeting that spring, Thompson's performance was called into question, some recognizing "the hopelessness of trying to run the business under present management, which has shown that it cannot change its old ways into new ones, and continually reaches for more expansion instead of taking care of what is already in such bad shape." It was the same old story: Thompson was far more interested in adding to than he was in improving upon. Simply put, he was a salesman, period. Although no one wanted to hurt Thompson, in Bell's view it was time to admit that he suffered from "a constitutional inability to attend closely to credit matters and routine affairs; he is never happy except when he is expanding."

At an informal stockholders' meeting in Tacoma on May 29, 1923, a resolution was passed instructing the executive committee to conduct a thorough inquiry, with John Kendall given the "authority to examine all books and records of each yard," after which he could "sell or trade each such yard or otherwise close the same, if in his judgment to the best interests of the company, and to do every other act or thing requisite to carry out the suggested program of liquidating unprofitable yards." Kendall, sales manager for Potlatch, had been similarly utilized in times past, so he was familiar with the territory. This assignment, however, was to be more "difficult and uncongenial," to quote from the resolution, than previously. Thompson Yards had grown "like Topsy," or so it seemed in retrospect. How to scale down equitably the huge debt of Thompson Yards? Long was content to await Kendall's report.

He didn't have long to wait. In fact, he and Horace Irvine met with Kendall on Saturday, June 23, in Seattle, after which he reported in detail to Bell: "What we did was to talk frankly with Mr. Kendall and I asked him what would be his recommendations." Kendall suggested decentralization, first separating the country operations from the city yards. Once that was done, he recommended a "weeding process of eliminating absolutely such yards as were worthless or which gave no promise of future value." Then he would organize the balance "into four or five, possibly more, units," the idea being that a good manager might supervise twenty-five to thirty yards. As Long noted, were this solution to be accepted, Thompson would have only the Twin City Yard left to manage, and "in passing [I] will say that this seems also to be the place where Mr. Thompson has made his most serious mistake, from a financial standpoint at least, and I refer to the credit extended in St. Paul & Minneapolis."

Kendall's advice seemed reasonable. In any case, it was necessary that they do something immediately. Bell received that message from Long: "While there is a great big loss ahead of us and while the task is not to our liking, yet it is really up to us to get in and straighten out matters and make a success of the venture instead of a failure and I think both yourself and Mr. Kendall have pointed out ways where that may possibly yet be done and is well worth the effort."

The executive committee met and recommended that Thompson continue at his present salary, $20,000; that he resign as president of Thompson Yards, Inc., and accept appointment as manager of the cities division; that the credit department of that division "be absolutely under the control of a man satisfactory to the executive committee"; that Ray Saberson, manager of the country yards, be terminated; and, in the most complicated of the provisions, that Thompson's "present indebtedness to the stockholders of Thompson Yards, Inc., incurred in connection with his agreement to purchase stock of the company, be cancelled and the collateral stock turned back to the companies now holding the same, with the further understanding, if this is done, that the contract entered into between Thompson Yards, Inc. and Mr. Thompson, whereby he was to receive a percentage of the profits, be cancelled."

The reason for the "undue haste," Bell explained, was that "somehow or other the whole purport of our action got noised about," and since Thompson had thus learned of the plans from other sources, it seemed better to proceed than to wait. Long, however, remained less than convinced that the new arrangements would solve their problems. Clearly Kendall felt better prepared to manage the country yards than the city operations, but, as Long saw matters, Thompson remained "in the same fix."

So much seemed to be happening at once. Allen and Long were making final plans for the meeting of the Western Forestry and Conservation Association's reforestation committee. The meeting was scheduled in Portland for July 13, which coincided with the availability of Oregon Senator Charles L. McNary, chairman of the Senate's Select Committee on Reforestation. McNary and his colleagues were in the process of developing legislation which would have a successful conclusion, with passage of the much-acclaimed Clarke-McNary Act on June 7, 1924. But such success was by no means assured in the summer of 1923, when Long and Allen agreed that it would be "pretty good diplomacy" to invite the senator to join them, providing evidence that "timber owners and the state officials and the Forest Service men that are represented on this committee, are trying to work out problems in a way that represents all the different interests that are involved in reforestation matters."

Long promised Long-Bell officials that he would be in Longview the day following, July 14, for the formal opening of the new hotel, the Monticello. In truth, of course, he was more interested in looking at and talking about the sawmill site, but the Monticello dedication gave him an opportunity to do so in an ostensibly offhand manner. And while all this was going on, Bill Boner was overseeing the loading of the SS *Hanley,* bound for Baltimore by way of San Francisco. As he explained to Louis Case of the Sales Company, "While it isn't such a task possibly to one who knows it, it is quite a little task to me," but things "seemed to work out fairly satisfactory."

The *Hanley* sailed the morning of July 13 with "a little over five million feet net of lumber," and with enough additional capacity "to take from 1,500 to 2,000 tons of general merchandise in San Francisco." The second ship, the *Pomona,* was expected to arrive July 23, and Boner inquired of Long whether the Shipping Board had been notified "to make this transfer direct to the Weyerhaeuser Lumber Company," presuming that "the procedure throughout will be practically a duplication of the *Hanley*." The Weyerhaeuser Timber Company was in the shipping business, no doubt about it.

Preparations for the Senate committee's visit went forward, and Snoqualmie Falls was to be featured in the itinerary. Long tried to orchestrate things. "Relative to the trip to Snoqualmie," he wrote Allen on August 1, "will say the committee could spend a day around the mills and in the woods getting their evening meal along about six o'clock," and then make the drive to Snoqualmie Pass before sunset, "giving them a view of a beautiful stand of timber and getting them back to Cedar Falls, a station on the C. M. & St. P. Ry," where they could board a train that would take them to Spokane the next morning. The enthusiasm over plans for hosting the senators suffered greatly the very next day, August 2, when Long and the world learned of the death of President Harding in San Francisco.

With all of the distractions, Long never forgot about his dreams for Klamath Falls. The summer of 1923 was noteworthy in that regard. First, he had to oppose, as respectfully as possible, those who continued to recommend sale of that property. But if Long was fighting a holding action with Bell and other directors, he was on the offensive down in Klamath itself. First he completed arrangements for the purchase of a mill site at a point near Klamath Falls where the Klamath River ran wide and relatively slowly, thus providing a place for log storage.

Next, he pushed negotiations for the extra tracts of Long-Bell timber intermixed with their own in Klamath. These deliberations, however, went more slowly, in part because of other complications, such as the Longview mill site that R. A. Long wanted to sell and George Long wanted to buy, and also some additional southwestern Washington timber that R. A. wanted to purchase and George wanted to sell. But, as usual, George seemed to be the one dealing from strength.

Long had last talked with Long-Bell officials at the time of the Hotel Monticello dedication, and now, in a personal letter to John Tennant dated August 13, he requested an update on those "preliminary suggestions of timber purchases and trades, mill sites, etc." Specifically, he asked what price Long-Bell placed upon its Klamath County timber, "which you wished to have us consider, in the event that you made additional purchases of fir from us." He further indicated that he was having the Weyerhaeuser fir timber examined and looked forward to making "a similar examination of the pine belonging to your company."

Next he inquired, almost casually, as to "what kind of a price you would want for a mill site." A related question involved "what kind of a traffic arrangement we would have with your railroad," the answer to which was crucial. Long concluded his letter full of questions, "wondering whether the time is ripe for discussing them."

Apparently it was. Within the week, Long reported on the possible trade of Long-Bell pine in Klamath County, Oregon, for Weyerhaeuser fir in southwestern Washington, informing J. P. Weyerhaeuser and F. S. Bell of the likely prices. In the fir district, theirs would be $4 per thousand feet for the green standing fir, cedar, and spruce and $1 per thousand for the dead fir and hemlock. The price for Long-Bell's pine was about what George had expected, but the quote on the intermixed was at least $1.50 more than its value. Compromise would be necessary.

On the question of the Longview mill site, no specific figures were forthcoming, although it was apparent that R. A. "was very anxious indeed to have us locate on the property and would fix a price that would be satisfactory to us." While many questions awaited answers before any decisions could be made, one thing was clear beyond a doubt: If a reasonable price was set for the Longview site and reasonable assurance provided as to the rail access, that was the place to be.

Bell found himself between the proverbial rock and a hard place. He did not want to oppose progress on a Longview mill site that seemed right from every consideration, but he also did not want to approve actions that might complicate the eventual disposal of the Weyerhaeuser Timber Company's Klamath holdings. It was evident that Long-Bell hoped to use the sale of its Klamath timberlands and the Longview mill site to offset the initial cost of additional fir purchases from Weyerhaeuser, deferring payments for several years. Bell sounded a warning, one that would prove out: "Mr. [R. A.] Long has a very heavy deferred obligation to us already and these next four or five years will be his critical years. I hope he is right in the optimism which plunges him into so great an adventure, but I think we are in it about far enough already unless we secure ourselves in future dealing against a series of happenings which may bring his scheme into embarrassment." Despite such warnings, the Weyerhaeuser brothers continued to support George Long's policies and positions.

Senator McNary's committee hearings were scheduled to begin September 6, and Long left nothing to chance, enlisting the participants, suggesting that they be certain to emphasize this or that, and explaining how their part would fit into the whole. He also indicated how he planned to convey his own taxation message: "I do not. think it is necessary to go into a long-winded discussion [about the burden of taxes on mature timber]. Neither do I believe that timber can reasonably escape its just proportion of the tax during its harvest, but I do think there should be some encouraging features adopted to grow a new crop of timber, and my idea . . . will be to outline how difficult it is under the present tax system in the State of Washington for an operator to be encouraged in the growth of a new forest."

All of the preparation for the hearings paid off, and the senators were surely full of the Pacific Northwest when they took their leave. Long was pleased with a job well done. As for his own remarks, they had clearly been effective. At the last minute, he had put aside what he had prepared and instead spoke extemporaneously—what he did best. His taxation sermon rang loud and clear. Some had tried to divert the discussion to fire prevention, but Long would have none of that. "Keeping the fire out simply solves the problem that timber will grow," he reminded. "It does not solve the economical problem of growing timber at a profit, and that is what you have got to solve if you ask private individuals to do it."

Possibly the best exchange followed Chief Forester Greeley's suggestion that taxation on cutover lands wasn't as large an obstacle to reforestation "as you seem to view it." Long queried, "Well, are you talking about today?" Greeley's affirmative response opened the door for Long. "I am talking about tomorrow," he began, "[and] the next day, 5 years from now, 10 years from now, 20 years from now, 30 years from now, when, during all that period, I can't withdraw from that property one single cent of revenue; I have got to pay the taxes every year, every year, every year, for 30 or 40 or 50 or 60 years; and I have got to take all the ups and downs that represents public whims about taxation—such 'a blind alley' that no man with one grain of common sense would ever start without some kind of a flashlight to at least tell him where he might get off. . . ." Long provided the committee graphic illustration of the problem, a year-by-year schedule of taxes paid by the Timber Company, beginning with 1900, on the same 342,154 western Washington acres. Over the period, taxes had increased every year but two: in 1900 they totaled $25,128, and in 1922, $583,490.

George Long was happy to return to his desk, only to find that he couldn't escape his two eternal irritants: Carl Hamilton's Weyerhaeuser Forest Products, the advertising arm; and Thompson Yards. Long once more attacked Hamilton for the "very slow development that has progressed in the house building line," again noting, "We have just passed through two of the best years the company has ever found for house building." And he closed expressing more unhappiness: "I will possess my soul with additional patience and hope that all will turn out well. In the meantime, hoping to see your books before a great while, although they have not yet put in appearance as per my request."

As for Thompson and his yards, there were some recent developments. Kendall had been called in from his inspection tour to meet with F. E. Weyerhaeuser and others in St. Paul on September 3. Thompson Yards seemed in chaos, the personnel well aware that something was in the air. A board of directors meeting had been held the following morning, and Kendall was appointed general manager of the company, "effective at once." Thompson had attended the September 4 meeting, and afterward advised Irvine that "he felt it would be a mistake for him to continue in the employ of the company after the first of the year." He received no argument on that score from Irvine.

Irvine proposed an immediate meeting, but, as Long explained, that wasn't possible. First, there were the touring senators to be hosted. "Then my son George has decided to get married on about the 25th of this month." It would be October before Long could get away.

Long spent the first week of that month going to and from St. Paul. As he subsequently observed, "It was rather a gruesome trip," reflecting more on business than travel. Thompson resigned and the board elected Irvine as the new president, with Kendall as general manager. The plan was as they had earlier discussed, "to split the company up into eight separate systems, with a manager in charge of each system and Mr. Kendall will have charge of the management and of the whole business." Kendall saw some hope for improvement, "after weeding out some twenty yards." At the very least, "We have a prospect of getting out of the hole or not getting in any deeper."

Sadly, it was soon discovered that Thompson had been more than just ineffective. Irvine reported a series of "unsavory transactions," which constituted "a great blow to all of us as we have felt that while Mr. Thompson's business judgment was poor, he at least was honest." But by this time, Long's mind was on other matters.

He hinted at some of these in a personal letter to his St. Paul banker friend, R. C. Lilly, in which he inquired into the financial health of Long-Bell. He explained his concerns: "We have large business dealings with them, all very excellently secured and taken care of and have no reason to be uneasy in any way as it may pertain to our own interests, but this concern is operating in such a broad way out here that we are indirectly interested in hoping that they have ample strength to walk away with the stupendous undertakings which they have on their hands."

The usual difficulties were being experienced between Bill Boner at Everett and Colonel James Long at Baltimore, between the supplier and the receiver. More often than not, Long sided with Boner. "Until you have been through the grind yourself," he admonished his brother, "you cannot appreciate what a regular cyclone of conflicting things come to a mill man on the Pacific Coast who at the same time is cutting special orders for China, for Japan, for the railway trade, for Baltimore, and at the same time keeping up an assortment in his own yard." Whatever the complaints, Baltimore did a booming business in the summer and fall of 1923, and if there had been any doubts about expanding the Atlantic Coast involvement, they had disappeared. With Dr. E. P. Clapp doing most of the negotiating, the company purchased a "magnificent Providence, Rhode Island, property at a very low price."

Events were shaping up nicely on the Atlantic Coast,

but Klamath Falls was never far from Long's mind. While admitting his interest, Long tried to assure Bell that no commitment had been made to build a plant. It may have been the best of times, but that didn't change Bell's opinion of Klamath. He repeated his concern about allowing Long-Bell to use its Klamath timber as a down payment on other exchanges. In Bell's words, "At this distance it does not seem to me they are getting forward very fast with their programme except that they are spending their money very fast."

At that moment, F. E. Weyerhaeuser was discussing with railroad officials plans for service to Klamath. The Southern Pacific seemed ready to proceed with construction of a line from Eugene, Oregon, connecting with Klamath Falls. Also, the Great Northern and the Northern Pacific appeared interested in providing service, although this would likely require trackage rights from the Southern Pacific. In F.E.'s view, "It would be advisable for us to push this project as hard as possible in the near future." So, while Bell and his group were wishing to be rid of Klamath, the Weyerhaeusers were working to make George Long's dream a reality. This wasn't the first time those lines had been drawn, nor would it be the last.

In any large business enterprise, some operations commonly are starting up as others are being phased out. In the late fall of 1923, this was the situation with the old Clarke County Timber Company. On November 16, a sale of 320,896,686 feet, "being the stumpage approximately on 7,444.00 acres," was consummated with the Murphy Timber Company. The terms were complicated, but amounted to $4 per thousand, or $1,283,586.74, with payments extending over the ten-year cutting period, and with the Weyerhaeuser Timber Company retaining ownership of the land. Long indicated that they would be watching Murphy "very closely, to see that the timber is cut in the right manner and payments are kept up promptly," but on the whole, he judged this a good arrangement. That would leave the Clarke County Timber Company with about a year's worth of logging left.

Long offered an assessment to the Clarke County shareholders: "On the assumption that the Murphy contract is good and we will get all of our money under it, the situation that we will enter upon the first of January, 1924, will probably mean that

Clarke County Timber Company will still have assets easily worth $1,500,000.00." Since its organization, Clarke County had distributed ninety-three dividends of 10 percent each, "so that realizing when we entered upon the task of logging the burned timber in 1903 that we were not very enthusiastic about the future, the results seemed rather to imply that Fortune has been rather kind to us after all in this venture." Most understood that the results owed at least as much to wise management as to good fortune.

TWENTY-FIVE

Good Times Can't Last Forever

JANUARY 1924 FOUND GEORGE LONG SITTING COMFORT-
ably in his office, enjoying a good cigar and the balance sheets of
the Weyerhaeuser Timber Company. The "nice mild cigar" was
from an expensive box, a Christmas gift from brother Jim in
Baltimore. Indeed, almost everything coming from Baltimore
recently had brought pleasure—the 1923 profits for that operation alone totaled $575,000, an amazing performance. But as
George was quick to point out, circumstances had worked to
their advantage, from "the extraordinary demand for lumber and
the increased price," to "the low cost of freighting." And as they
knew, such conditions would not continue indefinitely.

The year 1924 would prove to be an especially important one for the industry, largely due to final consideration of
the much-debated federal reforestation bill. Although success
was as yet far from certain, Long was optimistic, and when he
received the Chief Forester's annual report for 1923, he ap-
plauded Greeley's contributions: "You understand better than
anyone else how slow public opinion works and yet I feel that
remarkable progress has been made in calling attention to the
public to this matter [reforestation] and getting very warm en-
dorsement and happily I believe the lumber industry has responded more freely to this call during 1923 than ever before."

The task, however, was not quite complete, nor would it
ever be. A major concern at the state level continued to be tax
policy. University of Washington Dean Hugo Winkenwerder,
chairman of the Washington State Forestry Committee, was urging study of forest taxation. "The timberland owners," Winkenwerder noted, "fully appreciate the situation, and many of them

293

are interested in reforesting their holdings, but do not feel that they can undertake it under the present system of taxation." The dean recommended that they "meet the issue squarely and just as quickly as possible," and for that purpose proposed the creation of a forestry conference taxation committee, broken down into two working committees, a general committee and a subcommittee on research. Winkenwerder invited Long to become a member of the general committee, and he readily accepted.

As expected, business soon began to slow down. In early March, Long reported that logs had begun to pile up along the Columbia River, Grays Harbor, and Puget Sound. Though prices remained fairly firm, "the enthusiasm and ardor that was characteristic of logging last year" had cooled off considerably. The Japanese market and, more important, the California market had shown recent weakness. One didn't know how much credence to give recent disclosures of the Harding scandals, but as Long observed, "It looks like it is going to make of the presidential year not what we hope to see it, but one of considerable uncertainty and more or less disturbing to business."

Long was soon advising that they "trim sails," and brother Jim received a pessimistic report. "The fir business is shot all to pieces," Long noted, and "instead of having an easy time to sell lumber, we will have to get right down and scratch for business in an old fashioned way"—not by earning it, but with "lower prices and a whole lot of other things that are far from being agreeable."

Change was also afoot with the Western Forestry and Conservation Association. At the March 20 meeting, the trustees considered a plan for the association's foresters to "advise and assist timberland owners in such forest treatment as pertains to soil classification on logged off land, reproduction, and all topics having to do with reforestation, and possibly likewise to make studies and give advice as to the stand of mature forests, having to do especially with such subjects as healthy growth, maturity and decay."

The plan, presented by Allen, received a favorable response. Long worried that Allen had been too imprecise about what the service would cost, but he sent along the suggested $800 subscription in behalf of the Weyerhaeuser Timber Company, urging Allen "to have each one of the parties committed to a similar plan or obligation somewhat in advance of a positive employment of the man whom you have in mind, so as to make sure that this source of revenue is dependable." Allen was not the best of businessmen.

Long did some traveling that April, first to Baltimore and then on to Providence, where, in the company of Dr. E. P. Clapp and F. E. Weyerhaeuser, he wanted to inspect the newly acquired property. Trying to coordinate with F.E., Long was uncertain how many days would be required, "but I thought after looking over Baltimore pretty carefully, that we would go up to the Rhode Island site and have my brother go with us, so as to visualize all the physical conditions and map out the class and type of improvements and treatment of the lumber." Now, at least, they had some Atlantic Coast experience on which to build.

The Saturday following Long's departure, April 26, Hugh Stewart died of a stroke. Stewart had been the original colleague in the Timber Company office, and he would never be replaced. Long couldn't help but recall the reason behind Stewart's original appointment: he was to be Frederick Weyer-haeuser's watchdog. That, of course, was unnecessary, and over the years the watchdog became Long's trusted colleague. But it was Hugh Stewart, his friend, whom Long would miss the most. For him it was, as Stiles Burr understood, "a personal grief."

The trip itself went about as hoped, including the visit to the Rhode Island property. Everyone seemed well pleased with the prospects, and it appeared that satisfactory rail rates could be negotiated. Long subsequently informed Boner, "We have a very good piece of ground there; fully as much frontage and fully as much elbow room as we have at Baltimore." All agreed that they ought to proceed immediately with development of the site as the second Atlantic Coast terminal. This would be announced at the upcoming Timber Company annual meeting. Indeed, immediately upon his return to the Tacoma office, Long began preparing for that event.

In the truly important respects, this would be the easiest and most pleasant of all such meetings. As Long noted in his opening remarks, "The year 1923 was the most prosperous year the lumber industry of the Pacific Northwest has ever experienced." Everything seemed to be working in their favor, from weather to labor to the availability of railroad cars. Long reviewed the year's various acquisitions, both of timberland and of the steamships *Pomona* and *Hanley.* Then he talked of Portsmouth, calling attention to the contributions of Dr. Clapp in the acquisition of "a site that has a great many admirable qualities, so much so that we feel warranted in going ahead at once with the development."

And, finally, Long turned to what he considered the crucial unanswered question, what to do with Klamath. He had prepared well. Indeed, prior to the annual meeting, he had led a group of shareholders, family, and friends, on a tour of the region; or, more precisely, he had led them to Jack Kimball, who then served as their guide. Apparently, everyone had a good time. "One of the old stagers, Mr. William Musser," Long later informed Kimball, "came to me just as he was leaving, and said it was the finest trip he had ever had in his life, and he is considerable of a traveler." Whether all came away as enthusiastic over Klamath as George Long seems doubtful, but that was probably too much to expect. On the trip back from Klamath to Tacoma, some of the party stopped off at Longview to tour the Long-Bell facility and to see for themselves the mill site that had been proposed for a Weyerhaeuser plant.

After the annual meeting, Long looked forward with pleasure to the arrival of Chief Forester Greeley, who was celebrating the June 7 passage of the Clarke-McNary Act. The bill was passed, in his words, with "scarcely a ripple of opposition." Congressmen John Clarke had smuggled Greeley into "the sacred precincts of the House"—actually the Republican cloakroom—and he listened to the proceedings from there: "After four years of controversy, it was a great thrill to be in at the kill—even if the victory was bloodless." Section 3 of the law was of particular interest to Long, as it authorized a comprehensive study of "the effect of tax laws, methods, and practices upon forest perpetuation." Long forewarned Allen: "You no doubt will be able to discuss this matter quite fully with Mr. Greeley in the early future, and as you may know Mr. Chapman expects to be in Portland and at the same time Mr. Greeley is there, so I feel quite sure your further study of what to do first and how to go at it will be pretty thoroughly gone over in a preliminary way at least, laying a foundation for future activities.

George Long was not the only lumberman anticipating the consequences of the passage of the Clarke-McNary bill. J. J. Donovan, of Bloedel-Donovan Lumber Mills in Bellingham, forwarded a letter of inquiry to Long "from our mutual friend, Mr. E. C. Hole, manager of the *American Lumberman,* who wants to know what we are going to do out here in the way of cooperation with the National Government under the new McNary-Clarke Forestry Bill." Donovan reminded that the Washington legislature would be gathering in January, and added, "I think it behooves us to determine just what we want to do in a way of additional legislation."

Long was prompt and thorough in his response. By this time, Greeley had come and gone, and Long was able to enclose a copy of Allen's report of the Chief Forester's discussion with Oregon lumbermen. "In a general way," Long noted, "Mr. Greeley is preparing to work with the lumbermen in the matter of a study of the tax problem in each of the states and this [is] to be accompanied later by a recommendation that will have for its authority the full force of the Forest Service's viewpoint, as well as the affiliated lumber interests so far as possible." Their own Western Forestry and Conservation Association had appointed for each of the member states a committee "whose function is to work with the government officials on any problem that arises pertaining to forest problems." In addition, there was the Washington State Forestry Conference, chaired by Dean Winkenwerder, working under the auspices of the Seattle Chamber of Commerce. Long was convinced that "so far as machinery goes, there is one now organized and functioning which will help meet these questions as fast as they naturally arise and can be handled," and he concluded, optimistically, "I believe that after many years we are at least started on a program which promises some practical results." True enough, but the results would be a long time coming. The tax study, for example, would take a decade to complete, but as Donovan observed, "I suppose that as Rome was not made in a day, neither is a modern forestry policy built up in a few weeks." Or a few years, he might have added. Still, had it not been for these aims and accomplishments, the sort of yield tax Long envisioned would have been even further delayed.

Research of an entirely different sort was being considered a continent away. With the Portsmouth site decision approved, Long felt "quite an itching" to investigate similar opportunities "around the New York Harbor to see if there is not some economic nook or corner that we can acquire down there that would be of much benefit to us ultimately." But not all news from the East Coast was encouraging. Colonel Long reported that the Norton-Lilly officials had asked permission "to return *Hanley* in ballast," and there had been no choice but to agree. The message was obvious: "[It] shows just how rotten westbound tonnage is at the present time, and also I presume it is safe to say it might be considered as indicating the general business prosperity of the country."

Despite present discouragement, Long and Boner were seriously contemplating adding to their fleet. Prices for Shipping Board vessels seemed low indeed. Predicting that Weyerhaeuser might be moving as much as 150 million feet within a year, Long wired Norton-Lilly to ask whether the company would be willing to "continue an arrangement involving more ships than the *Pomona* and *Hanley,* in so far as the west bound tonnage was concerned."

It was not a propitious time to expect an encouraging response. Indeed, Long soon had second thoughts himself. On July 12, he wrote again to Joseph T. Lilly, indicating, "For the time being we have decided not to make this purchase," but, he continued, "we would like to have you express your views as to the subject, so that if an opportunity presented itself for quick action, we could know what to do." Regardless, it was apparent that the Atlantic Coast would assume greater importance as a market for Weyerhaeuser timber.

As the summer wore on, there were some serious fires, most notably in the area of the Clarke County Timber Company operations. Almost as worrisome was the issue of declining demand. Long was philosophical as he looked over the Everett midyear statement, observing for the benefit of Boner that the figures were "not very pleasant to reflect upon, but so it runs in life." Although he thought he had sensed some improvement in conditions over the past month or so, the future was uncertain: "This thing of running on a slow bell is deadly in its effect so far as profits are concerned and yet I cannot see where there is anything to warrant running any stronger than you are, in other words, I think it is absolutely dangerous to pile up more lumber than you are selling monthly." The Everett statement showed business volume at about 80 percent of the previous year, with prices ranging from $6 to $7 lower.

As Long admitted to William Carson, "There is not much money, if any, in a lumber operation right now." Everett was currently operating Mill B both day and night, but working only a daytime shift at Mill A, and both plants were on a four-day-a-week schedule. Baltimore was faring somewhat better, with sales up to July 1 exceeding those for 1923 by some four million feet. The average price, however, was lower by about $4 per thousand. Even though the market had been weak, Long seemed reasonably pleased with the work of the Sales Company, and went so far as to suggest that it had been "functioning very well." Its special merit, he noted, was that it could cover such a large territory, "picking up an order here and there," an advantage over smaller organizations.

One most welcome interlude occurred in late July 1924— George Long visited Snoqualmie Falls to dedicate the new community center. It was a grand affair, the program opening with several rousing Sousa marches played by a sixteen-piece band. There were the usual speeches by plant officials and dignitaries, and some more entertainment, including Harry Wood, a shed foreman, who, in the words of a young journalist, Stewart Holbrook, "raised one of the sweetest Irish tenors I've ever heard in 'Isle of Dreams,' and the house turned bedlam at the finish." When order was finally restored, Holbrook saw "a tall, spare figure step from the wings. I already knew him by sight. He was George S. Long, Sr. He was dressed in somber clothes, and a black tie with large knot circled the snow-white old fashioned stand-up linen collar. He walked to the middle of the stage, cleared his throat, grasped a coat lapel with one hand, and started to speak. His voice was calm, mellow. I do not recall his exact words, but his duty was to dedicate this new community hall, and he did it with grace and humor, and with dignity that had nothing of the stuffed-shirt quality which has ruined the efforts of so many other wise good men when they rise to pass a few remarks." What young Holbrook didn't realize at the time was how much George Long enjoyed such opportunities. In many ways, he was a showman.

But in most situations Long avoided center stage, proceeding quietly, preparing for the future. Others watched and wondered. Southwestern Washington seemed on the verge of something big. What was Weyerhaeuser up to in these parts? To one who inquired, Long replied matter-of-factly: "The Cowlitz Development Company started a year and a half ago, and is simply the building of a railroad eight or ten miles in length by ourselves and the Ostrander Railway & Timber Company, over and across a logged off area, to reach timber where ourselves and Ostrander Railway & Timber Company had uncut timber." Nothing more, nothing less—old news in every respect. Only it wasn't. Where were those logs going? That was the question awaiting an answer.

Although most of Long's thoughts were focused on Western properties and opportunities, he was planning another East Coast swing. They needed to make some final decision as to the Portsmouth development, and Long also had to get together with lawyer Gus Clapp to iron out the legalities of the Snoqualmie Falls and Cherry Valley merger. Clapp wanted to know what the proportion of merger would be, and Irvine had indicated that "we roughly estimated" that Cherry Valley's share would be 25 percent. Now Irvine asked for the "final figures," and hoped that Long could bring those with him when he came east.

Another subject for legal consideration involved logged-off lands. As noted earlier, this subject touched upon one of Long's pet peeves, the suggestion by many promoters that such lands were appropriate for standard agricultural uses. As he told lawyer Clapp, "I do not like to use the word 'Colonization' in our logged-off land department out here, because the word 'Colonization' as applied to western Washington logged off land has been very greatly abused by a lot of fake treatment." Accordingly, he wanted such lands as belonged to Weyerhaeuser to be managed by a separate corporation, the Weyerhaeuser Logged-Off Land Company. He acknowledged that the designation was "somewhat cumbersome," but he wanted the name Weyerhaeuser attached because it implied a policy that he wanted to preserve.

And that is how the Weyerhaeuser Logged-Off Land Company came to be.

Long was accompanied on the eastern trip by his wife and a daughter. Carrie planned to visit a sick sister in Detroit, and the three of them subsequently hoped to combine a bit of pleasure with business. But as always, business came first. They proceeded to Baltimore and while there, Long received important news from F.E. concerning railroad developments at both Klamath Falls and Longview.

Charles Donnelly, president of the Northern Pacific, had paid F.E. a visit on the afternoon of October 2, inquiring, among other things, "point blank whether we had in mind any immediate development of our Klamath Falls timber." F.E. was not being evasive when he replied that they certainly wouldn't build without the assurance of "an outlet over the northern roads," and further that he couldn't answer "definitely until the Trustees of the Weyerhaeuser Timber Company had considered it." Donnelly also wished to discuss the Longview situation, but again F.E. demurred, suggesting that they wait until George Long arrived in St. Paul, at which time he could "inform them more definitely than I am able to, both as to the situation at Klamath Falls and what we shall need in event we build a mill at Longview."

The St. Paul conferences with railroad officials and representatives of the Long-Bell company came off as planned. Negotiations with Long-Bell were clearly reaching a critical stage. R. A. Long, President M. B. Nelson, Vice-President John Tennant, and the chief counsel and traffic manager met on October 23: "The most important question was the tonnage or traffic arrangement for hauling logs and lumber to and from the plant." The premise was simple enough: "That we wanted to be placed in just as good position as they were in every respect." A suggestion was offered by R. A. Long, "that maps would be prepared showing the location of the tracks which were to be used by the Weyerhaeuser Company railway, and this trackage, before being finally determined upon, would be fully approved by Mr. Geo. S. Long of the Weyerhaeuser Timber Company." An additional understanding was reached, George agreeing that should a "so-called Weyerhaeuser railroad" be constructed, they would not "solicit or carry any

business or tonnage out of Longview except what might origi-
nate at their plant and for their own use in logging." In short,
they would not compete as a common carrier with any Long-
Bell road.

The situation at Klamath Falls remained confused, what
with the different railway companies uncertain about how or
even whether to proceed with construction. As he expressed in
a letter of October 30 to R.A. Booth of Booth-Kelly, Long
preferred that "all three lines build jointly from Bend around
Summer Lake and heading towards Lakeview," but no commitments were forthcoming. He was sure of one thing, that the
Great Northern was "quite anxious to get into that territory . . . for they need lumber tonnage."

But some matters were resolved, among them the arrangements to merge the Cherry Valley Logging Company
with Snoqualmie Falls. Long wrote to Snoqualmie manager
Rod Titcomb upon his return, noting that it was "deemed quite
desirable to have our final consolidation made as of December
31st, 1924," which meant, of course, that they had to have their
figures available very soon. "All this means," he continued,
"that your complete inventory of logs, lumber and everything
else will have to be taken" so as to be ready for a December 16
meeting with Horace Irvine in Tacoma.

As always, one of the foremost subjects for any St. Paul
discussion was Thompson Yards. This time there was progress,
or at least there was change. As announced in a letter dated
November 1, 1924, the Weyerhaeuser Timber Company had
acquired the 25 percent interest in Thompson Yards (formerly
owned by Potlatch Lumber Company, Edward Rutledge Timber Company, and Humbird Lumber Company), and "now
holds all of the stock of Thompson Yards, Incorporated." The
stockholders of the Timber Company were thus given the option to purchase "certain stock" at a price of $25 per share,
utilizing "a special cash dividend to all of the stockholders of the
Company, payable December 1st, 1924." It was further noted
that "while the price named is probably greater than the whole
or any part of the stock could be sold for to investors or outsiders, nevertheless we believe that it is to our best interests to
accept the offer made by the Company to sell the stock. . . . We
hope and believe that the changes which have been made in the

management and policies of Thompson Yards, Incorporated, are going to result in placing the affairs of the Company on a sound basis."

Calvin Coolidge had easily won the presidential election, much to the relief of George Long and most of his friends and colleagues. But local elections were another matter. Jack Kimball had plenty to worry about down in Klamath Falls, particularly the strength shown by supporters of the Ku Klux Klan. Long tried to assuage Kimball's fears. "It is, of course, quite disappointing and disgusting to have such manifest expressions of ill will reflected," he acknowledged, but he reminded that occasionally "the fellow who is loudest in his denunciations proves to be rather sensible and reasonable when he gets in office and gets responsibility." Kimball was also concerned about the likelihood of threatened recruising of timberlands for tax assessment. Again, Long counseled patience, advising against any concerted opposition. "Be reasonable in pointing out the lack of necessity," he advised, "and then if you find you are licked on the issue, try to see that the job is done, not by enemies but by people who have a fair sense of justice." And he concluded, "I feel like encouraging you to use your best diplomacy to handle the situation without the possibility of stirring up too much trouble with the fellows who happened to win out in this case." The advice was vintage George Long. Cooperate. Confrontation should be the last resort.

Another ingrained Long trait was secretiveness. He saw no benefit to advertising either success or intention. Even major shareholders were forced to listen carefully to discern Long's designs. In November 1924, Long was upset to learn that Paul Kendall, manager of the publicity department for Long-Bell, had apparently gone public concerning recent negotiations between his company and Weyerhaeuser. Long wrote directly to R. A. Long to complain. In this instance, he indicated that "our people . . . do not feel very much like passing out to the public very much that might be called publicity." There might come a time when an announcement would be appropriate, but not yet. "Already," he continued, "I have had an indirect suggestion that the close affiliation of our two companies was apparent and indicated something which the radical people call 'Combination' and which you know is a word that all demagogues like to play with."

Long's displeasure was well-founded. On November 13 an article appeared in the *Oregon Journal,* headlined, "Weyerhaeuser Co. May Build Lumber Plant Near Kelso," accurately noting "the presence here during the past few weeks of parties of Weyerhaeuser engineers and officials, who are making detailed examinations of sites in this vicinity." Long did his best to put a stopper in the rumor bottle. To one Kelso inquiry, he told the truth and misled at the same time: "The fact that we are building a logging railway near Ostrander, with the idea of marketing these logs in [the] Columbia River, seems to be the starting point for many wild rumors, none of which are wilder than that we contemplate building a mill at any point on the Cowlitz River or at Kalama." Long's statement was literally true. Longview was situated at the confluence of the Cowlitz and the Columbia, and the Long-Bell property under consideration was on the Columbia.

Plans were proceeding, but not as fast as George Long wished. As he explained to Dr. E. P. Clapp, "Between various other important matters that are coming up, like the consolidation of the Cherry Valley Logging Company and Snoqualmie Falls Lumber Company, the cleaning up of the Thompson Yards matter and fussing with the Portsmouth situation, I have, with current affairs, had my hands pretty full for the last two or three weeks, so I have not been pushing the Long-Bell matter." But he was certain of one thing: "If we take the property immediately below Oregon Way, I am going to try to get the lands all the way down to Coffin Rock." The Longview project would illustrate once more Long's commitment to unimpeded development.

If one remains in a responsibility over time, one invariably encounters some all-too-familiar problems. So it was with George Long and F. E.'s accounting arm, Frank Poole. F. E. once again had received a report from Poole that some of Long's office managers were not playing by the rules. "Do you think," F. E. subsequently inquired of Long, "I am asking too much in insisting that the Managers comply with this request?" Actually, it was Baltimore that had caused the tempest, Poole complaining, among other things, that expense accounts were not being itemized.

Long had to wish that it was unnecessary to admonish his brother. He tried to be diplomatic, allowing that it was indeed a nuisance "to keep track of all the little petty expenses that are involved in travel . . . and all kind of little things that confront a man every half hour in the way of expenditures when he is around a busy town like New York or elsewhere." But, he quickly added, "it is a rule which we follow quite carefully in our office. Every expense account that is passed upon in our office, except the writer's, contains an itemized account of everything. My own expense account is rendered monthly in sum totals, but I do keep it item by item in a little book and these little books are kept on file so that at any time the monthly expense account can be checked up against the book." So George Long ordered James to do the same—keep a record.

This was a relatively minor irritation, as Long realized upon reading a mid-December letter from Boner. "My physical condition," Boner began, "is such that it seems to be impossible for me to do anything like justice to the position of Manager here . . . and it is an absolute detriment to my health to continue to make a pretense of so doing." He closed requesting that he be relieved of his responsibility, "all of which has been very pleasant, and I regret it very much, but it is the only thing to do." Long was taken aback. "Somehow I cannot get reconciled to the idea suggested by your letter," he wrote in reply. Boner, however, wasn't exaggerating. And Long had to reconcile himself to that fact.

What to do? The news from the Portsmouth dredging site was discouraging, and now word of Boner's heart problems. On December 19, George sent a wire to Colonel Long: "Am writing you today suggesting that we defer immediate activities on the Portsmouth plant." The letter that followed explained some but not all of the reasons: "I really feel chagrined that I have slipped around to a point where I am a little at sea about the Portsmouth proposition, but such is my mental attitude this morning, and several other facts have been presented to me, involving somewhat the question of our doing a little further looking around before we finally decide on this particular site." Most of all, Long just needed some time to reassure himself that the pieces could be put together in Portsmouth without the benefit of Boner's assistance.

Edward B. Wight tried to assume his father-in-law's responsibilities, but his reports were not encouraging on any score. According to Wight, the doctors could not help Boner, other than alleviating "pain with the use of opiates." As for the mills, they had hoped to reopen on Monday, December 29, but Everett was experiencing a serious water shortage, and no plants using city water were allowed to start before January 2. It was, as Wight noted, "quite a disappointment as we are all ready to start and need to."

Nor had December disappointments been limited to Everett and Portsmouth. Progress toward settlement of the Thompson Yards ownership and management questions was slow and difficult. Most of the Thompson Yards stock had been sold to Weyerhaeuser Timber Company shareholders. These totaled 56,935 shares, transacted at $25 per share, bringing in $1,423,375 and resulting in a loss to the company of approximately $3.7 million. The remaining shares awaited purchase or assignment, Long suggesting that the new corporation, the Yards Securities Company, agree to purchase the balance, 3,065, at $25 a share, submitting a note to the Timber Company in payment. That was making the best out of very bad circumstances.

Long was hoping for better results regarding the proposed merger of the Cherry Valley Logging Company properties with those of the Snoqualmie Falls Lumber Company. There was no question about the desirability of the merger. There could and would be questions as to the details. In fact, Long could hardly have been surprised when Irvine, on behalf of the Cherry Valley ownership, expressed some last-minute reservations. In mid-December Long was still optimistic, scheduling a stockholders' meeting for December 15, "it being an all-around conference simply to pave the way for positive action at the formal meetings which will be held later." He asked the two managers, T. M. Williams of Cherry Valley and Rod Titcomb of Snoqualmie Falls, to be present, bringing with them details, "inventories, etc., as well as your cruising records, so that we will have before us all kinds of information that may possibly be needed." The meeting was held, but little progress was made.

Long tried to explain the problem to friend O. D. Fisher: "Mr. Irvine had some little different ideas about arriving at the percentage which the Cherry Valley Logging Company would be entitled to," and added, "He did not get the thing thrashed out just to his own satisfaction, and certainly not to mine when he left us yesterday afternoon [December 17]." Fisher thought they had compromised more than enough with Irvine. "I hope," he wrote in response, "nothing will be injected into the negotiations that will require a complete revamping of the whole basis of consolidation. I cannot see how it could be worked out to be more favorable to the Cherry Valley company than the line-up arrived at at our conference on Monday and Tuesday." With the opposing sides apparently firm in their positions, Long advised Titcomb on December 29 "that the amalgamation with Cherry Valley Logging Company will not be made during the year 1924, and likewise it looks as though it would not be made at all." The ramifications were immediate. They meant that Titcomb had to "revamp" his plans "pertaining to logging, railway development, etc."

The difference between the two sides was not large, or it didn't seem so on a percentage basis. Irvine now argued that Cherry Valley ought to receive a 22 percent share of Snoqualmie in return for its merged assets. Long and Fisher maintained that the 20 percent earlier stipulated was more than fair, and perhaps it was the fairness aspect that stiffened their backs. Long summarized the situation for Irvine: "As much as I would like to see the amalgamation made, I would not want to see it made if it was not entirely satisfactory to everybody." So it seemed that each would go its separate way. For those outsiders who assumed that the so-called Weyerhaeuser affiliates operated as one, this was another piece of evidence to the contrary.

TWENTY-SIX

The Busiest of Years

THE YEAR 1925 BEGAN INAUSPICIOUSLY. ONE OF THE first letters to cross George Long's desk came from Horace Irvine, remarking on Long's "unfailing fairness" as concerned the unsuccessful merger effort between the Cherry Valley Logging Company and Snoqualmie Falls. George could only bite his lip, given the frustrations of dealing with Irvine over the years. But of greater immediate worry was the management question at Everett. Long had no idea who could replace the ailing Bill Boner. He admitted as much to F. S. Bell: "I have taken no steps yet to arrange for his [Boner's] successor, and in the meantime am giving to that business more of my personal attention."

Then there was Longview. Dr. E. P. Clapp was to have submitted a report, complete with recommendations, on the subject of purchasing a mill site from Long-Bell. But Clapp had been ill, and the report was delayed. If Clapp's paper was ready in time for the annual winter meetings in St. Paul, then perhaps Long would attend. If not, as he explained to F.E., "I may feel under obligations to remain at home."

F.E. had been trying to move the Klamath Falls rail negotiations forward, a task made somewhat easier by reason of his membership on the Great Northern board. Recently he had consulted with Great Northern's president, Ralph Budd. According to F.E., Budd wanted assurance that if his railroad were to proceed with construction, the Timber Company would proceed with its plans, assurance that "as railroad director I cannot make." Long could, almost. He wired Budd indicating the obvious, that Weyerhaeuser had "large quantities of timber tributary

308

to these sites," but he stopped short of making any commitment, "until additional railroads are made available for wider and more economic distribution of lumber." It was a question of which assurance came first.

To F.E., Long restated his own position: "We have a whole lot of money tied up in that proposition, the timber is ripe and regardless of whatever developments we may make in the fir region of our property, I do think that we should get busy when conditions are right for transportation in developing our property at Klamath Falls." Those who had opposed operations at Klamath were clearly on the verge of final defeat. But if Long's Klamath plum seemed ready to drop in 1925, the appearance was deceiving. An option to the Southern Pacific would not be available for another three years, and there was nothing to do but wait. And hope.

Waiting was not the order of the day elsewhere, and certainly not at Portsmouth, Rhode Island. There, a special town meeting unanimously approved an exemption of taxes on buildings to be constructed on the Weyerhaeuser property. As the *Providence Journal* reported, steady employment for about two hundred workers was promised at the site, "whose location in the town will mean much to Portsmouth." Clearly the time had come to unveil plans and solicit help. Long, however, concerned that they might run into serious engineering problems, was not quite ready to make his move.

Too often, or so it must have seemed, Long's plans were frustrated by causes beyond his control. Life had been simpler when his only responsibility was to manage timberlands. But even that was complicated by constant fluctuation in the value of trees and lands. When a potential investor in western Washington timber inquired as to the value of an "average stand," Long replied that it was "a good deal like asking, what is a horse worth?" When buying a horse, one "would first want to know how old he was, how heavy he was, whether or not he was blind, or lame, or had the heaves, all of which had to do with the value of the horse." So it was with timber. Age, density, quality, location, all affected the value. And, "if this timber happened to be hemlock it would have but very little value; if it was young growth fir, it would have more value, but far less than large old growth fir."

One element appeared to be in flux. Weyerhaeuser owned a large amount of hemlock, and although it may have had small relative value, as was indicated by a recent sale to Crown-Willamette, it had value nonetheless. No longer was it ignored by those who cruised the western woods. And George Long looked to increase its value. He had no choice.

Word of Bill Boner's death came in early February. Long had lost not only a valued friend but his most important lieutenant. There would be no replacing Boner, at least not in the foreseeable future. Other sad if not surprising news came from Tom Humbird, who announced his intention to resign as president of the Weyerhaeuser Sales Company for reasons of health. He shared his decision with Long, writing "I feel particularly responsible to you, and am confiding in no one else this proposed action." The two had been allies for many years on most issues.

And there were concerns closer to home—in fact, at home. As he informed brother James, Long's late winter travel plans to St. Paul now seemed in doubt. George explained that "in the last two or three days my wife has had quite a serious illness," and that unless she recovered quickly he would not be able to leave her. She failed to improve, and George canceled the trip. Apparently Carrie had suffered a slight stroke. Although she never lost her ability to speak and seemed otherwise only slightly affected, she did experience difficulty in breathing, and indeed would be bedridden for the remainder of her life, a matter of months.

Even though Long wasn't able to attend the St. Paul gathering, his presence was felt, at times strongly. That was the case with the publicity executive committee meeting, called by Chairman George Lindsay. Long offered to share his thoughts on the agenda, doubting, as usual, the efficacy of the program.

Though Long agreed that Weyerhaeuser should "keep lumber on the map as something that should be used," he also contended that the burden ought to be borne equitably by those most likely to share the benefits. When they went alone, as with "occasional advertisements" in the *Saturday Evening Post* and *Literary Digest,* education ought to be the objective, "letting the world know that the lumber is not yet all exhausted." But for the most part, advertising should concentrate on retailers and the industry as a whole. "We are in business to make money," he asserted, "and not to put out a propaganda . . . unless it accomplishes the end of being a profitable venture. . . . I am aware that . . . what I have said is a bomb shell in the very magnificent program which has been carried on in the past and which is contemplated in the future, and most certainly do I hate to be the one that throws the bomb." But throw he did.

To a considerable degree, the issue bothered George Lindsay far more than it bothered George Long. Long's objection to advertising was simple: He hated to spend money for what he considered nonessential matters. And so he continued to swat at Lindsay and his program. In this instance, Lindsay took large exception and fought back, sharing the debate far and wide. This surprised Long, and perhaps disappointed him as well. The disagreement had become personal. "Above all things," Long explained to Dr. Clapp, "I do not want to hurt Mr. Lindsay's feelings." But it was too late for that, and Long would soon be trying to salve the person as he assailed the program.

There was so much else of greater importance awaiting attention. Atop the list were the rail negotiations with Long-Bell, as well as the final determination of a mill site. Long had been champing at the bit. As he explained to F. E. Weyerhaeuser in a February 2 letter, "I really think it is unfortunate that we are making such slow progress," adding, "My anxiety about the Long-Bell matter is not wholly on account of their restlessness, but really because we would like to get started on that development just as soon as we have our feet on the ground in the right way."

Long had been studying the Longview choices for many months now, noting that the location below Coffin Rock had the easier access to deep water, while the upper location provided "somewhat better ground for our improvements, is quite a little nearer the town, and affords a very excellent opportunity to make log storage in the Columbia River that will be ample for all of our possible requirements." But the Coffin Rock site was finally selected, and the only question was how much land to acquire. To Dr. Clapp he admitted that "sometimes I am so grasping that I feel like buying not only the area between Coffin Rock and Oregon Way, but also to dip down 1000 ft. below Coffin Rock." This, he agreed, "would mean a rather enormous investment for a site, but we have enormous resources tributary to this location, and no one can tell what subsequent developments may follow in this day and age of high products and the full utilization of forest growth, much of which is not and cannot be converted into lumber profitably." And he closed his argument in typical style: "I am, as you know, from Indiana, and as the old lady said in The Hoosier Schoolmaster, when she was told that her husband had too much land . . . 'Maybe so, but when the gitting is good, git a plenty.' " Apparently Long always kept that dictum in mind. At least there are very few instances where one could charge that he should have built bigger. Generally his plans appeared overly grand, but within a few years any who doubted their correctness were eventually convinced.

Although the nation's economy in the mid-1920s seemed to be strong and getting stronger in most sectors, such was not the case with the lumber industry of the Pacific Northwest. The owner-operators understood that they could do little or nothing individually to improve conditions: It was the old story of too many manufactures producing more lumber than the market could absorb. And once again some thought aloud about limiting the output. One of these, Charles D. Johnson, president of the Pacific Spruce Corporation of Portland, dusted off the old idea and shared it with Kansas City operator, Charles S. Keith, president of the Central Coal & Coke Company. "When the 'muckraking' period of the lumber industry is over with," Johnson predicted, "[people who were] now looking seriously into the question of the conservation and perpetuation of the forests [might begin] to develop some really practical ideas." The most practical of those ideas, according to Johnson, involved "some method of regulating the production in accordance with the demand."

When Johnson appealed to George Long, inquiring "whether or not you think there is any merit in this proposition," Long offered little encouragement, although he did allow, considering the expansive atmosphere of the 1920s, that one "should hesitate to predict anything is impossible." Still, he noted the industry's repeated failures in curtailment schemes, and concluded that any attempt at combination presented "so many impossible features that I do not think it can or will be done." Curtailment might be achieved in other industries, but

"I do not believe that this can be done in lumber." Johnson wasn't persuaded, but he should have been. Long was right.

Another proposition caught the attention of F. E. Weyerhaeuser, and nothing could have pleased George Long more, if only for the moment. F. E. told of plans to meet with representatives of the Shevlin-Hixon lumber interests regarding their holdings in the Klamath Falls district and the possibility of a merger of properties. In F. E.'s words, "I look on the timber west of Klamath as the best operating proposition we have—one of the very best in the country—and so I shall be sorry to see it sold." It would not be sold. Both Long and F. E. may have been wrong in their optimistic assessment, but together they made a majority and their decision would carry the day.

Indeed, Long was so sure of the eventual outcome that he was soon advising Kimball of plans. His words, as usual, cloaked the certainty in uncertain terms: "We have been giving some thought to our mill site property down in Klamath Falls, and for the lack of something else to think about have been wondering what kind of development we might want to make there some day, and with that object in view, I have had our mill engineer up here make some sketches for the kind of things that one has to have about him in a large manufacturing plant, and to figure it out to see how it would fit the ground which we have." Long went on to advise Kimball that, "in order that we may not unduly arouse any local excitement over the matter, we thought best to send down one of our own engineers to stake off the ground, to see how it would fit with the plan which has been suggested." Thus Lloyd Crosby arrived to get the lay of the land, Long assuring Kimball "with perfect truth and frankness" that nothing definite was in the works, that this was "simply a try out to see what can be done in this particular location."

Suddenly, it was Longview's turn in the spotlight. R. A. Long announced that he would be arriving in late March, at which time he hoped that George Long, F. E. Weyerhaeuser, and possibly others might meet with him and other Long-Bell officials. As it turned out, F. E. was not available, and George Long couldn't leave Tacoma. "I have not been away from home over night since you were last in our office," he explained in a March 25 letter to John Tennant, this due to "the rather serious continued illness of Mrs. Long." Fortunately, however, St. Paul lawyer Gus Clapp was in town, and R. A. Long, Tennant, and associates agreed to gather in George Long's Tacoma office on March 28. There, the decision would be made.

Not unexpectedly, some opposed any action. Dr. Clapp, for example, wired his preference for a delay, at least until the annual meeting. George Long allowed that Clapp's was "a very safe and natural" position, "yet this matter has been gone over very thoroughly . . . and in so far as my own mind is concerned, it is made up, unless we find something in the negotiation that disturbs the situation." Long was anticipating no such disturbance. It was time to proceed.

Although the formal announcement was delayed, the decision was reached during that Saturday meeting. Long was embarrassed that the agreement with Long-Bell made the pages of the press prior to any official statement from his office. "R. A. Long and his usual coterie of officials met in conference with us in our office," he informed the directors in an April 3 memorandum, and "we had with us our attorney Mr. A. W. Clapp, together with Mr. J. P. Weyerhaeuser and W. L. McCormick and the writer." The primary point of negotiation involved the rail arrangements, specifically whether Weyerhaeuser would have its own line or would use the existing Long-Bell trackage. Previously, R. A. Long had opposed allowing "us an exclusive track of our own," but now seemed willing to compromise, either through granting outright ownership or long-term lease. In any case, by the end of the conference, "there was no reason why we should not go ahead and agree to the proposition as outlined above, and a written agreement to that effect was duly drawn and executed by both parties." Long concluded, "This commits us to the idea of going there with our plant."

Not everyone received this news happily, but F. E. wasn't among the detractors. And it was his opinion that mattered most. "You certainly have been given full authority," he observed upon receiving the report, "and your contract merely carries out what really was agreed to at our St. Paul conference with the Kansas City people."

Now that they were firmly committed to Klamath, F. E. and others in St. Paul were discussing cooperation in developing the manufacturing plant. They decided against it. Included in those talks were Rudolph Weyerhaeuser, Horace Irvine, and Harold Richardson. As F.E. reported, "All three agreed that we should not do so, that we ought to manufacture this tract of timber ourselves, and that they would prefer to hold the timber for a considerable length of time in the thought of our ultimately building our own mills, rather than make a combination with the Shevlin people."

F.E. also replied to a recent indication that Long was considering moving Titcomb from Snoqualmie Falls to the Tacoma office, presumably to provide George some much-needed assistance. According to F.E., the St. Paul contingent agreed entirely, everyone approving of such a change. He continued, "I never cease to marvel at your ability to take care of such a vast amount of detail work and, at the same time, not neglect the bigger problems which are constantly at your desk. I trust you will pardon me, however, in saying that we all feel that you should . . . conserve your strength for working out the bigger problems of the Weyerhaeuser Timber Company. You have a genius for this thing and are the only one who has the vision to carry out a program of development that is in proportion to the possibilities of the timber assets of the company."

Long's vision was being thoroughly tested in 1925, and he was proving equal to the challenge. Even as logging ceased and timber sales diminished, wherever he looked, he saw opportunity. "We find we have a half dozen cruisers on our hands without very much to do," he explained to Axel Hanson of the White River Lumber Company, and so Long recommended that they "spend the next two or three months up in the White River country examining our land and your land." If nothing came of this, he assured Hanson that there would be no expense, and "if the expected does happen, which I am sure it will, to-wit, that we will get together, we will be willing to make such adjustment of this cruising expense as meets with your full approval." In short, what amounted to dull times for some merely meant different work for Long.

And if ever the Tacoma office grew quiet, Long could renew his debate with George Lindsay on the subject of advertising. At the moment, Lindsay was interested in reserving time during the annual meeting of the Timber Company in May. Long thought it more critical that the directors spend any available hours in the field, touring the Longview site and the Everett and Snoqualmie operations. Though Long knew that the directors would do as he scheduled, the general subject of advertising continued to rankle. In Long's view, the best a lumber advertising agency could hope for was to assure the public that "there is still an abundance of it so as to prevent all manner of substitutes from supplanting it." But, he asked again, wasn't that a responsibility of the industry rather than of Weyerhaeuser? "I do not think that our crowd should bear the burden of this kind of propaganda single-handed, unless after an earnest effort to have the industry do it they refuse."

Besides, the Weyerhaeuser name was already well known: "No traveling salesman connected with any other company can walk into the office of a retail man, or of an industrial buyer of lumber, and present to him so many varieties of lumber to fit so many uses and purposes as the Weyerhaeuser representative." He wasn't finished, far from it. "We least of all other lumber companies need advertising," and, "Our great asset is our resources of timber supply, our wide range of woods and our facilities for serving the trade who buys it. . . . I most unhesitatingly say that the amount of money contributed by the Weyerhaeuser Timber Company for the advertising program up to date has not yielded on the returns anything at all to compare with the expenditure. Our contributions to this fund during the four years of active advertising represents an amount fully 50% of what it had cost us to sell our lumber."

Accompanying the letter were memoranda of "financial expenditures, etc." of Weyerhaeuser Forest Products, the advertising arm. In the four years since September 1, 1920, advertising expenses totaled $1,213,728.85, and during those same four years, the Timber Company manufactured 3,336,824,817 feet of lumber. Thus the average advertising cost per thousand feet of lumber amounted to 36 cents. But Long contended that those figures deserved refining, for they included local sales, which were not "affected a particle by any advertising program." Subtracting such sales, the costs of advertising increased to 48 cents per thousand feet. He concluded, "This is rather a narrow view to take of the subject but I submit it for what slight bearing it has on the question." That was George Long at his most caustic.

There was a hint of good news concerning Klamath Falls, specifically that officials of the Northern Pacific and Great Northern were seriously considering plans to extend service in that direction. Long wasn't certain of the results, but he had been asked to help in the survey of the projected line from Bend to Klamath Falls. Thus he informed agent Kimball to allow the railroad officials to "count on you in an unlimited way."

The annual meeting was upon them, and, as usual, the directors sat patiently while George Long read his report. But if the scene was familiar, the details were different, many unpleasantly so. Long began by noting the "startling changes" that had come to the lumber industry, particularly to the fir district of the Pacific Northwest. "In 1923," Long recalled, "every condition was favorable," with the result that the year "was the best of all the years that the fir manufacture has ever experienced." But the good times had ended as quickly as they had begun. First, the Japanese demand for lumber, "which during 1923 had reached a total of 750,000,000 feet," dropped by more than half in 1924. Then the California market "was thrown into a semi-panic early in 1924 from various reasons," and its demand for lumber declined by some 500 million feet. The Atlantic Coast markets held reasonably firm, "but the demands of the agricultural districts of our country still were dormant and not to exceed 50% of its normal consumption."

Perhaps the most significant discussion at the annual meeting, at least in the longer term, concerned logged-off lands. The directors reviewed the decision of the previous fall, to organize a separate company "to take over, own, control and manage our logged-off land." What next? Although there were no final answers, the most desirable course seemed clear. "In so far as it is practical or possible," Long announced, "some attention is being given to the possibilities of reforestation on these areas." Although this presented "the usual conundrum," Long promised to forge ahead, studying "the possibilities of successfully reforesting them, realizing at the same time that the task is hopeless from a financial standpoint, until such time as the State of Washington sees fit to enact legislation that will make it possible to hold lands for that purpose." In the meantime, they had only a couple of choices: "Either to hang on to this property and do the best we can with it under existing conditions, or let it go for taxes, and we have chosen to do the former through the medium of the Weyerhaeuser Logged-Off Land Company."

That amounted to another crucial decision, born more of faith than present fact. Regarding the retention of the logged-off lands, George Long might have said, as Frederick Weyerhaeuser was purported to have said a quarter of a century earlier: "This is not for us, nor for our children; it is for our grandchildren."

Twenty-five years. The directors excused George Long in order that they might officially honor his tenure. Upon a motion offered by F. E. Weyerhaeuser and seconded by Dr. Clapp, they unanimously adopted this resolution:

> W H E R E A S , George S. Long, the Vice President and General Manager of the company, has completed a quarter of a century in the service of the company; during most of that period he has been the responsible head of the company's business; he has conducted its affairs with singular ability and tact; under his direction its properties have been built up and its business has prospered, and at the same time there has been established for the company a reputation for fair and equitable conduct in private and public relations for which his personality and character are largely responsible, and which is an unmeasurable asset; Mr. Long has devoted the best years of his life to the single purpose of advancing the interests of the owners of the company, and in doing so has himself continually grown in the esteem, the confidence and the affections of the Trustees who have cooperated with him, and of all of the owners; for his valuable services he has been paid; the esteem and affection of his associates, the Trustees and the owners, and their appreciation of his warm-hearted friendship and his geniality and tact in all personal relations cannot be measured in money, and yet there is a universal feeling among the owners that they as a group would like to at this time give him some concrete evidence of their esteem and of their admiration for and appreciation of him as a man, an associate and a friend, and that that can best be accomplished by a gift from the owners through corporate action; Now therefore, be it

R E S O L V E D , that there be paid to Mr. Long from the Treasury of the Company the sum of One Hundred Thousand Dollars; and that Mr. Long be informed that this is the gift of all of the owners as a group, and is made as a token of their personal esteem, admiration and affection.

Long was appreciative, of course, but the recognition was not something he would take too seriously. He did his best because he could do no less. It just so happened that his best was very good indeed, and those with whom he worked were forever impressed by the fact. And forever grateful.

The major business decision made at the annual meeting involved Longview and the rail arrangements. On June 4, by telegram and letter, George Long confirmed an earlier agreement between lawyers Clapp and Andrews, indicating the land option selected for the mill site and the accompanying decision to use an independent railroad. Within the week, Long advised R. F. Morse, Long-Bell general manager, that Al Raught would be representing Weyerhaeuser in Longview. Raught had been employed by the Clarke County Timber Company for several years, first as an accountant and then in charge of logging. Now that Clarke County had ceased operations, "We are taking advantage of his idleness to throw him into the breach temporarily as our representative at Longview until our affairs drift along a little faster when we may want to perfect a more permanent organization." Within the week, Raught was on the job, already having rented an office in the Ross Building, about a block south of the Long-Bell office. Decades would pass before Raught left his Longview responsibility.

It was good that George Long kept busy and found pleasure in doing so. There had been no joy at home in recent months, what with Carrie's prolonged illness. She passed away on June 22, the primary cause of death officially listed as a "cerebral hemorrhage," but in her last few days pneumonia was also a factor. While George could tell himself and others that it was a blessing, given her suffering, he couldn't fill the void. And he never would.

During the summer, sales matters from Spokane to Minneapolis to the East Coast kept Long busy. On the East Coast, George was pushing brother James to pursue a Newark opportunity as a third Atlantic terminal. Once the Longview plant was in operation, coupled with the Everett production, they were going to have "a whole lot of lumber and I am rather ambitious to have the Atlantic Coast so safeguarded that we can put through Baltimore, Portsmouth and another yard in the metropolitan district (and the Newark yard suggested suits me) a total of at least 250,000,000 feet when things are running smoothly." Within a week, he was seeking the approval of the directors to lease a portion of the government dock facilities in Newark.

Long knew that it would seem an unnecessarily hasty decision to some. "I fully realize that we are going pretty fast in this Atlantic Coast program," he acknowledged, "but I am beginning to comprehend and realize more and more that the marketing of a large quantity of lumber is made much more stable and profitable if you can create facilities that will bring you business."

As it developed, approval was nearly unanimous. William Carson's reply was typical: "Of course the rental seems high to a lumber jack." But he wasn't about to disagree with Long's recommendation. In Carson's words, "My conclusion of the conditions of the manufacturing and selling of lumber by the Weyerhaeuser interests is that the Fir end excels all our competitors in manufacturing and marketing." And later he observed, "In the Pine end of the game [Idaho], we are not so favorably situated." There was no need to note that the pine end lacked a George Long.

The extent of Long's 1925 summer involvements is impressive. He was simultaneously studying the system of grading Douglas-fir lumber, the management of the Midway wholesale yard, plans for the Longview facility, the construction of the Portsmouth plant, whether to purchase or to lease a Newark site, and what to do at Klamath Falls. On July 30, George Long, J. P. Weyerhaeuser, F. E. Weyerhaeuser, and attorney Gus Clapp met in Portland with representatives of the Shevlin-Hixon interests to discuss a possible merger. Offers and counteroffers were made and discussed. In fact, the larger problem was not with the figures themselves—they came close, no doubt uncomfortably close, in Long's eyes. He sent a lengthy letter to F. S. Bell in which he included a "Query. Shall we drop the matter as it is and advise them that the deal is off, or shall we make a counter proposition?"

Bell must have known that Long already was formulating an answer. And if he didn't know, he was so informed: "I think one of the ideas that had something to do with our considering the amalgamation was that we would be relieved of the executive part of the work and now that we are somewhat committed in our minds at least, to the necessity of realizing profit on this timber, and are likewise reasonably convinced that we cannot sell it for what we think it is worth, that our best policy is to quit thinking about the amalgamation and make up our minds to handle the proposition ourselves, and unless you have some different ideas from this conclusion, I think that will be my answer to the Shevlin-Hixon people."

E. L. Carpenter, president of Shevlin, Carpenter & Clarke Company, expressed understandable frustration with the dealings, in reply to Long's August 8 telegram indicating that an agreement was not possible. Allowing that "unless the question is raised again by your people, we will call the incident closed," Carpenter could not refrain from reciting the history of the discussions, that the original suggestion "to combine our interests in the Klamath Falls district came from Mr. [F. E.] Weyerhaeuser." And, he continued, "Excepting the tentative suggesting of price from Mr. Long during our recent conference at Portland, which was satisfactory to us and was accepted, we have, at no time since the negotiations opened, had a price from you on the timber west of Klamath Falls. It is impossible to conclude a negotiation of this nature unless each party is willing to advance definite proposals. In no other way can 'minds meet,' which they must do if a negotiation is to be successfully concluded." Carpenter was right. George Long had gone through the motions, confident that the figures would suffice to defeat the effort. This confidence proved wrong, and he had no choice than to stop talking without explaining why. In any case, there would be no joint undertaking at Klamath Falls.

Long spent much of August traveling around the country visiting various new operations with Onstad and Titcomb, "to study conditions that are new and novel in mill building," all with a view toward designing the best possible plant for

Longview. By the end of August, the land clearing at the Longview site was under way, to be completed by December 1. That news pleased George Long, sure evidence of progress at last.

Less pleasing was the fact that with so much going on, individuals were stretched to and beyond their limits. That was certainly the case with Onstad. Coincident with the Longview effort, young Phil Weyerhaeuser, manager of the Clearwater Timber Company, was planning a big mill at Lewiston, Idaho. Phil hoped that Onstad might be available to help. Long admired Phil greatly and genuinely wished that he could comply. But he couldn't, explaining that "the [Longview] job we have got for him [Onstad] is the biggest one he ever had; is full of a great many new and unexpected things and that in connection with other developments which we are sure to have in other places, makes it seem to me that we need and ought to have at least 100% of his time and energy."

One of those other places would be, sooner or later, Newark. Long planned to go there personally in mid-October, at which time they might complete negotiations on a site. But that was one of many matters awaiting decision. "Talk about moving fast," he wrote to Dr. Clapp, "we have about concluded to purchase two additional ships, similar to the two which we have. The two which we now own, the *Pomona* and *Hanley,* which cost $350,000.00 cash and we can buy two just like them, made by the same people—one for $263,000.00 and the other for $262,000.00, less 2% for cash and we have decided to buy them." The final terms proved even more favorable. The *Hegira* and the *Heffron* were purchased on September 22 for $500,000.

The decision to purchase two more ships must have surprised some of the directors, to say the least. Long explained matters to Bell: "The situation developed where I had to act quickly and the only one I could confer with was Will McCormick, who O.K.'d it." The fact that McCormick was a rubber stamp seemed immaterial. No one could really argue the case. If two ships were required to supply Baltimore, surely at least two additional would be needed to meet the demands of the Portsmouth and Newark yards. Will Carson's reaction to the purchase was again typical: "If you approve of it, I do too." Then he added, "If you could raise the price of Fir about $4.00 a thousand, I would probably advise buying a few flying machines."

Meanwhile, work was going forward on the Portsmouth facility. The side tracks had been laid; foundations for the sheds were being constructed; dredging was under way, to be completed within two months; and a vessel was due to arrive in Providence on September 15, containing 500,000 feet of structural timbers, to be followed within two weeks by another with an identical load, "as well as a consignment of fir piling from this Coast, so we expect to be very active there in the immediate future." Across the continent at Longview, the foundations for the bridge across the Cowlitz River began to take shape, and a contractor was grading the right-of-way from the bridge to the mill site. Long hoped the work would be completed by May 1, 1926. As for construction of the mills, he wanted the land cleared first "so as to get a full view of the expanse which we have to work with."

Still undecided was whether the shipping arrangements would continue as before, although Long believed that they had "reasonable assurance" that "the Norton-Lilly people" would accept responsibility for the westbound cargoes for the new boats as well as the old. "This arrangement," he reminded, "means we come West with our boats at a slight profit over the cost of operation." All in all, "The boats have proved to be magnificent carriers for our business."

On October 8, George Long left for points east aboard the Oriental Limited. He was accompanied by his daughters, Margaret and Helen, and also Titcomb. Tom Humbird joined the party in Spokane, headed for the October 12 Weyerhaeuser Sales Company meeting in St. Paul. Long had to forgo that pleasure; his destination was New York City for negotiations on the Newark port lease. Following the usual bureaucratic delays, the lease was finally granted. All that remained was approval by the mayor's office.

From New York City, Long and his daughters traveled leisurely southward, visiting brother Jim and family in Baltimore, and then continuing on to Monroe, Louisiana, where his nephews, the Kellogg boys, sons of sister Eva, lived. From Monroe, the party went to St. Paul for a meeting of the Timber

Company trustees on November 16, at which time Long was authorized to organize two new steamship companies.

There was, however, a much larger subject under consideration, one that was discussed in hushed tones and often indirectly. Thus it was that Long subsequently contacted Dr. Clapp: "Mr. [William] Carson also spoke to me privately about another matter which probably has to do with his desire to brush up a little on Weyerhaeuser Timber Company resources and he said he had met you somewhere and spoken to you on the same subject." What Clapp, Carson, and unknown others were contemplating was the possible sale of the Weyerhaeuser Timber Company properties to a group of eastern bankers.

Once again, and for a lot of reasons, it would come to nothing. In the process, however, the prospective sale created hard feelings and placed Long in a touchy and tenuous position. Evidently, Dr. Clapp had assumed that Long would favor such a sale. He assured George that "the parties are most responsible, and . . . stand ready to pay the price named, conditional only on being shown that the 1924 balance sheet correctly represented the assets." He then explained his reasons for wanting to push such a sale. "It would seem that this is perhaps a psychological time to dispose of the property—stocks and bonds are at an abnormally high level, and there seems to be an enormous demand for something in which to invest the vast sums of money which have accumulated in this country."

George Long was nearing the end of his working career—he had celebrated his seventy-second birthday on December 3—and he couldn't help thinking in terms of his legacy. As for himself, Long admitted that the idea had appeal, that "looking at the matter selfishly, I should undoubtedly favor it from my own personal standpoint, but on the other hand I would not even make any overtures pertaining to the matter unless I was dead sure in advance that the trade would be made." And he wasn't. Indeed, his assessment was otherwise. "Personally," he observed, "I have never taken the matter very seriously." For Long understood what Dr. Clapp failed to understand: the Weyerhaeuser brothers would never approve the sale, however profitable. To do so would be tantamount to betrayal of their family and their father's memory.

Following his return to Tacoma, Long turned his attention to matters of management. The first step was to bring Titcomb to the Tacoma office. Replacing Titcomb as general manager at Snoqualmie Falls was Charlie Ingram. In his letter of appointment, Long had little advice to offer young Ingram, although he did remind him that "it has been our hope, ever since we started the Snoqualmie Plant, to build it up in a way that would have an appeal to those employed by the company and to have a good working family organization that would work harmoniously together in an effective way." Charlie was a good choice in almost every respect.

Matters were less clear at Everett. Dr. Clapp hoped his son James would be promoted, but George Long had other ideas. As he explained, "Until the right man is thought out for General Manager [I] do not deem it advisable to disturb present situation." That present situation had E. B. Wight, Bill Boner's son-in-law, in charge of Mill A, and Bill Peabody running Mill B. As yet, neither had proved himself to Long. "My thought is that we need a real big general manager," he wrote to Dr. Clapp, "and that no one on the ground is competent for that position." He further explained that he wanted to do something for James, "but I don't want to do it until conditions are right, and I don't think they are right now."

The year ended with the usual reflections. This time, however, looking back took on a special significance, for 1925 had marked the end of a long and fruitful marriage, as well as the quarter-century anniversary of the Weyerhaeuser Timber Company and of Long's service managing it. On Christmas day, old friend Fred Conant wrote a thoughtful note to his boss, recalling what they had seen and what they had done over the years. The elderly cruiser expressed the wish that, starting New Year's day, they might commence "another 25 years of work together, this time to renew the timber that has been destroyed since our association began. Many times I have stopped in the woods, alone, and taken my hat off to a fine tree—the only thing that gives me a genuine impulse to do that." Conant remarked that from boyhood he had always wanted to be in the woods, "and I have had my wish." But, given his love of the trees, he would have preferred to be in the business of growing and not just cutting: "It seems that all I have done was preparatory, and had to be done before the work could start that I was best fitted for and would have enjoyed most."

George Long appreciated such thoughts, and in his reply spoke as much to himself as to his friend, regretting not that forests were being logged, but that so much of the trees remained behind, unused: "To look at the waste around a logger gives you a weird, dreary feeling; to look at the homes and barns and shelters that the lumber makes, that come from these same trees, give you a sense of satisfaction, so that after all we are not engaged in a very ruthless business when we cut down the forests." Finally, and of greatest importance, "I think we have all been mindful of this waste and all been anxious to correct it and really have more men in our midst who are thinking of forest problems than almost any other region where timber is found in the United States." Long surely hoped that was the case. If not, he was a failure on his own terms.

TWENTY-SEVEN

Too Much Lumber

THE CALENDAR TURNED OVER, BUT MANY PROBLEMS lingered. For George Long, a constant charge for the 1920s was trying to determine the appropriate role for Weyerhaeuser in advertising. In general, he believed that the responsibility lay not with any individual company, but with the industry as a whole. After all, most of the advertising benefited all producers of lumber.

At a meeting of the directors of the National Lumber Manufacturers Association in December 1925, one resolution adopted focused on the increasing dangers confronting the lumber industry through use of substitutes in building materials. The resolution called for an early February convening of the trade-extension committees of all "regional manufacturers' associations and wood-using industries," in the hope of adopting a cooperative program that would benefit the entire industry.

There was no difficulty eliciting Long's support for such an effort. As Long noted in a letter to F. E. Weyerhaeuser, "The custom of regional advertising very largely is to suggest the merits of one wood as against another wood," when the real danger was to all woods. He cited recent advances by those in the cement, brick, and steel industries: "I believe we should do whatever we can to promote this [cooperative] effort, and to lend to it our deep interest, our best efforts and a liberal financial support." There were reasons to launch the effort now, including "attacks that were being made upon the industry by more or less irresponsible parties, who appealed to prejudice rather than to reason." Bell and Dr. Clapp added their support to the idea of an industrywide campaign.

But there was little unanimity on a more crucial matter—machinations aimed at selling the Weyerhaeuser Timber Company. Long learned of the latest in a January 7 telegram from William Carson: "[Horace] Rand notified me Dr. Clapp anxious to have Timber Company conference St. Paul soon as possible. Thinks you and [Bill] McCormick should be present. I believe New York bank should make cash offer. My opinion sales best interest of stockholders if conditions satisfactory. At any rate believe directors should have chance to consider proposition. Will be in Burlington [Iowa] Monday."

Long, in an unhurried reply, mentioned a recent discussion with J. P. Weyerhaeuser, whom Long quoted as saying "unqualifiedly" that he "thought it would be a very foolish thing to do; that the property ought to be kept together and afford an outlet for the future activities of the coming generations, etc." That was about as predictable as a response could be. Long himself seemed ambivalent: "For a great many reasons I would prefer to not take a very active part in this matter, other than to express my opinion that if I had a chance to sell my stock in the company at the price suggested, I would sell it, but I would hardly feel like urging the matter to anyone who might entertain a different notion." He also noted that the question had "never been before any of us in what might be called even a semi-official way," suggesting that such a consideration seemed the proper first step.

Carson agreed to discuss the matter "with a number of our people" in St. Paul on January 21, at the Wood Conversion Company's annual meeting. He also reported that the New York bank involved was Goldman, Sachs & Company, "one of the oldest and most responsible banks in the United States." And, in postscript, he shared the latest from banker Russell Hawkins: "My New York friends advise conditions for purchase are in wonderful shape due both to proposition and general conditions. The hope is expressed your associates act promptly as possible."

Things went just about as Long assumed they would. When Dr. Clapp submitted a proposal to sell the Weyerhaeuser Timber Company assets to the available directors, "nothing was done on the subject, and very little said." The silence must have been deafening.

Still, the thoughts lingered on in the minds of some, including Carson. Six weeks later, after remarking that no one had "a higher regard for the value of W.T. stock than I," in the next breath he allowed, "If anybody offers me $1200 [a share] . . . the gentleman would find himself in possession of it as soon as I could deliver it."

Long gladly turned his attention back to other matters, such as soliciting cooperation in advertising within the lumber fraternity. A general meeting of Douglas-fir manufacturers was scheduled for Seattle, with Long serving as chairman. He tried, in his inimitable way, to encourage attendance. To one, Long described the situation as he saw it: "I personally entertain the notion that there are a very great many men engaged in the lumber business who ought to be engaged in something [else], and this applies to manufacturers as well as it does to any other branch of the business, but they are here and will always be with us, and the main point is to help them solve their own problems."

To another, he emphasized the need "to boost fir lumber, probably to advertise it quite extensively and to ask all fir manufacturers to contribute to this fund." For that purpose, John Tennant, chairman of the publicity committee of the West Coast Lumbermen's Association, established an executive committee to oversee the effort. The executive committee, however, had trouble recruiting members. A typical reply ran thusly: "We have always advertised our own lumber and our own grades and spent considerable money in developing a personal name for our stock in territories that we wished to sell in." That, in general, was the same argument that Lindsay and others had consistently mounted regarding the Weyerhaeuser publicity campaign. As Long described it, "Mr. Lindsay has always felt that the industry would never do the job, that whatever was the job of everybody soon became the job that nobody attended to, and he may be right, but I hope not."

At the Seattle meeting, Long and Tennant had a slight disagreement over the proposed effort of the National Lumber Manufacturers Association to promote wood without regard to region or species—simply wood as opposed to the feared substitutes. Tennant wasn't so sure the time was right. Long, however, felt so strongly about the need for a national campaign that he volunteered to go to the upcoming Chicago conference on the subject in mid-February. And he encouraged others to accompany him.

Despite the generally dismal business outlook for Pacific Northwest operators, there were bits of encouraging news for Weyerhaeuser Timber Company. Colonel Long reported that dredging of the Portsmouth channel was complete. His son, Allen, had been placed in charge there, and George requested that Allen provide a "weekly resume" of progress in order that they might know when to arrange "for cargo movement, so as to arrive just about as soon as you are ready to receive it." And the news from Newark was also encouraging. On January 26 the long-awaited word came via wire that the mayor had finally signed the lease to the port site. George sent his brother "express congratulations at the final conclusion of this very important event."

Long did as he had promised, boarding the train to Chicago for the gathering of "the lumber fraternity," where "all lumber regions will take part in preaching the gospel of wood." The assembled proved faithful, but to no one's surprise they also proved better at stating problems than formulating solutions. Still, Long believed that his colleagues were "being awakened" to the dangers of wood substitutes. The delegates departed, charged with raising at least $500,000 a year to advertise wood. Long summed up the situation with a query: "Will the lumber industry talk about it or do it?"

Though Long was committed to furthering the industrial cause, there were nonetheless limits on how much he could manage personally. And he knew how to say no. He did just that in response to notice of his election as a trustee of the National Lumber Manufacturers Association: "It is absolutely impossible for me to take on the work suggested," he informed Secretary Compton, adding, "I regret this, but I have full knowledge of what is ahead of me in the way of work, etc., and know that instead of increasing the load I should lighten it."

But some assignments were not as easily sidestepped. In the fall of 1925, President Coolidge had established within the Department of Commerce the National Committee on Wood Utilization, ostensibly "to work for a closer wood utilization and better manufacture and distribution practices." Secretary of

Commerce Herbert Hoover was designated chairman, and Col. William B. Greeley, Chief Forester, vice-chairman. In mid-May, Long received an invitation to be "one of five or six to represent the Pacific Northwest industry on this committee." One can almost hear George Long's sigh.

A happier aspect of federal relations involved monies made available as a result of the Clarke-McNary law. Specific benefits were listed in the annual report of the Oregon Forest Fire Association. Its secretary noted that the first allotment to Oregon under the law was "somewhat disappointing, but it represented approximately a 50 per cent increase over the Weeks law allotment of the previous year." Also, an amendment to the law recently passed promised an easing of fiscal difficulties.

The annual meeting convened on May 27, featuring the by-now-familiar, even comforting figure of George Long standing before the group, cigar in hand, reading aloud. "There was no profit accruing to those who buy logs in the leading log markets of Puget Sound, Grays Harbor and Columbia River and confine their operations to raw materials thus obtained," he reported. Mills that owned their own timber fared better, at least on paper, but that was an accounting profit rather than an actual one. Long did use the occasion to express his concern about wood substitutes in construction: "The lumbermen are beginning to realize that their competition is not only amongst themselves." Then there was another factor in the equation, the public: "The thought has been expressed that the country at large entertains the opinion that forest growth and lumber is just about exhausted, and that it is high time to turn to substitutes."

There was much new to report. First on the list was the Columbia & Cowlitz Railway Company, "a line about eight or nine miles in length, which runs from our proposed mill site at Longview to connect up with the Transcontinental Lines [the Northern Pacific, the Great Northern, and the Union Pacific] at Rocky Point, a few miles north of Kelso, and likewise to connect up with our railroad known as the Cowlitz Development Company which swings back some eighteen miles into the timber which we propose to log to the Longview mills."

As for the Longview mill site, it covered more than 600 acres, with a frontage on the Columbia of nearly two miles. Long admitted, with a smile, that he had been "rather liberal" as to the extent of the purchase, but reasoned that they should have "ample room for any further developments that might ultimately be required in the manufacture of our timber tributary to this plant, which conservatively speaking is at least twelve billion feet with the probability of a larger amount."

Down in Klamath Falls, they were attempting to "protect" the mill site by means of a Klamath River Boom Company. There had been a hitch in those negotiations, but Long expected things to sort themselves out shortly. He noted the purchase of the two additional steamships, the *Hegira* and the *Heffron,* the start of construction at the Portsmouth terminal facility, and the twenty-year lease of the Newark port property. Looking ahead, he predicted that the three Atlantic terminals—Baltimore, Portsmouth, and Newark—"should easily give us an outlet for at least two hundred and fifty million feet per year and in time more than that." That must have seemed like a lot of lumber to those who had grown up as Mississippi River millmen, but George Long understood the Pacific Northwest situation, and it was of an entirely different scope.

In the background, almost as an underlying theme, was the dream of growing a new forest. As everyone knew, the first problem was that cutover lands were so scattered and so widely owned. Long had put Allen to work surveying the lands in the Grays Harbor district, suggesting plans and procedures for reforestation. They differed somewhat on what should be done with the results of that survey. As Allen explained to Minot Davis, he had initially wanted to define the "joint problem," and then "the next stage would be an individual, selfish, confidential ownership study by a few progressive concerns, to see wherein their interest lies; and thirdly, probably the realization that not study, but solution policy, would draw them together again into some sort of merger to deal with the situation." Long, however, felt that there was no room for an individual effort anywhere in the Grays Harbor region, "because of interlocking ownerships, and that the strategy lies in making everybody feel that way." (Twenty-five years hence, the very lands that Allen and Long were studying would be included within the Clemons Tree Farm, later to be designated the nation's first official tree farm. In 1926, Allen talked of their "joint problem"; in 1941, Phil Weyerhaeuser proudly spoke about their "joint undertaking.")

F. E. Weyerhaeuser had been following the reforestation efforts closely. In a letter to Long, he observed that if Allen was "within a 'gun shot' of the truth," reforestation in western Washington showed promise. In any case, F. E. saw it as "our duty to leave our cut-over lands in such condition that nature will have a chance to reproduce the forest, providing, of course, the expense of doing so does not become unreasonable." That, of course, was the critical factor.

But F. E. had little doubt that George Long would keep a hand on what was crucial in any transaction: "After an examination of the properties of the Weyerhaeuser Timber Company, I never fail to come home with greater enthusiasm over its future possibilities, and, if you will pardon my saying so, I think the stockholders of the Weyerhaeuser Timber Company should have a tremendous appreciation of what you have done for the company in protecting and developing its properties. I do not mean to infer that other stockholders have not this appreciation. I am sure that they have. I merely wish to tell you that I have great admiration for the breadth of vision you have shown in working out the enormous problems that have arisen in the development of these vast timber holdings."

F. E. appeared to be giving scant attention to the merger question, but in truth he had been fuming over the behind-the-scenes activities of Dr. Clapp and William Carson for several months. Treasonable—that is how such activities seemed to him. The situation was made all the more trying by the fact that Clapp and F. E. were, at least in the winter months, Pasadena neighbors. Some weeks earlier, brother Charlie, by nature a peacemaker, had encouraged F. E. to forgive, if not forget: "Considering the great number of partners we have and the large properties we have to care for and the many responsibilities we are called on to assume, I think we are to be congratulated in receiving as little criticism as we do from our stockholders, and, at present, I am in a frame of mind to be willing to make friends with Dr. Clapp."

F. E., however, was not in a forgiving mood. Months later he had cooled off a bit, but not completely. As he explained to his sister Margaret, "I am not very much surprised at anything Dr. Clapp does, for he has no long-established points of contact which might have brought about a finer sense of loyalty than he has displayed," adding, as a reminder to himself, "I think it is wise for me not to talk business very much with Dr. Clapp."

In the meantime, Long was listening to offers from Long-Bell officials to make immediate settlement on their so-called second purchase, changing the terms of the original time contract. Long wasn't interested in any new arrangement. Others agreed. C. R. Musser wrote, "I am rather glad to know that you have advised the Long-Bell people that we do not care to make any concession in terms on our deal No. 2."

If Long and his colleagues at Long-Bell were at odds over that contract, elsewhere there was cooperation. In mid-summer 1926, fires raged through the pinelands west of Klamath Falls. On July 1, Long informed Long-Bell's John Tennant of plans to sent Titcomb and Davis to Klamath to "make a study of the situation and decide what is best to do. . . . As our timber is so badly intermingled, it would seem that we have somewhat of a common problem down there where it might be possible to work together." Specifically, Long urged Tennant to send his representative down with Davis and Titcomb, noting "that we will have to do something in a very active way during the remainder of this year in order to salvage it to any advantage." Tennant responded immediately, agreeing that J. B. Woods, Long-Bell's forest engineer in the district, would meet Davis and Titcomb in Klamath Falls on the evening of July 6, "to assist in the proposed investigation and formulate recommendations as to the best way to salvage the damaged timber."

One thing did work out as expected, although its resolution had taken more than twenty years. This was the sale of most of the Cherry Valley Logging Company's timberlands to the Snoqualmie Falls Lumber Company. Long had been trying to accomplish this forever, or so it must have seemed, but Horace Irvine's position was uncompromising. It was always something of a tender subject, with Long caught in the middle, endeavoring to make the best business decision negotiating with a friend and major Weyerhaeuser shareholder. Irvine proved difficult to the very end. The final terms stipulated a price of $5.50 per thousand feet for all fir, cedar, and spruce, Irvine insisting upon payment over six years at 6 percent. Long would have preferred to pay "on or before" the six-year schedule, but he agreed, relieved to have the deal at last concluded.

Much was happening in and around St. Paul. A meeting of the directors of Thompson Yards and the Yards Securities Company was held on July 15, at which time it was voted to reduce the capital stock of Thompson Yards by $600,000. The reason was clear enough: Thompson Yards had that amount in unneeded cash, and the easiest way to return it to the shareholders, "without the question arising whether it represents income or not, is to actually reduce the capital stock by the amount of the payment to the stockholders."

Following that action, several of the Weyerhaeuser Timber Company directors spent a couple of hours discussing the desirability of issuing preferred stock. No decision was reached. Last, they talked about the continuing depressed state of the Pacific Coast lumber markets, "and particularly the danger of some of the fairly well-known companies getting into financial difficulties." According to F.E., there was a consensus that the Weyerhaeuser Timber Company should "retain all of its present funds, including any payments that may come from the Long-Bell Lumber Company, investing them in high-grade easily convertible securities, so that, in an emergency, the company might have a large cash reserve, to be in position to take advantage of any very attractive purchase that might come up in the next year or two."

There was considerable cash on hand—$13 million in mid-July—and another Long-Bell payment due within thirty days. Several projects were under way, of course, including development of the Newark plant, construction of mills and railways at Longview, and building of a logging railroad and camp in Thurston County. Still, as Long acknowledged, "We will be receiving money all the while, possibly enough to take care of this development, so that any way we look at it we will be accumulating quite a large amount of money for which we have no immediate use."

One decision, long delayed and much debated, was formally announced on August 12, when William H. Peabody was named manager of the Everett Branch. Peabody was bright and had performed ably. Still, there was a roughness about him that would anger a few and intimidate many. In any case, Everett now had its leader.

Various industry meetings were held in late August, including one in Vancouver, British Columbia. Long urged his friends to attend and to bring their friends. One rather unexpected result of the discussions was a restatement of the old grand merger scheme. Surprisingly, for once Long seemed interested: "It goes without saying that an undertaking of this size would call for the very highest ability in its management and the very magnitude of the undertaking justifies the belief that the best financial executive and businesslike efficiency in the lumber industry would be in charge of the enterprise. Assuming that this would be the result, I do not hesitate to express the opinion that the undertaking would be financially successful."

The plan was soon made public. Long forwarded an article on the subject from the *Portland Oregonian* to F. E. Weyerhaeuser, indicating "that a great many of the leading mills out here are looking favorably upon the project," and concluding, "If the merger is put together right and ably handled thereafter, it will be a great boon to the entire lumber industry and especially to the fir industry, and undoubtedly very beneficial to our company." Long knew, however, that the old questions remained: What will we do with the money, and what will our children do with their lives?

Back at the office, changes were afoot. Titcomb was gradually assuming more responsibility, now officially Long's assistant general manager. One of his assignments was to oversee the salvage-logging efforts in the Klamath district. He advised Kimball of plans to visit the area, including a tour of Weed Lumber Company operations, in order to watch how they cut and graded, after which he hoped to bring Weed's superintendent "up to the Aspen Lake operation and go over that with yourself, Mr. [Bert] Peterson, and Mr. [Charles] Chapman." They were all beginners when it came to western pine.

At Everett, things weren't running very smoothly. Wight was unable to accept the appointment of Peabody as his father-in-law's successor, and on October 30 submitted his letter of resignation in order to accept a position with the Walton Lumber Company. Wight was gracious in departing, asking only that "there may go with me the good wishes of yourself and of all the officers and employees of the Weyerhaeuser Timber Company." "If there is anything at all in sentiment," he added, "it was nineteen years ago today that I entered your office," and it would be "nineteen years ago next Monday, November first, that I started work for the Weyerhaeuser Timber Company."

One upcoming event held promise. A special train carrying Ralph Budd, president of the Great Northern Railway, "and a party of friends" was scheduled to stop at Everett, and President Budd had requested a tour of Plant B. More important, the special train would then head south, and Long arranged for Kimball to get aboard when it stopped at Eugene. Kimball would thereby be afforded ample time to discuss the desirability of service to Klamath Falls. Long was clearly doing some courting for a change.

He was considering possibilities of yet another sort, citing several of the Puget Sound's newly established pulp and paper operations. Long specifically noted the Zellerbach Paper Company, with "two or three enterprises . . . some of them in operation, the one which I have in mind being at Port Angeles where they use hemlock for their raw material." He was particularly familiar with that pulp mill because it was purchasing hemlock slab wood from Everett. Still, Long lacked confidence in the future of pulp and paper: "How much money there is in the paper business out here I do not know, but I do know there is quite an epidemic right now of paper mill talk. . . . The query is whether or not they will overdo it." What an eternal question.

Shortly after Thanksgiving, Long headed east, arriving in Chicago for a gathering of the National Lumber Manufacturers Association to consider the trade extension movement. He had preached so loud and long on the subject that his absence would have taken on added significance. But there was another reason for the trip: a chance to tour the Atlantic terminals at Baltimore, Portsmouth, and Newark. Long found trade at Baltimore disappointing, where only 7 million feet was shipped in

November, some 2 million less than projected. Portsmouth was another story. There 3.5 million feet had been sold, "which is doing pretty well, considering we got started late in the season and are just a beginner in that country." As for Newark, it was still a few months away from operation. But construction was proceeding on schedule, and Long was certain they would "have a dandy plant at that point, and while I think at all of these Atlantic Coast depots we are building, we are a little ahead of the times, yet I am sure we will want these avenues for our future distribution."

Upon returning to Tacoma, he first had to consider the size of the next dividend. Thanks largely to the Long-Bell sales and the maturing of government securities, the Timber Company till was full to overflowing. Long had about decided upon a distribution of 30 percent in January. "The facts seem to be," he indicated in a letter to directors, "that our profits for 1926 promise to be close to $5,000,000, and of this amount there will be about $2,500,000 of undivided profits, so that if we should make a distribution of 30% in January, $2,500,000 of such a distribution would be from our undivided profits of 1926, and 10% of the proposed 30% distribution would be a distribution from our capital assets and consequently would not be subject to tax to the recipient of the dividend."

Looking ahead to 1927, he worried over several large developments, starting with the building of a railroad in Washington (the Columbia & Cowlitz) serving the Longview mill, "which will cost more than $1,000,000 we think." In addition, there was the Longview plant construction program, to begin seriously in the spring, "and the probabilities of our making good with our promise to build a mill at Klamath Falls if the Northern Lines, i.e., the Northern Pacific and Great Northern build from Bend, Oregon to Klamath Falls." Three major undertakings were on the horizon, and still Long felt confident that they had sufficient resources to pay a 30 percent dividend.

No shareholders complained about the large dividend, but a few expressed reservations concerning plans for 1927. Dr. Clapp, for example, continued to be less than enthusiastic about development at Klamath Falls. He and others, including the Weyerhaeusers, were heavily involved in Idaho operations, and

Klamath would be offering direct competition in western pine. "With the hard time Boise [Boise Payette Lumber Company] has been having for several years, there is not much inducement toward developing at Klamath a competitor both in kind of timber and territory supplied," he said, adding, "I was in hopes someone would want to purchase the Klamath timber."

But the hard times were everywhere, and no purchasers for Klamath had offered themselves. Not that George Long would have listened. No one, and certainly not Dr. Clapp, was going to talk Long out of Klamath Falls.

TWENTY-EIGHT

Longview,
Klamath Falls, and 4-Square

TACOMA'S WELCOME TO THE NEW YEAR, 1927, WAS damp and depressing; so too was the lumber market. There seemed little reason for optimism. George Long worried, and so did everyone else in the lumber fraternity. He noted that many of the Northwest mills were "again attempting a curtailment project," closing down for two weeks over the holidays instead of one, and thereafter running five days a week instead of six. "But," he continued, "it is beginning to dawn on us that this curtailment propaganda is not going across very well and that there is a whole lot of lumber being made." In other words, curtailment was working about as well as usual.

It must have seemed a bit incongruous, talking curtailment while simultaneously overseeing the start of construction at Longview and thinking about doing the same at Klamath Falls. The latter project was at a critical point: the northern lines, which is to say the Northern Pacific and Great Northern, were again considering the rail connection, but officials of the railroads continued to be at an impasse. F.E., a member of the Great Northern board, reported that while his group was "unanimous" in favor of extending to Klamath Falls, some board members from the Northern Pacific "are very much opposed to the extension." It must have been a bit touchy for the family, what with brother Rudolph Weyerhaeuser serving on the Northern Pacific board. In any case, a joint meeting of the two boards was scheduled for January 19, and F.E. predicted that even if they decided against the extension, "In a few years the Southern Pacific will have so thoroughly developed the timber resources that a later application to the Commission on the part of the Northern lines could scarcely receive a favorable decision."

Construction at the new Newark plant was ahead of schedule. One fly in the ointment, however, was old nemesis Carl Hamilton of Weyerhaeuser Forest Products. Hamilton was eager to announce the Newark promise to the world at large. Colonel Long judged that a bit premature, and so did brother George. The latter allowed that Hamilton "must think we are running something like Barnum's Show." What should be done, Long advised, was to wait until things were running smoothly before calling attention to the operation, "then we can do at Newark just what we did at Baltimore once upon a time and that is, invite the whole lumber fraternity to see your real stunt at a time when you can put it on with a certainty of its being a worthwhile performance."

In the meantime, feeble efforts at curtailment continued. Though many realized that they were simply going through the motions, somehow the motions nonetheless seemed important. On February 24, there was a meeting of lumbermen in Seattle, poorly attended, but as Long reported, "It was put up to me [that should Weyerhaeuser resume a six-day workweek] we would topple the whole program over and render useless the meeting which is to be held on Thursday next week in Tacoma." In a personal note to Charlie Ingram at Snoqualmie, Long admitted, "quite against my judgment," that he had reluctantly agreed to continue the five-day shift for another week.

The big problem, as always, was "that these curtailment efforts are not entered into heartily enough by all of the lumbermen to justify us in being good when the other fellows are not." Still, that didn't lessen Weyerhaeuser's responsibility: "It is a very unsatisfactory situation that exists, but one of the penalties of being big fellows is that we always have to do what we promise and anytime when something is attempted for the betterment of the industry, we are expected to be pacemakers and quite a crowd of the lumbermen are still of the opinion that if we could hold down the cut a little longer, it would ease off one of the most dangerous features to the market situation, and that is, of course, overproduction." The curtailment effort would continue through March.

Keeping to the terms of the curtailment agreement was inconvenient for Weyerhaeuser in early 1927 because of the constant problem of supplying the Atlantic Coast terminals. Production was one thing; shipping was another. No final resolution had been reached with Norton-Lilly, and Long seemed ready to admit that no arrangement could "always be depended upon and will always work smoothly." The *Heffron* had recently completed an eastbound voyage carrying somebody else's lumber, but Long hoped that such trips, "which came about from dire necessity," would not become a practice.

Meanwhile, the lumber market continued to be slow. There seemed no end to it. As Gus Clapp commented in St. Paul, "The condition . . . is bound, under present merchandising methods, to be more or less permanent." And he added, "Anything within reason which can be done to better the situation should be done."

One possible remedy, proposed by lumbermen such as Charles Keith, with several colleagues' support, was "whole hearted cooperation between merchandisers of lumber." The key to the problem, as everyone knew, and "why the situation with lumber differs so from other basic commodities such as cement and steel, is the great number of separate merchandising units." Thus, if lumbermen were somehow able "to improve their merchandising methods, to cut out useless and expensive duplications of selling endeavor, to make themselves independent of middlemen, or at least give them adequate control over them, to be able to furnish stock desired as and when it is desired by the buyer, and generally to effect selling economies," the market might turn around. In short, they envisioned a super sales company.

This was not quite the revolutionary solution it may have seemed. Clearly the Weyerhaeuser Sales Company had been organized for the same reasons. And as Gus Clapp observed, "If the Weyerhaeuser mills were to come to me today without any sales organization and ask me to suggest a plan for one, I am free to say that the plan that I would suggest would provide for fewer trustees, less frequent meeting of trustees and a great centralization of power in the responsible officers of the sales organization."

But Long, who had never been a strong proponent of

the Weyerhaeuser Sales Company, was far from convinced that Weyerhaeuser should be a party to consolidation efforts within the industry. As he subsequently explained, "We do not want to take part in any price fixing or price agreement plan, and I am pretty well satisfied . . . it behooves all of us to comply strictly with what we understand to be the law in such matters."

Long's position remained constant. Cooperation should not be in sales; it should be in advertising. He continued to solicit funds in support of the industrywide campaign extolling the virtues of wood. Long reminded his colleagues of the objectives—"to combat the inroads of substitutes upon the use of wood and establish structural engineers and other skilled expert men to show up what wood is really worth; to employ a corps of men to see that adverse legislation is headed off, as far as possible, pertaining to the use of wood in cities and towns for building purposes; and to employ chemical experts to study the better utilization of wood waste."

To further the cause, Long agreed to attend another Chicago meeting of the National Lumber Manufacturers Association, all the while wishing it wasn't necessary. Before his departure, there was a hint of better news. To brother Jim, Long noted "a stiffening in the selling price of lumber going to the Atlantic Coast." At the same time, he warned that it wasn't likely they could meet the needs of Portsmouth, Newark, and Baltimore, amounting to some 5 million feet per month. As he explained the situation, "We would be running our mills night and day if it were not for the sentimental reason that out here there is a very serious and determined effort being put forth to curtail and prevent overproduction."

The annual meeting, scheduled for May 28, was nearly at hand, and Long was busy putting together his report. There were other preparations, chief among which were legal questions involving rail construction for the Longview operation. For that purpose, Gus Clapp would be available. They would be acquiring rights-of-way over state lands, thus entailing certain obligations, and would have to determine just what the requirements would be for "our common carrier road, the Columbia & Cowlitz Railway Company."

Another important matter awaited resolution—purchase from Long-Bell of some 40,000 acres of pine timberland, "adjacent and intermingled with our holdings on Klamath Lake." The Long-Bell officials had proposed to turn over the pine timberlands for credit on the original contract, reducing the balance due. In conjunction, they wanted to sell some fir lands adjacent to Weyerhaeuser's other fir holdings "on something like a long-term transaction, somewhat similar to our first and second deals." George Long was willing to negotiate, but he made no promises. He was sure of one thing: "Personally, I believe very strongly that we should acquire their pine." None could have been surprised.

The annual meeting passed quietly, almost dully. Although there were plenty of significant items on the agenda, there were no major points of contention. J. P. Weyerhaeuser's reaction was typical: "Considering the large outlay of money that will be necessary to make the contemplated addition to the manufacturing capacity, and the 30% annual dividend looked for . . . the situation is fine." And he closed, "I congratulate you." Of course, much unfinished business remained, but clearly they were progressing. If the markets continued to disappoint, it was the same for every producer, and when the good times returned, Weyerhaeuser would be in an advantageous position.

It seemed that the Klamath impasse might soon end, the northern lines having decided to build the extension. The long-awaited word arrived in a June 11 wire from F. E. Weyerhaeuser: "[President Ralph] Budd advises of making agreement with Southern Pacific covering Natron Cutoff and Strahorn Lines and that they [the Great Northern] will be operating into Klamath within eight months." Long immediately responded: "We have to get busy at once and I fully realize this and am beginning to think seriously of all of the questions involved in this undertaking which, at the same time, evidently calls for careful studying of conditions and quite a good deal of haste."

Long's earlier optimism regarding the lumber market evaporated as spring gave way to summer, and he found himself worrying again about advertising, particularly Weyerhaeuser's own. Carl Hamilton of Weyerhaeuser Forest Products had come up with a new idea to answer the complaint that their advertising costs ought especially to benefit their own products. The program was labeled 4-Square, ostensibly referring both to square dealing and to an exactness of lumber dimensions. The plan envisioned trade- and grade-marking of all select grades, regardless of species. Special attention would be required to ensure exact lengths and ends precisely squared and wrapped. Great care would also be exercised in lumber delivery, so that the appearance of Weyerhaeuser's retail product would clearly be superior to all others. Long needed a lot of persuading. Initially he envisioned increased production cost and little else.

The matter had been discussed during the annual meeting. Tom Humbird was among those who listened to the various points of view, and, after returning to Spokane, he surprised friend Long with an expression of support for 4-Square. "I am wondering if the total amount of the budget proposed was not largely the cause of your hesitancy to endorse," he inquired. He was right about that. Humbird reasoned, "Our need of advertising to overcome the sales resistance occasioned by overproduction of lumber is very great," adding, "This is the first time that it appears to have been possible to connect up our advertising with our particular production." Although estimates were that 4-Square would probably double current advertising costs, Humbird nonetheless favored giving the program a try. "I firmly believe we are warranted in making this expenditure and I am very hopeful that after giving the matter further thought you have arrived at the same conclusion, and will so advise Mr. Lindsay."

A few years earlier, Long would probably have fought 4-Square with greater spirit. But facing an agenda already full to overflowing, he simply didn't have the energy to combat what the majority so clearly favored. In responding to Humbird, Long suggested that they "pick out some locality and try it out on them and see how it works." In the end, he promised to support the will of the majority, "with full vigorous effort." But if his surrender was gracious, it was hardly unconditional: "I am more in favor of this idea than any other one we have yet attempted, but did think there was some danger in the rather broad and sweeping advertising matter that was presented to us for our consideration at the Tacoma meeting."

It was with relief that Long turned his attention from 4-Square to Klamath Falls, where things were looking up. F.E. had lunched with Budd on June 14, and reported that the president felt "very happy over the outcome of his efforts to get the

Great Northern into Klamath Falls." As F.E. explained the plans, since the Great Northern had purchased "the Shevlin Road south from Bend and an agreement with the Southern Pacific for running rights from Paulina to Klamath, there will be very little new construction work," and what was required went through level country. Budd had asked about the Timber Company plans, and F.E. had responded that he was quite certain "you would build on the west side of the river." Long immediately reassured him of that.

Some wondered about the logic of Weyerhaeuser's expansion in the face of overproduction. One was Charles S. Elms, a San Francisco lumber dealer. When plans for construction of a Klamath Falls sawmill caught his eye, Elms stated the obvious, that "new mills increase the production but do not increase the consumption," adding, "I believe it would be a great mistake for your company to add to the production of lumber." Better, he advised, to take over some of the present operations in Klamath Falls and manage them more efficiently.

Long, of course, saw the matter differently. He viewed Elms's position as theoretically appealing but practically flawed; postponing manufacturing could not be justified by any expectation that future prices would be high enough to recompense the carrying charges on the timber. Like it or not, in the short term the only prudent policy argued in favor of liquidating Weyerhaeuser's timber asset. Further, a Klamath Falls operation had been Long's dream from the first day he had walked through those pine forests, and at the moment he was more concerned about finding a manager for the branch than selling lumber that branch would produce.

He broached the subject of a manager to F. E. Weyerhaeuser, admitting that he himself had no one in mind. Whomever they selected, Long wanted to get him in place "in the very early future—the sooner the better—to get acquainted with all the various things that are to be learned." As for designing the plant, the approach was familiar: Long, Titcomb, and Onstad would "visit some of the best mills in the upper California and Southern Oregon District, who will saw timber more or less similar to ours, and learn what we can of the type of a mill that is best adapted for our use at Klamath Falls."

In the meantime, work continued at the Atlantic terminals. Though business was booming on the East Coast, Long preached caution, particularly in the area of credit policies. In studying the May statement from Baltimore, for example, he thought he saw "somewhat of an increase of past due accounts," and expressed his "usual concern in matters relative to past due accounts and credits generally." He then offered a general economic assessment: "For the past two or three years the American business has been so prosperous and money so abundant and credit so easy; financing everything from the purchase of a paper of pins to the building of a skyscraper, that I think there is going to be a reaction some day and there will be lines drawn a little tighter, so it is well for us to keep our sails set in pretty good shape, anticipating a possible squall some time when dealers who are not in real good condition may not prove to be able to pay their debts."

The Newark yard welcomed its first shipload of lumber on July 13. Colonel Long was on hand for the occasion, subsequently reporting that in the first two days, some 1.2 million feet were unloaded, which he thought "was fairly good when you take into consideration a brand new crew, not only in the stevedoring but amongst our own men as they had to be broken in to our method of unloading." It also happened to be the hottest weather of the season. The machinery performed well, with "no stopping of the sorter chain or break down in the lumber storage shed and all of our crane hooks worked in fine shape."

Long coordinated a visit to Klamath Falls and vicinity with a San Francisco meeting of his Trade Extension Committee on August 2, followed by a regular National Lumber Manufacturers Association confab. He came away from his tour of the pine sawmills in northern California impressed with "the quality of the sugar pine and of the pondosa pine [*sic*] . . . and even more impressed than ever before with the very high quality of their lumber, its soft texture and the remarkable width they could get in their uppers and in their factory stock." As he forewarned Humbird, the competition from ponderosa pine was only going to increase, making "it all the more difficult to dispose of the Idaho white pine at anything like a higher differential in price."

The decision regarding a Klamath Falls manager was

still pending. Long had his eye on young Ralph Macartney, then employed in Cloquet, but he worried about the possibility of making those at Cloquet unhappy, thinking particularly of Rudolph Weyerhaeuser and Harry Hornby. Accordingly, he "asked a favor" of F.E.: "namely, that if there is no doubt about the attitude of the management at Cloquet as to Mr. Macartney leaving them, that you discuss the matter with Mr. Macartney and see if it will be agreeable to him to consider this opening and opportunity."

Lawyer Clapp arrived in Tacoma as requested, and a good deal of progress resulted, not only with the Long-Bell negotiations, but also in drafting a contract between the Columbia & Cowlitz Railway Company and the Weyerhaeuser Timber Company. Long was also negotiating a small deal with Mark Reed of the Simpson Logging Company. At question was a Weyerhaeuser tract that included, according to Long, "certain odds and ends that we have on our hands and which may not offer a very attractive logging situation." This tract had never been logged, but much of it had been burned in the fires of 1902. The Simpson and Weyerhaeuser cruises were surprisingly close on green fir, spruce, and cedar, but Simpson listed 4.5 million less of dead fir, "and for once," Long indicated, "we are going to assume that you know more about that than we do." His original price on the tract had been $85,000, and on the basis of the 4.5 million difference, he lowered it to $72,000.

Reed must have smiled. "To be real frank," he responded, "the dead timber which is cruised in this scattered stand is in fact a liability rather than an asset," explaining that one could "hardly imagine, without having the experience of operating in this timber which has been dead for twenty-five years, the amount of waste thru breakage." He enclosed a check for the $72,000.

Negotiations were not always so straightforward. The joint cruises of the Long-Bell pine timber in Klamath County, Oregon, and Weyerhaeuser fir timber in Lewis County, Washington, were completed, and the results were interesting if not surprising. As usual, the figures differed, depending on the basis of who was selling and who was buying. For three sections of pine timber, for example, the Weyerhaeuser estimate was 17,107,000 feet, and Long-Bell's 23,696,000 feet. And on three sections of fir, it was Weyerhaeuser, 124,800,000 feet, and Long-Bell, 74,000,000 feet. Check cruises in both instances fell somewhere between. In reporting the estimates to his boss, Minot Davis admitted that he was impressed with the accuracy of the Long-Bell cruise of its pine, and suggested, "We will at least be justified in splitting the difference between the Long-Bell cruise and our own as to fir, because it is manifest that on several sections at least the Long-Bell cruise was so low as to be out of all reason." Long agreed, and on October 31, Davis met with John Tennant in the latter's Longview office. The two reached an easy accord, accepting an "average of your cruise and of our cruise on each variety of timber."

Meanwhile, things were really moving at the Weyerhaeuser Longview mill site. Titcomb inquired of his brother-in-law Phil Weyerhaeuser whether Phil had any materials "left over from your construction [of the Clearwater Timber Company's new plant] which you will not need and of which you are desirous of disposing." He also wondered whether Phil could make any recommendation regarding the parties "who operated your boarding house or cafeteria during your construction work." They had made "a very favorable proposition" to Al Raught, and Titcomb—ever cautious—sought assurance. Onstad was busy drawing blueprints of Mills 1 and 2.

Long had left on an eastern junket in early October. Nearly two years had passed since his last visit to the other coast, and it was time for a personal inspection. When he returned to Tacoma in early November, he turned his thoughts to how and when they might expand into the Two Harbors district of western Washington. He had discussed this possibility informally while in St. Paul, and apparently received support. So he now announced plans to survey "the situation down there and find out a little more than I now know, as to the possible physical conditions of some of the plants that are suggested, and as to whether or not the idea would be agreeable to those whom we would want to consider as desirable associates."

One thing didn't change during Long's eastern travels: the lumber market remained depressed. There seemed to be no solution to the problem. Long wondered "whether or not we can get manufacturers to voluntarily reduce their cut, or whether everybody will keep on plugging away until we all run up against the stone wall." He hated to think that the industry would simply continue in its "bull-headed way, which means the survival of the fittest, for that is a hard word to use, and I hope some remedy will be thought out and followed that will obviate that extreme performance."

His fears were not misplaced, although the timing of their expression was awkward. Several took notice of this. C. R. Musser's reaction was representative: "Under the conditions, it rather strengthens my belief that it would be good policy for us to utilize some of the present production facilities rather than increase production." John S. Owen, his old friend from Eau Claire, took Long to task on a number of counts: "Your wages are too high, $4.00 a day as against $1.75 in the South and $3.00 in this section. Your business is run by the 4-Ls, your commissary is extravagant (no one of the owners have such food)." And he wasn't through: "It breaks our hearts to see the Weyerhaeusers building big new mills all over the West—mammoth ones. Is it right, and what will be the result? I think you will, if you keep on, put one-third of the mills in the West out of business—bust them higher than a kite. I don't think you mean to. You will put a lot of banks in the same 'pew' and put thousands out of work. Anyway, it will all be laid at your door, and the name 'Weyerhaeuser' will be made the goat. I know there are no fairer people in the whole United States, but you have got a situation to consider, and one that no one can handle but the Weyerhaeuser Company. It is a great responsibility."

Long surely agreed about the responsibility part. As for the rest, he suggested that "some of these matters are not so easily handled as one might think," and noted that it was "rather a big order when one company on account of its size is looked upon by everybody else to set the pace and tell the other people what to do." On the subject of expansion, he acknowledged that the activity "may look absurd to the outsider, and certainly comes at an unfortunate time for market conditions, and yet there are some deepseated reasons for it." Explaining that to others would take some doing.

Long still held some hope in that national advertising campaign, but the fact was that the national campaign was experiencing serious problems, chief among which was the industry's inability to adhere to standard qualities and grades.

Long indicated as much to Wilson Compton: "While I dislike to admit it, yet I am forced to, by the fact that the industry, especially on the West Coast, is not and cannot at the present time prepare its lumber up to the highest standards that are recommended in this proposed advertising." Given that, going ahead with the program "might be a boomerang that would do us a great deal of harm."

The management problems neared resolution. Ralph Macartney and Long spent a day together in Klamath Falls, after which Macartney stayed on, to be shown about by Jack Kimball. Long warned "that the wintry winds will be blowing all through the western pine country quite soon," and suggested that Macartney might want to arrive in advance of that "to soften up." Little by little, the new leadership team was taking shape: Peabody in Everett; Charlie Ingram at Snoqualmie Falls; Macartney in Klamath Falls; Harry Morgan and Al Raught at Longview; Colonel Long with Atlantic terminals; and Minot Davis and Rod Titcomb in Tacoma. All in all, it was a strong crew, and one that would largely stay in place for years to come.

But strength in management did not diminish the seriousness of the problems challenging them. Long shared his concerns with his lieutenants, suggesting that given "the prospects of the future, we would have to lay awake nights to study ways and means to cut down our expenses and get more efficient results out of our organizations." During good times, the tendency was not "to watch corners quite so close unless you are so inclined," but now, and in the foreseeable future, "it is the man who has the best organization and has the least expense that can be obviated that is going to live and last." Accordingly, Long was asking his managers to prepare a list of employees, "stating what the employment is, the character of the work they do and what the wages are." Review processes are a necessary if unpleasant part of doing business. In this instance, a knife was clearly being readied, and Long wanted to make sure that the carvers-to-be were well informed.

Bill Peabody was considering a special problem, involving a substantial change in his source of logs and a likely increase in costs. The nearby Cherry Valley logging operation was closing down, and although Everett would continue to receive logs from independent companies, its major supply would come from timberlands at the extreme southern end of the Sound, in what was then known as the Skookumchuck Camp, years later to be the Vail-McDonald Tree Farm. Long knew that Peabody would try to obtain every advantage for Everett, which was as it should be, but he wanted it understood that the new circumstances dictated new rules. For example, to serve Skookumchuck they had purchased terminal sites on the Sound at considerable expense. They also had to buy a thirty-mile right-of-way, "and while we did a good job and got it reasonably cheap, the chances are that the cost of the railroad from the headquarters camp to the terminal will show an investment of at least $1,000,000 [and] possibly more." Cherry Valley, he reminded, had carried no such burdens. Also, for Skookumchuck, they would have between $200,000 and $300,000 invested in logging cars, plus $50,000 in locomotives. Somebody was going to have to be charged for this, and since Everett was the beneficiary, why not Everett?

Furthermore, they would need to purchase at least 4 billion feet of Skookumchuck timber to provide for its future supply. A going price for that timber would be between $3.50 and $4.00 per thousand. Just suppose Everett made such a purchase, Long continued, "then they would have an investment, together with all of its other kinds of property everywhere, close to $25,000,000 and you can readily see how difficult it would be for the Everett Mills to make even 5% on that kind of an investment."

Long was trying to alleviate the competitive Peabody's fears of suffering by direct comparison with the likes of Snoqualmie Falls. Peabody knew that Snoqualmie would enjoy a great advantage in such a contest. But, of course, George Long knew it too. Indeed, he cited some of the differences between the two operations, noting that when Snoqualmie was organized, "Two different interests merged their timber in one corporation and each one took stock in proportion to the amount of timber it contributed." Given those circumstances, the value placed on the timber didn't much matter, and Long remembered that at the time there "did seem to be some reasons why we should assert a conservative value, largely because we did not want to reflect in our capital organization a very large investment." At the time, the company was trying to avoid attracting the attention of the tax assessor. Accordingly, they had agreed upon an average value of $2.50 per thousand. This decision had not been altogether beneficial:

> We were not wise enough to foresee that there would be a revenue law and if we had we no doubt would have asserted the value of our timber at [at] least $3.50 or $4.00 when we entered it on our books, for I think it was worth it, but having it on our books in 1915 as a purchase, we could not get the [Internal Revenue] Department to allow us to change that figure, so that in the cost of logs of the Snoqualmie Falls Lumber Company on that original purchase, which embraced 50,000 acres of land, we have to report it every year on the basis of $2.50 per thousand cost. We made some large purchases since then, where the timber really cost us from $4.00 to $5.00 per thousand and merging all of these properties together now makes the depletion rate something like $3.00, which is still very low.

Thus Snoqualmie was assured the luxury of a cheap timber supply.

Nor were the "purchase figures" the only edge enjoyed by Snoqualmie. The facility managed with a relatively short railroad, had "magnificent ground on which to log, and still further, the block of timber which they first entered was at least 25% to 30% better than the average of the entire stand." As a result, Snoqualmie Falls showed abnormal profits. Long assured Peabody that "none of us on the inside are fooled at all by the magnificent showing which the books give of the Snoqualmie Falls profit." Indeed, he preached a warning at every annual meeting that soon they would be faced with higher logging costs while harvesting timber of a poorer quality.

But even with all of its advantages, when one figured the investment at Snoqualmie, the value of timber, mills, town, lumber, and accounts, the earning interest on its entire property would be little more than 6 percent over the most favorable years of operation. And what was true at Snoqualmie was generally true of the Timber Company as a whole. Long noted, for example, that for 1926, "the Weyerhaeuser Timber Company in its lumber operations, including its boats, including its eastern yards and including its stocks of lumber and accounts earned something over 4%."

Furthermore, largely because of its resource base, Weyerhaeuser was doing better than the average. For most others, it seemed to be as Long's Canadian friend James McCormick had earlier concluded: "Our industry is a good one to keep labor employed, keep business active for the supply and material houses, but not a good business to give back adequate returns to the capital that is invested in it."

Progress was being made on a variety of fronts—in Long's opinion, too rapidly on some, and not fast enough on others. Among the former was the Weyerhaeuser Forest Products, whose executive committee scheduled a December 13 meeting in Spokane, the day before Weyerhaeuser Sales Company trustees were to assemble. The main subject for consideration was the 4-Square program. Tom Humbird had warned Long that it was full speed ahead, and Harry Payzant, the engineer, had recommended placing an order with the Sumner Iron Works "immediately," so as to expedite delivery of machinery to implement 4-Square. Humbird had given his approval, half hoping for Long's support.

One question still unresolved was what grades 4-Square ought to include. Long warned that it was "dangerous to adopt the practice of making an especially good grade for a special purpose," and argued that the lumber packaged "should be our regular standard grade." Many, if not most, felt otherwise. Humbird had assumed that 4-Square meant something better than standard. "If you are satisfied to have the regular grades as now made go into these packages," he wrote to Long, "there is nothing further to be said." But, he continued, "This would not accord with my idea, my thought being particularly in the matter of dressing we are going to have to be more careful."

But standardizing the product was easier said than done. Long attempted to explain a part of the problem, noting a difference in the thickness of some items manufactured at Snoqualmie Falls and Everett, "especially in one point where Snoqualmie surfaces lumber two sides clean, i.e., its flooring and drop siding, while Everett does not do this in all cases." Additionally, Snoqualmie produced its rough inch-lumber a little thicker than Everett, Long adding, "I would have to write you a whole page or two to explain why, but there are good reasons for what each concern does." What bothered Long most, however, was Humbird's apparent intention "to make something a little better than our regular grades." Now, if Humbird meant only to ensure uniformity in thickness, width, and dressing, that was one thing, "but if your thought is that we should make a better grade of lumber than the stock which we propose to put in the 4-Square category, I certainly do not agree with you. . . . When you begin to do this you shoot to pieces our regular well-established grades, and when an organization gets an idea of that kind into their head there is no telling how much it costs you in dollars and cents. I think it would mean a great deal more damage than any benefit you would get."

Payzant seemed to be everywhere at once, designing the necessary machinery for branding and packaging for the many scattered operations. Long would later observe, "Mr. Payzant makes his headquarters under his hat." He forewarned brother Jim that at least one of the Atlantic terminals would be a part of the "experiment," and the question was which one, Baltimore, Newark, or Portsmouth? George Long thought Newark might be most suitable because of its available space in the planing mill, Payzant having indicated that the equipment required an area twenty by eighty feet. It was to be installed, he warned, "at our expense." Regarding the project as a whole, he hoped for the best but expected considerably less.

He was much more enthusiastic about Klamath Falls, where he was eager to begin the work, which meant getting Lloyd Crosby on the railway construction job. At the moment, however, Crosby had his hands full putting the finishing touches on the Skookumchuck network. Crosby was willing, and almost ready. They had completed some sixty miles of railway at Skookumchuck and logger Ronald McDonald was already on the scene, with Walter Ryan expected shortly to relieve Crosby. Looking ahead, Crosby couldn't imagine "anything that should bother Mr. Ryan or Mr. McDonald after January 1st," adding, "I feel certain that Mr. McDonald will be able to start logging in volume very shortly." As for his next assignment, "I will be very pleased to go to Klamath Falls, or anywhere else you might want to send me."

In view of the number of decisions required of him, it is

not surprising that Long made an error now and then. In one such instance, he accused the Midway Yard of purchasing only 20 percent of its fir lumber from "the so-called Weyerhaeuser Coast mills." Horace Irvine immediately corrected this estimate, assuring that it had been a much higher percentage. Long readily admitted his mistake, allowing that he was "very glad to see that the Coast Mills have been so handsomely treated." Still, he couldn't avoid teasing Irvine: "It would have been a little bit nicer if you had not given me such a positive call down. You must know, if you give it a second thought, that it is not considered polite for a youngster like yourself to dispute any fact that is uttered by an old man, even if as often occurs, the older he gets the more he gets twisted with his facts."

Irvine, within two weeks of his fiftieth birthday, answered in kind: "Flaming Youth extends his best wishes to the 'old man,' and in closing admits in twenty-five years of association this is about the first time he has been able to get anything on him." One thing that George S. Long had learned over the years was that it was the friendships that mattered most. And in that regard he had been richly blessed.

TWENTY-NINE

The Return of Good Times?

EARLY 1928 SEEMED A TIME FOR NONSTOP PLANNING. There were plans for Longview and plans for Klamath Falls. Then there were plans for the 4-Square campaign, which would eventually involve every Weyerhaeuser-affiliated plant. And, in addition to plans, there were rumors. James Eddy of the Siler Logging Company inquired about the truth of one, that the Weyerhaeuser Timber Company intended to acquire Long-Bell. This rumor was doubtless fueled by hopes: "Lumbermen feel such would be the beginning of the end to their troubles in over-production." Long replied matter-of-factly. The Tacoma office was accustomed to rumors, he said: "The public is always having us do something here or there." As for buying out Long-Bell, "the rumor is all bunk."

There certainly had been negotiations between Weyerhaeuser and Long-Bell. But those negotiations had to do with land exchanges and were nearly complete. Long announced that the Timber Company had assumed an $800,000 mortgage in the purchase of Long-Bell's timberlands in the Klamath region, a mortgage previously held by "the Western Pacific crowd, otherwise known as the Shevlin-Hixon outfit." Long was eager to pay the balance due, thereby avoiding the 6 percent interest payment.

All this was on the eve of another trip to St. Paul, where Long intended to announce that Colonel William B. Greeley would be the next director of the West Coast Lumbermen's Association. Recruiting the Chief Forester constituted a coup of coups, ensuring strong and progressive leadership in the years ahead. Humbird arranged to share the ride east with Long so as

357

to have plenty of time to discuss 4-Square. Peabody was already scheduled to accompany his boss, and Long now invited Charlie Ingram to join the party.

But soon all of these plans came apart at the seams. First, Long's sister Rhoda died on January 14, while he was en route. He thus proceeded directly to Indianapolis, where he spent nearly a week arranging for her funeral. Then brother Jim's mother-in-law passed away, and her funeral was in Eau Claire, Wisconsin. So plans were further revised, and as a result, little in the way of business was accomplished. During a brief stopover in St. Paul, there was some consideration given the 4-Square program, but nothing substantial resulted.

Of greater significance was a meeting to review the continuing scheme of a grand merger. F. S. Bell, F. E. Weyerhaeuser, and Long constituted "the committee representing Weyerhaeuser stockholders and directors," and they agreed upon their position, which was fortunate, for the merger advocates subsequently pursued Long to Tacoma, meeting with him there on February 14. What they sought was a counterproposition, but Long was having none of that, "simply because I did not want to step outside the boundaries of the negotiations as our committee left it."

In fact, thoughts of a grand merger would not fade unless the market changed for the better. Once more, Long imagined he could see a glimmer of hope, assuming that American farmers enjoyed another year comparable to 1927. Farmers, he reasoned, had been paying off their indebtedness, "and consequently we have not had any rebound of improved conditions in the agricultural district." At last perhaps things were about to turn around. The annual meeting of the West Coast Lumbermen's Association was set for February 17, and Long predicted that "there will be more or less talk about curtailment and other methods of improving market conditions." The Everett mills had been operating on a five-day work schedule since December 1, with two mills on day shift and one mill on a day and night shift.

But if the market moved slowly, nothing else did so that winter for George Long. "As it is now," he explained in early February, "I am expected to spend at least two days at Longview, where we have 200 men at work building a mill and where we have to decide almost at once on some important plans connected with construction; and expected to do the same thing in Klamath Falls within the next two weeks; and expected to be at our Snoqualmie plant and our Everett plant for at least two days each place, to pass upon pressing problems that must be handled soon, all coupled with the usual routine daily business, plus some West Coast Lumbermen's Association activities and West Coast Lumber Trade Extension Bureau activities which will absorb quite a little of my time in the next two weeks." Long didn't even mention a pending trip to Carmel, California, in connection with settling his sister's estate.

The reference to planned visits to Everett and Snoqualmie—visits that never materialized—had largely to do with 4-Square fallout. When in St. Paul, at Humbird's request, Long had met with Hamilton and Payzant to try to anticipate the practical problems likely to be encountered at the mills. Everything wasn't settled, of course. Charlie Ingram subsequently raised questions regarding his instructions, some of which appeared to call for procedures "at variance with the common practice."

One specific complaint concerned the recommendation that all items should be trimmed to exact lengths. Ingram thought that reasonable in terms of fir and hemlock, but "when it comes to Cedar Siding there is an established trade custom to rebut one-half to one inch short of the given length," and, he continued, "I feel rather doubtful that the added cost of trimming the Siding to a longer length could be offset by any sales advantage that might be gained in view of the established custom." Long could only suggest that Ingram write directly to Hamilton: "I took the liberty when in St. Paul at a conference in January, to say to Mr. Hamilton and Mr. Lindsay that we would do everything that was possible to do to carry out to the letter the entire program of this 4-Square work, and of course meant it or I would not have said it, but if there is anything so very radically wrong in doing this as it pertains to the cedar, why let's take the matter up frankly and as quickly as possible with Mr. Hamilton and see if we can adjust it."

Pushing other questions aside, at least temporarily, Long turned again to association matters. A joint session of the West Coast Trade Extension Bureau and the West Coast Lumbermen's Association met in Seattle on February 17, and all went according to plan. An amalgamation of the two organizations was accomplished—henceforth there would be a single West Coast Lumbermen's Association—and Colonel Greeley was elected as its secretary-manager. Long forwarded the news to Greeley, observing that everything "went off very smoothly without any opposition and apparently with entire satisfaction." He also reported that John Tennant had been elected president, with "a very strong board of directors. . . . So the setting is all made for a real effective machine."

Now everyone was anticipating Greeley's arrival to spearhead the effort to increase membership and get "the whole industry into this new revised association." The initial plan, as Long described it, envisioned a "flying squadron," composed "of the leading men of the industry to travel around among the various lumber centers to tell the whole story in a more effective way than we have ever been able to."

The Chief Forester intended to resign on or about May 1, allowing sufficient time "to carry through the important appropriations and legislative matters coming up at the present session of Congress." He thanked Long for his "interest and efforts in bringing this new organization to pass," adding, "I hope that I will not disappoint you." Tennant had hoped that Greeley would be available sooner, explaining, "Our work here is rather marking time until you are able to be with us." Perhaps, he suggested, Greeley might join him and Long at the mid-March meeting of the National Trade Extension Committee in Chicago, which "would give us an excellent opportunity of discussing some of the preliminary matters with you and would permit of your making your plans in the East a little more definitely."

Long departed for California on the evening of February 21, where his stay was extended as a result of meetings on the old merger question. Although it was not yet a dead issue, Long wasn't offering any encouragement, certainly not without further discussions with F. S. Bell, F. E. Weyerhaeuser, and others who might be available in southern California. For F.E.'s benefit, Long observed: "I am not at all enthusiastic about this particular crowd having a combination that will be ultimately successful or that will bring about any great improvements in the lumber market."

In the early spring of 1928, important hearings were taking place in Washington, D.C., on the Revenue Bill of 1928, and it was section 115 of that bill, "Distributions by Corporations," that precipitated the greatest fears. This would replace section 201 of the Revenue Act of 1926, and would eliminate tax-free distribution to stockholders of the earnings or profits accumulated or increase in value of property accrued prior to March 1, 1913. The effects of such legislation to Weyerhaeuser shareholders and to many others would be devastating. Mark Reed and Alex Polson, with a hard push from Long, went to Washington, D.C., in late February to speak against the provision. Stiles Burr was in the audience, and came away delighted with Reed's contribution. "I do not believe anyone else could have done just what he did," Burr reported to Long, after noting that Reed "blames me for the suggestion which brought him here, although admitting it was at your command."

In the wake of Reed's testimony, Burr seemed confident that the "accursed dividend" provision would be defeated, but allowed, "I am a little too old to feel sure of anything," and promised to keep on working. It was good that Burr and others did keep on working, for the battle wasn't yet won. Within a month, Frank Wismer, chairman of the Advisory Tax Committee for the National Lumber Manufacturers Association, wrote to Long of the latest attack, an attempt to limit the tax-free distribution "to an amount not in excess of the value of the shareholder's stock, as of March 1, 1913." What may have seemed a compromise to some was of no help to such as Weyerhaeuser, where comparatively little of the increases had been realized or distributed. Wismer had closed with a postscript, urging Long to get his friend R. A. Booth to contact Oregon Congressman Willis Hawley, asking him to seek a restoration of section 201, "without any change or modification." In mid-April, with the issue still undecided, Burr remained cautiously confident: "Nothing, however, is settled in Washington until it is settled."

Closer to home, Long turned to a public relations matter in Klamath Falls. Ralph Macartney had inquired as to what kind of advertisement Weyerhaeuser should purchase in the "congratulatory" issue of the *Evening Herald* on the commencement of Great Northern's rail service to the community. Long advised that they "be rather modest in our appearance," and instead of taking a whole page, "as though we were the biggest people down there," take a quarter-page, or half at most, "even if we should pay the full price of a page."

What was expected, at least from the editor of the *Evening Herald,* was an announcement of Weyerhaeuser's plans for Klamath Falls, but Long wasn't ready for that. As he explained to both the editor and Macartney, President Budd of the Great Northern had invited him to attend the celebration in early May, suggesting that this would be "a good time for us to make an announcement of our plans," and Long agreed, making a commitment to do just that. He sympathized with the editor's position, recognizing that "this may not satisfy the real red-blooded instincts of a news gatherer who would like to be the first to make the announcement," and he promised to send him the details, providing the information was not made public until the official celebration.

It seemed to Long that he was in constant motion. In April, he paused to take note of recent travels. "So far this year I have been a wanderer, twice in Chicago and three times in California, and see ahead of me for the next ten days or two weeks some more of this wanderlust." There was the new camp at Vail, "full of interest and full of a lot of things and I have not yet seen it. We are building two big saw mills at Longview and I have only been down there once in two months. I have not been to our Everett mills during 1928 and only once at our Snoqualmie mills, and that was after night." It was too much for one man, even a young man, to manage. And Long wasn't a young man. Still he found himself trying "to do a hundred things that have been left undone, between now and early in May," and the annual meeting was only a month away.

As if that weren't enough, F. E. Weyerhaeuser had begun to take a personal interest in the subject of selective logging. In part this was because his nephew Phil was experimenting with just such a program in Idaho, adopting a plan designed by consulting forester Carl Stevens. Indeed, while Long was in California, F.E. was meeting with Stevens in Portland, and came away impressed. He would soon recommend that Stevens's outfit "look over our Klamath County country before we go into it with our logging development." Long agreed that Mason & Stevens understood "these matters very thoroughly and know how to get up very illuminating reports," but he wasn't quite ready to commit to them, explaining, "I want to have our own organization give it a close going over, as the background for any subsequent study." It wasn't foot dragging. F.E. doubtless assumed that what worked in one pine district would work in another; George Long knew there were a great many variables to be studied.

At the moment, however, Long was concentrating on the possibility of making the timber resource more valuable, not by selective logging but by increasing the value of the finished product, lumber. There seemed only one way to manage that: manufacture less of it. And so the eternal interest in curtailment continued, only this time there was briefly a hint of serious compliance. Lumbermen gathered in the various producing centers of the Pacific Northwest to consider their options. And, for a change and for the moment, results were forthcoming.

When Long received a report that 95 percent of the Everett mills had agreed to cease Saturday operations, he immediately wired back, "Congratulations. You are all fine fellows." In Seattle, some 60 percent of mill managers agreed to a five-day workweek. The Chehalis and Centralia district advised that they would likely get 100 percent cooperation, and Grays Harbor promised a 20 percent reduction for the remainder of the year. In Tacoma, Long found no great enthusiasm for the program, but expected that most of his colleagues would fall in line, noting that the St. Paul & Tacoma Lumber Company had already shut down the night shift at both of its mills. He now advised Charlie Ingram to follow suit at Snoqualmie Falls, indicating, "I would be glad to have you do this commencing with Saturday of this week."

At the same time, a committee began collecting data to determine each mill's "proper quota of reduced output." The procedures were detailed. The chairman of each district would appoint a subcommittee of four or five, and they in turn would be responsible for specific mills, completing forms prepared by the Davis Statistical Bureau. Included were figures for the past three years for total sales, orders, prices, and inventories, along with current weekly reports of sales, orders, and prices.

In the midst of this feverish activity, Long went to Klamath Falls for the Chamber of Commerce celebration dinner on May 10. There, as promised, he outlined just what Weyerhaeuser had in mind for that community. It was impressive: a sawmill with four headrigs "of the usual band saw type," and all of the necessary accoutrements, "such as edgers and trimmers, to completely finish the manufacture of lumber." In addition, there would probably be a large gang saw, to saw the fir logs. The capacity would be somewhere between 250,000 and 275,000 feet per shift, Long indicating that "the condition of the lumber market, of course, will determine whether we run on one or two shifts." There would also be a planing mill, at least twenty dry kilns, and a large box factory "sufficient to manufacture into boxes that part of our lumber products which is better adapted to box manufacture than for other purposes." As for the logging, that would commence immediately to the west of Klamath Lake, and would require the construction of a logging railroad "some 12 or 15 miles in length . . . and this logging road will be extended from time to time to ultimately reach all of our holdings in that vicinity." If all went according to plan, the whistle would blow on July 1, 1929.

By the end of May, Colonel Greeley was settled in a Seattle office, beginning his work as secretary-manager for the West Coast Lumbermen's Association. Initially this involved a series of meetings in communities throughout the Northwest, including Vancouver, B.C. Long described the purpose as being "largely to introduce the new Manager to the industry and at the same time to stimulate amongst the operators a little more sympathy and loyalty to the West Coast Association." Long would have enjoyed accompanying Greeley, but, as he explained, other responsibilities intruded: for "the next ten days the writer is in jail, so to speak, having on his hands the annual meeting of the Weyerhaeuser Timber Company with a host of visitors and a great many little details." He did, however, write letters of introduction for Greeley to a variety of friends, and promised to join the tour as soon as possible.

The annual meeting included a visit to the Longview site, where the directors witnessed a veritable beehive of activity, most prominently the framing of the huge sawmills. The annual report contained, as usual, good news and bad. Net profits for 1927 amounted to more than $3.1 million, but nearly a million of this was deferred payments on outstanding contracts negotiated in previous years. In addition, $1.3 million resulted from interest on securities and other bills receivable. It was thus clear that a very small fraction of earnings came "from the actual operations of our saw mills," and, Long warned, 1928 was likely to be even more disappointing. Even where there seemed to be a profit, such as with the Snoqualmie Falls Lumber Company, it was "more or less fictitious, as the value of the stumpage charged against the operation was the March 1, 1913 valuation, and not the real value of the timber during 1927."

Long spent plenty of time discussing business in general, noting that "the lumber industry as a whole has been giving more attention to its markets and making endeavors to stimulate the demand for lumber." He referred specifically to the national campaign, "raising a fund of $1,000,000 per year for five years, all with the object of demonstrating that wood still has its proper place for all kinds of construction, and to put up a fight to hold the markets which we now have." Weyerhaeuser was contributing $50,000 each year to this effort. He also mentioned such regional programs as theirs on the West Coast, raising $400,000 annually "to advertise its woods and establish the worth of its product."

As to the possibility of a super merger, Long reported the current level of interest to be "luke warm," most assuming that "the matter had been dropped and that in all probability it would not be again presented." Long, surprisingly, seemed somewhat disappointed: "In the Pacific Northwest there have been several efforts made to organize several large corporations, generally referred to as mergers, with the hope that by the control of a large amount of the production, more economies could be practiced in management, better results obtained in marketing and a more stabilized condition generally developed, but the efforts so far made to bring about these combinations or mergers have not been successful, but it does seem apparent that something of this kind may ultimately be necessary to bring about more stability and better earning capacity in the industry." No formal consideration, however, was given the subject.

When the last of his shareholders had headed homeward, Long hastened to join Greeley's traveling show. That he devoted nearly two weeks to the tour evidences the importance he attached to it. As Long and Greeley wended their way, mixed news from the markets followed them. After the March lows, lumber prices had increased by an average of between $1.50 and $2.00 per thousand, and Long thought if they could hold on to that or possibly improve it a little during the remainder of the year, it would be a better year than any since 1924. The advances were not the result of curtailment efforts, although those seemed now fairly well in place, encouraging the thought that such gains might be sustained.

While the five-day workweek was generally being honored, statistics were slow in coming. Indeed, at the end of June, A. B. Davis, of the Davis Statistical Bureau, reported that his agency had yet to receive "from any member of the committee any expressions as to the normal capacity of the mills in the respective districts." Long called for a committee meeting on July 10, and in the meantime tried to "stir up this question." He also recommended that Greeley's West Coast Lumbermen's Association assume responsibility at an early date for putting together all of the production facts presumably being assembled by the Davis Statistical Bureau.

Even with Greeley now on the job, the various campaigns were proving difficult to manage. It was like driving a twenty-horse hitch. Secretary Compton, in behalf of the National Trade Extension movement, was proposing a program guaranteeing that lumber bearing the trademark of the National Association would adhere to established standards for width and thickness, and would also be well seasoned. Long was asked to comment. For himself, he saw little difficulty in manufacturing lumber to the standards and "proper grades of the regional association to which a member belongs, but I do see quite a large amount of possible trouble in the guaranteeing of lumber well seasoned." Then there was all of the branding, this while he was still agonizing over Weyerhaeuser's own 4-Square.

The problem of seasoned lumber was most acute in the Atlantic terminal operations. Though they had initially tried to supply Baltimore with dry lumber, under the current practice, Long acknowledged, "We could not use the national brand on our shipments out of Baltimore, Portsmouth and Newark, if the national brand specified the lumber should be well seasoned." Perhaps they ought to be working toward the exclusive use of dry lumber, but that amounted to "a Herculean task." Shipping green lumber was economical, and the manufacturer got a quicker return. It was one of those "perplexing situations."

Another perplexing situation occurred as the curtailment campaign began to dissipate. J. S. Magladry, president of the Willamette Valley Lumbermen's Association, reported little interest among small operators in his locale. "Because of their method of operating," he explained, "they lost a lot of time anyway." In short, they were already within the prescribed limitations by reason of their own inefficiency. Further, why should they curtail when they looked about them and saw many of the larger mills running double shifts. But by far the most discouraging news came via Peabody in Everett, who reported that the Eclipse Mill Company was resuming a six-day workweek. The problem was beyond easy resolution: as prices began to improve, however slightly, the reasons for curtailment simply seemed less compelling.

Though the unraveling of the curtailment efforts was disappointing, no single disappointment caused George Long to reflect on his future. It was a combination of things, especially the realization that he was finally feeling his years. On July 20, he sat down and wrote to F. S. Bell, the new president of Weyerhaeuser Timber Company, advising of his intention "to withdraw from the active management of the company in the near future":

> I will be seventy-five years old in December next, and while the years have rested lightly on me, until recently, I am conscious now that I am slipping physically, and that my remaining years should not be loaded with weighty matters, neither for my own comfort and plans, nor for the best interests of the Company. We are broadening out in our activities, by our mill developments, that have a great appeal to me but they call for the best efforts of a younger and stronger man than I am now, and most certainly than I will be in the nearby future. Then again there are a few things I want to do, that will take my time; I have had so much enjoyment in my work that

> I have not sought for much elsewhere, but after all, this in a
> sense is selfish, when one considers what one owes his family;
> I am derelict in this particular and should make some amends.
> I want to do this.

In some respects, it was a sad commentary. Even George Long
had his limits, and, as he now acknowledged, it was his family
whom he had neglected.

Bell responded sympathetically, but expressed the
hope—"if consonant with your purposes and your best interests"—that action could be deferred until they had a chance to be
together. Bell was planning a trip to Tacoma in September, "and
personally I would like to talk with you as to arrangements for
the future."

Colonel Greeley already seemed at home. The physical
part came easily, Greeley leasing a suite of offices on the third
floor of the Stewart Building in Seattle, where the combined
staffs of the West Coast Lumbermen's Association and the Lumber Trade Extension Bureau would be housed. Other questions
proved rather more difficult to answer, among which was the
line Greeley must walk between industry cooperation and possible violation of the Sherman Anti-Trust Law.

The trustees of the West Coast Lumbermen's Association gathered in Longview on July 31, at which time Greeley
was directed to draft a resolution "covering the general informational service of the Association regarding market and other
industrial conditions and meetings of sales managers as a part of
such informational service." The trustees also instructed Greeley to seek legal advice concerning the resolution, advice subsequently provided by Charles H. Paul of the Long-Bell Lumber
Company. Judge Paul cited two 1925 Supreme Court cases,
Maple Flooring Manufacturers' Association v. U.S., and *Cement
Manufacturers' Protective Association v. U.S.* In essence, the court
had found that the mere gathering and distribution of information was not in itself a violation of the law, "unless the prosecu-
tion shall sustain its burden by further evidence." Judge Paul
warned, however, that "the present legal situation certainly warrants the close attention to detail and manner of handling these
matters."

As influential as Greeley was, not everyone immediately

enlisted in the Association. J. H. Bloedel and J. J. Donovan were two of the more important recalcitrants. John Blodgett recounted for Long a recent discussion with his friend, Mr. Donovan. Although both Bloedel and Donovan "have great faith in Colonel Greeley and Dr. Compton," he noted that Bloedel found the West Coast dues to be excessive, and also held that it was "just as important that Long-Bell join the export company [Douglas Fir Exploitation Company] as that Bloedel-Donovan join the West Coast Association." Blodgett assumed that this was no news to Long, but only a "recent confirmation." As for himself, he was delighted to see some industry initiative finally being taken: "We have usually waited for an earthquake in Japan, fire in San Francisco or the digging of the Panama Canal to temporarily restore the balance between supply and demand."

Long was indeed familiar with the position taken by Bloedel and Donovan, and regretted it very much. He considered it "doubly unfortunate because these two gentlemen are, in my opinion, rather head and shoulders above almost any of the lumbermen out here in their ability to solve problems and handle big questions." Still, he had hopes that with Greeley in charge, Bloedel and Donovan would soon be won over, giving "us the benefit of their financial support and ability to direct things."

In fact, however, Long was getting a bit discouraged. Perhaps he was simply tired of trying to pull others along. At such times, he sought the refuge of the woods, in this instance going out to visit Axel Hanson. The two were scratching their heads over the future of the White River Lumber Company, Hanson wondering whether the time was right to undertake changes earlier discussed, making the sawmill and planing mill into a single unit, thereby increasing the output considerably. The two finally agreed to go ahead. Long did allow that "the bane of this increased output is what has been troubling us," but thought that for those "of us who have large timber holdings . . . the time has just about gotten ripe when we should run our business the way it seems best."

Longview continued to be an exciting place, and Long delighted in watching that facility grow. He also didn't mind the decision making that was required. On one occasion, Manager Al Raught, his assistant, Harry Morgan, and purchasing agent

C. L. McPhail came to Tacoma to determine whether they should buy some of the new machinery or manufacture it in their own machine shop, at a probable saving in cost. Long, Onstad, and Titcomb met them at noon in the old Tacoma Hotel. Onstad opened the discussion, which, as McPhail recalled, "waxed hot and heavy." Long quietly ate his lunch, making no comment, listening as Onstad's position—that they should build a substantial machine shop of their own—seemed on the verge of prevailing. At that point, Long, by then finished with his meal, stood up and summarily announced, "Gentlemen, we are in the business of manufacturing lumber; we know nothing about manufacturing machinery. We will buy the machinery. Let us return to the office." And so they did.

Bell was clearly taking seriously his responsibility as president of the Timber Company. As promised, he arrived in Tacoma in early September and set out to learn as much as he could about business, recent developments, and plans, most prominently those involving Long's retirement. Long took genuine pleasure in hosting Bell. The two headed immediately down to Longview, and upon returning to Tacoma they were joined by F. E. Weyerhaeuser and Horace Irvine, providing "a large enough crowd of our directors to discuss various questions that are important." The weather was perfect, the onset of the rainy season seeming to stall until the visitors had departed. Bell summed up the general outcome: "From now on the cooperation of elective officers of the company should be counted upon to a greater and greater extent." Though no one actually said so, apparently George Long had consented to stay on into the new year. He had a big job before him—initiating his successor, Rod Titcomb, as general manager of the Weyerhaeuser Timber Company.

THIRTY

It's Time to Retire

ANY DEAL AS LARGE AS THE ONE BETWEEN THE WEYER-haeuser Timber Company and the Long-Bell Lumber Company was bound to be followed by some problems sooner or later. Thus George Long was not particularly bothered when Long-Bell called upon him to reconsider the terms of payment. Although there had been several separate timberland purchases, Long-Bell naturally preferred to consolidate the financing when it seemed advantageous to do so. The first purchase had been consummated on February 2, 1920; the second, September 1, 1920; and the third, October 1, 1927. This third purchase consolidated the debt remaining from the initial sale with the debt for the purchase of other parcels in various townships. Long-Bell had paid the purchase price on the second contract in full by means of a general mortgage bond. That may have settled the matter so far as Weyerhaeuser was concerned, but Long-Bell now was left worrying about repayment of the bond.

Long-Bell had commenced logging on the first tract immediately after purchase, and through 1928 it paid off that principal ahead of schedule. But the second tract was another matter. Though Long-Bell had not yet begun cutting that timber, the terms of the mortgage bond required the company to pay into a sinking fund annually. Long summarized the problem for F. S. Bell: "They have to pay us full price [for what they are now cutting] and in addition pay into the sinking fund considerable sums of money." This was proving difficult, and Long-Bell had come to Long with a proposition for relief. The proposition suggested a release from the third contract, with Weyerhaeuser accepting "in lieu thereof timber of equal value and quantity that we have heretofore conveyed to them" in the second trade.

Long needed Bell's counsel and, in the end, consent. He therefore inquired, "If by such a realignment we could still preserve our full security, should we entertain the proposition?" His own inclination, clearly, was to assist, even if it meant "we will be taking in less money annually for some little time to come." He would soon recommend just that, a recommendation that was accepted by the Timber Company directors.

Cooperation of a larger sort—production curtailment—was less easily managed. The efforts sputtered along. There had been a preliminary get-together of lumbermen on December 16, 1928, poorly attended and "not at all enthusiastic." They gathered again in Tacoma on January 9, where both attendance and enthusiasm were better. The group approved "a continuance of curtailment in a radical way at least until the first of April." On the surface, that seemed encouraging, but Long was not reassured. Still, he had no choice but "to subscribe to the plan," which meant, as he forewarned brother James, that Everett would fall short of supplying the requirements of its Atlantic terminals.

When George Long wasn't thinking about supply and demand and Long-Bell's problems, he could always think about the future of management in his own organization. He wanted to resign just as soon as possible. Titcomb was to be his successor. That much was settled. But Longview and Klamath Falls were due to get under way during 1929, two new operations which would present the new general manager with special challenges. As Long explained to Bell, Weyerhaeuser's activities with logging, production, and sales would then represent a "more than 100% increase over what it has been doing in the past ten years." Even if Titcomb proved a genius of a manager, he would need lots of help, and Long recommended that Charlie Ingram be transferred from Snoqualmie, "as an assistant to Mr. Titcomb to look after all manufacturing problems." Perhaps they could replace Ingram with "Tip" O'Neil, Peabody's assistant at Everett. Long also foresaw an increase in Minot Davis's responsibilities, "having him assume general charge of the logging operations, general charge of all cruising and handling most of the problems that have to do with our timber."

Not all were quite ready to accept the changes, at least those at the top. F.E. soon entered the discussion. "I trust," he wrote on January 29, "[you do not] contemplate giving up your connection with the Weyerhaeuser Timber Company. I think you will never know how great is our appreciation of the service you have given the company over these many years, and, while we fully sympathize with your desire to be relieved of some of the tremendous amount of detail work that is passing over your desk, we hope that the real direction of the affairs of the Weyerhaeuser Timber Company will still remain in your hands." Although Long appreciated the thought, it didn't change his mind. Anyway, he never intended to cut the ties completely. He couldn't. Weyerhaeuser was his life.

In that connection, what worried him most these days was production and marketing, in large measure because the Timber Company seemed to be at the mercy of others, including Weyerhaeuser's own Sales Company. In the past several years, about three-fourths of total production from the Everett plant was destined for water shipment, and the Sales Company in Spokane had little responsibility for its eventual disposal. Even with Longview operating, Long presumed that upward of 50 percent of the combined Everett-Longview output would continue to be shipped by water. Under the circumstances, Long argued in favor of locating a sales company branch office on the West Coast, "exclusively for the purpose of handling fir." But what seemed so logical from George Long's Tacoma desk seemed less so from other vantages.

On another front, the management of Klamath Falls would be different in an important way. This wasn't a fir forest, at least not primarily. It was pine. At long last, Charles Chapman was preparing to be what he had been trained to be—a forester. Specifically, he was instructed to develop a reforestation plan, one that would be in place prior to the start of Klamath's logging operations. The times were changing, ever so slowly, but changing nonetheless.

Some of the changes presented Long with a continual source of irritation. Consider, for example, lumber branding, a practice that Long still regarded with a jaundiced eye. First, there was an unfavorable federal court decision on patent rights. That, of course, could be managed. But was the branding itself effective? Long still didn't think so. In his words, "The manner of our imprint and the symbol which we use does not convey very much to the trade, as it is usually pretty badly smeared over the end of the lumber and does not give us an outstanding distinctive mark that is easily recognized." Moreover, he foresaw a problem of too many markings. The National Lumber Manufacturers Association was urging adoption of an association brand or trademark, supposedly indicating "guaranteed lumber," and regional associations had similar programs in mind. Furthermore, most lumber would carry a grade mark, resulting in so many marks "that it would not only be confusing to the retailers, but be expensive to apply, and would take on altogether very much of a Chinese puzzle to the consumer."

When Long began worrying about lumber branding, he knew it must be time for a vacation. He had been looking forward to an extended absence from the office. It was well earned, and he hadn't been feeling very strong since a bout with the flu early in the year. It was also good that Titcomb begin assuming more responsibility, something that could never be managed with his mentor around to lean on. Long had his doubts about Titcomb, but if not Rod, then who? There wasn't anybody else.

Despite the presence of Titcomb, George Long found his desk piled high when he returned to Tacoma from Arizona after nearly a month's absence. Much of the correspondence pertained to the continuing industry and association efforts to combat the depressed market conditions. Tennant, president of the West Coast Lumbermen's Association, asked if Long would continue to serve on the Trade Extension Committee. George finally consented, "with the understanding that if my plans are such that it will be impossible later on for me to serve, that I will not be looked upon as a quitter if I should have to ask you to name a successor." Colonel Greeley urged Long to attend a general meeting of West Coast lumbermen and loggers at the Winthrop Hotel in Tacoma on March 21 "to consider the vital facts as to production, inventories and market conditions in the West Coast lumber industry for 1929 to date."

The popular solution proposed at the meeting once again was curtailment. It was decided, finally, to limit production until June 1. Night-shift manufacturers were instructed to reduce by a quarter of normal production, and those operating a single shift were to reduce by a sixth. Long considered this to be a "rather unfair division" for the night-shift plants, but given the circumstances, the market would probably "go all to pieces if we could not announce to the world that curtailment of production was still the program." And so they had consented. But as always, it was more easily announced than accomplished.

Despite his sincere promise to comply, Long, pressured by Peabody at Everett and his brother at Baltimore, soon felt compelled to seek some slight relief from the association's curtailment formula. Central to Weyerhaeuser's problems was its need to supply the Atlantic terminals; they had to manufacture lumber now in order to fill orders months down the road. "The question is simply this," Long explained to Greeley and Tennant, "what would be the effect if we were to turn our mills over to a six day per week production during the months of April and May and then agree to pull down the production in June to the level of our allotment by closing down the mills for a specific time, and in going to this program announce and have it understood that we will later on live up to our schedule of the amount of reduction?"

The responses were as expected. Tennant sympathized, but stated what Long understood all too well: "The very fact that your company, as well as our own, is looked upon to take the lead in matters of this kind regardless of the hardship that it may work on our respective operations, causes the other operators to expect us to do it." Greeley's message was the same: "It would be far preferable if the plants at Everett could continue on the course previously adopted; and I hope very much that this can be done."

Long wasn't going to insist on special treatment for Everett. He had tried to explain Weyerhaeuser's particular problems with compliance, but he wasn't willing to go beyond that. A major factor in his reluctance to press the issue, beyond his awareness that success depended on a collective approach, was Weyerhaeuser's expansion at Klamath Falls and Longview. The last thing Long wanted was to call attention to the new plant. "A great many shafts have already reached me," he admitted, "about the unusual interest I am taking in trying to get other people to reduce production and at the same time go ahead and build three big saw mills which will go into production in 1929." Further in this vein, he mentioned that following the March 21 meeting, "in which I very strongly urged the reduction of output, the evening papers announced that the Clearwater Timber Company had just put on a night shift and that the Potlatch Company [Weyerhaeuser-affiliated operations in Idaho] expected to do the same thing in a few weeks." At the very least, it was embarrassing.

As annual meeting time approached, Long naturally began to think about the sorts of questions most likely to be discussed. One that nearly always arose was the size of the dividend, and with Bell now serving as the president, this took on added significance. Bell would certainly come down on the side of distribution of earnings, a fact that he soon made abundantly clear. Responding to Long's expressed concern over amounts then being expended, particularly at Longview and Klamath Falls, and the likelihood that part of their investment in government securities would be needed, Bell saw nothing "sacred about the accumulation which you have made in government bonds and certificates," adding, "I believe the accumulation lay in your mind in the first place as a safety fund against needing money for just such things as the Longview and Klamath Falls operations; and now in your mind the accumulation lies as something which must not be disturbed." Bell recommended that a 10 percent dividend be made prior to the annual meeting, to be followed by another of like size "very soon after that." "You know I have always had a tendency to consider those among our stockholders who are not interested so much in the long-range policy of the company as in getting some money out of it as they go along." Yes, Long knew that.

As for other agenda items, Bell promised to bring up the matter of preferred stock again, still convinced that "we ought to do something to relieve the stockholder who is under pressure from the necessity of selling his present stock at a sacrifice, or perhaps give him something which he can hypothecate, when needed, without writing his name on the back of his Timber Company stock." He also forewarned Long of plans to arrive in Tacoma several days in advance of the meeting so as to visit as many of the properties as possible. Long looked forward to that; he enjoyed nothing more than showing visitors around.

Among those Long hoped to see in Tacoma was lawyer

Gus Clapp. For one thing, he thought Clapp might have a copy of what had been discussed the previous fall regarding bonuses, another item that Bell wanted on the May 30 agenda. In general, what had been proposed was that operating profits from the manufacture and sale of lumber, to the extent that they exceeded 5 percent annually "upon the investment in the operation," would be available for distribution to selected salaried employees. Each branch would be treated separately—Everett, Longview, and Klamath Falls; all of the distributing yards; and the four steamship companies. If Everett made a profit, the Everett employees would share in the earnings, even while their colleagues at Klamath Falls endured a poor year. As was stated in the "Purpose," the company benefited from "a number of salaried employees of more than usual ability and efficiency," and it was desirable that these "should not only have such incentive as can reasonably be furnished to give their best efforts in the company's interests, but that they should be made to feel that successful efforts reap commensurate rewards."

Normally, common stock would be available, but that wasn't true with the Weyerhaeuser Timber Company. So alternatives had been considered, and the profit-sharing (bonus) plan was the result. Clapp took more than a lawyer/consultant's interest in the proposal, and was disappointed that he was unable to attend the annual meeting "to sit in conferences in this matter and have my share in whipping some plan into shape." He hoped the thing would be approved, "that in treating with tried and true employees, enlightened selfishness loosens the purse-strings." Long was subsequently delighted to report that Clapp's plan was "adopted in toto."

Another item on which Clapp had been working was an amendment to the Timber Company by-laws creating an Executive Committee, to be composed of F. S. Bell, F. E. Weyerhaeuser, George S. Long, Horace Irvine, and Bill McCormick. The purpose was obvious: to assist Rod Titcomb following Long's resignation.

This was to be a very different annual meeting. A nervous Titcomb became the first other than Long to present the report. But the words were, as before, Long's. He noted the continuing depression in the industry: "The last year when there was real prosperity in the Northwest was the year, 1923, when all conditions seemed favorable, and since 1923 the trend has been downward." He also noted the effort to curtail production: "It was the unanimous thought of all that one possible remedy was to make less lumber." But "unanimous thought" had not been translated into unanimity of action. There had been some reduction, and prices had "advanced by a slow steady process at least $1.50 per thousand, but even at that, there was no profit in the business [other than] on the part of some of the mills owning their own timber and having economic operations." Weyerhaeuser was one of these, reporting net earnings for 1928 of more than $3.5 million, with an additional $570,000 net profit at Snoqualmie Falls. What lay ahead? How could they be optimistic in the face of disappointing returns during a period in which the nation generally enjoyed prosperity?

"Better utilization of by-products is apparently one way in which we can improve our conditions and a continually closer revisement of our products will help quite a little," Titcomb read. "But after all, two basic facts stand out very plainly, one of which is that our timber . . . is 2,000 miles from the center of population, making expensive costs for delivery and high prices to the retailer and to the consumer. The other basic fact is the constantly increasing use of substitutes, which to a large degree can be produced near the location where the material is used."

But the subject of greatest interest involved management changes. Titcomb was officially elected general manager, with Ingram as his assistant, and Davis continuing as manager of the timber and logging departments. Long would be a vice-president of the company and chairman of the Executive Committee. Gus Clapp subsequently assessed the effects of these changes:

> From the practical standpoint, if you will just remember the powers which Mr. Long has exercised, either by direct delegation of authority or by tacit consent, I think that you will appreciate [the power of] the newly constituted Executive Committee. . . . It is all nonsense to say that the Board of Directors cannot delegate authority; they can and have delegated authority many times to the officers or to a single offi-

cer such as Mr. Long. There is no reason why authority should not be delegated to an Executive Committee. An Executive Committee probably would not have been provided for at the present time were it not for the fact of Mr. Long's gradual withdrawal from executive supervision of the company's affairs, and the realization that this meant the advisability perhaps for all time in the future of a small committee which could consider, pass upon and carry out policies and plans and transactions which in the past have been largely, although not entirely, left to Mr. Long.

Change was in the air, but life at Weyerhaeuser still featured George Long. Indeed, to most in Tacoma it must have seemed that nothing was different, that it was business as usual, with Long coming to the office as before and dealing with many of those old enduring problems.

The lumber business revived somewhat in the spring of 1929, but not everywhere. In the Pacific Northwest, log prices were at least $1 higher than a year earlier, with lumber for rail shipment averaging between $1 and $2 over 1928 prices. At the same time, hemlock was "getting a terrific wallop, there being a surplus of hemlock logs," and prices at the Atlantic terminals remained generally soft. Still, Long advised brother Jim to meet the market in a conservative way. In other words, they should not "be the leaders in making any decline," but it would "be folly not to keep our lumber moving at the very best possible price you can get for it."

Meanwhile, the curtailment efforts dragged on, to the dissatisfaction of many, not the least of whom were the employees. Working five days a week reduced their wages, which was distressing in itself, but worse was watching others, occasionally within their own organization, continue on a six-day schedule. For example, at Snoqualmie Falls, the sawmill crew was limited to a five-day week, while those in the planing mill and loading crews worked six days. Long had no ready solution to this problem, although he allowed that they could probably increase the hourly rate by some figure, or simply establish the five-day week as basic, with pay equivalent to six days. What he hoped was that the industry would soon sort itself out, that there would be fewer manufacturers competing. But what he hoped wasn't what he was predicting. In fact, he foresaw a "much demoralized lumber market."

Not unexpectedly, considering labor's complaints, the old 4-L leadership—Colonel Disque's World War I brainchild—was being challenged by new organizers. Long called attention to recent developments in La Grande, Oregon, where the American Federation of Labor was actively recruiting: "The general feeling over here is it would be very unfortunate if the lumber industry and its workers should become a part of the A. F. of L." Long knew only that action was required, and soon, regarding wages, and at the moment there seemed no alternative to the 4-L. As he explained the matter to Long-Bell's president, John Tennant, "[Since] we have so generally decided to continue the curtailment program . . . I believe if something is not done about the wage question soon we are stirring up trouble that may be very annoying." He then suggested that immediately following the July Fourth shutdown, Long-Bell and Weyerhaeuser announce a new wage schedule, "and if at the same time we will feel inclined to join the 4-L movement with our two companies the combination of these two events would have a very healthy and stimulating effect."

But Long wasn't ready to make any promises to W. C. Ruegnitz, 4-L's leader. Indeed, when he considered the matter privately, he acknowledged that the "company union," as its detractors labeled it, had become an anachronism, if not an albatross. Thus, in responding to 4-L calls for support, he equivocated, noting that Reugnitz's "conclusions should command every thoughtful man's serious consideration." Then he added, "At the same time many of us who are supposed to sit in the seats of the mighty, are not always in the position to make the final determination on various lines of policy." The lines of battle weren't even clear.

In the meantime, the merger to end all mergers was again being discussed. Heading the effort was William E. Boeing, already building airplanes, but not yet famous for it. Boeing's proposal had been mentioned at the Weyerhaeuser annual meeting, but it had aroused little interest. Central to the scheme was inclusion of a number of major producers, including Merrill & Ring, Bloedel-Donovan, the St. Paul & Tacoma Lumber

Company, the Simpson Logging Company, Weyerhaeuser, and a good many others. Boeing envisioned a gigantic consolidated company that would purchase the several properties, "making of each transaction a business one where the values would be mutually agreeable." The new company might also buy other properties whenever such purchases would "strengthen its position." Gradually, it would assume "the management of the business including manufacturing, selling, etc."

Boeing and a friend, Valentine H. May, visited Long in his Tacoma office on June 19, and, as Long noted, "It so happened that Mr. [William] Carson, who was passing through Tacoma, remained over at my request to be present at the interview with Mr. Boeing." Carson had been one of the early merger enthusiasts, but it was Long who sat as judge and jury, and, as he subsequently reported to F.E., he wasn't persuaded: "I had to say to Mr. Boeing that I thought that the first important thing was to definitely get all of these various properties . . . in the ownership of the new proposed company and have no uncertainty about this, for I could imagine a situation where the prominent men connected with each of these companies might be willing to take stock in the new company, but might hesitate selling their properties to the new company . . . [that one or another] might think that he might fare better in timber and logging facilities to keep it under his own personal ownership and control—I still think so very firmly." Boeing, however, seemed unworried, apparently convinced that once the individuals were in the new organization, they would naturally bring their properties with them.

F.E. fully agreed with Long. Only if Boeing first got "Merrill & Ring, Bloedel, Butler and the other successful operators into agreement as to the value of their properties" could they hope to proceed, and F.E. added, "I do not believe Mr. Boeing nor anyone else can buy these properties at a reasonable price after the cash for purchase has been provided."

The merger mania seemed to spread to all things. Some, for example, were suggesting that the three Pacific Coast associations should become one, but Long felt otherwise. "My thought," he countered, "is that I would rather have three associations working intensely on their particular problems and thus enlisting the earnest zeal of its membership." He held a similar view toward a national lumber program: "I think the whole world is now to a certain extent oversold on the idea of combinations and mergers, some of them are worth while and some of them I think [are] fabrics put together to enable people to sell stock at a profit instead of running the business on a more scientific basis."

The forestry committee of the West Coast Lumbermen's Association met as scheduled in Tacoma on July 2. Colonel Greeley made a presentation, noting the likelihood "of renewed discussion and agitation of further national legislation dealing with forestry," and suggested that it would be preferable were the lumber industry to put forward "a constructive program of its own rather than be placed in a defensive position under public attack and criticism." No strategic decision was made, but the members agreed "to formulate a statement" and then leave "it to Mr. Long's judgment as to how such a statement should be placed before the industry."

The statement adopted was, by and large, the one proposed by Greeley. It included reference to promoting "industrial forestry to the fullest extent that the economic conditions controlling its operations will permit," and it specifically recommended two practices that deserved special emphasis and study: "First, the possibility of operating existing stands of timber on the basis of selective logging, whether by size of trees, species, timber types or areas classified with respect to accessibility and logging costs, with a view to harvesting such timber in the order of its actual economic value, and reserving for future utilization the portions of such stands whose present cutting actually yields a low or negative return to the operator. Second, the practicability in many cases of organizing forest properties and manufacturing plants on a basis of sustained yield in cutting and manufacture, either in lumber or other forest products, or as nearly a sustained yield basis of cutting as local conditions will permit."

On the subject of curtailment, Long was in a defensive position, forced there by the paradox of Weyerhaeuser's current expansion in the face of soft market conditions. Several again took note of the unfortunate timing of the start-up of the Weyerhaeuser operation at Longview. The comments of W. B. Nettleton of Seattle's Nettleton Lumber Company were typical: "We cannot conscientiously congratulate any one starting up a new sawmill under the present conditions, but if new plants must be started we prefer that they be operated under the direction of the Weyerhaeuser Timber Company rather than any one else."

There was no lack of coverage of the start-up. The *West Coast Lumberman,* for example, devoted an entire thirty-two-page illustrated article to the event, which its editor described as "the most complete and comprehensive sawmill feature ever published." In that issue, Long was given opportunity to defend Weyerhaeuser development at both Longview and Klamath Falls. In his essay, "The Ideal," he emphasized the responsibility of lumbermen to use the resource to the best possible advantage: "The great God-given Western forests are too valuable to be whittled away." The point, of course, was that these new operations were the result of careful and farsighted planning: "There is a particular purpose for which every piece of lumber is best adapted and it will be our aim to find what this use is and cut our timber accordingly."

Other trade journals used the occasion to announce the recent personnel changes at Weyerhaeuser. Sounding much like a eulogy, the *Mississippi Valley Lumberman* article of July 12 detailed Long's professional career: "Mr. Long has been identified with the promulgation of ideas which have been adopted on an important scale by the industry." These included "deep water transportation of rough lumber" to Atlantic Coast terminals, and "fire protection of timber under the auspices of producers associations or by private enterprise when that is required, reforestation on a commercial basis by private interests and the intensive utilization of lumber through the development of by products."

No longer general manager, Long was trying hard to slow down, but he couldn't seem to. The fact was that work had been his whole world. Perhaps he arrived at the office a few minutes later, and didn't always hurry back from lunch, but the changes were hardly noticeable.

Even the newly formulated Executive Committee brought little easing of Long's responsibilities. Bell had hoped to visit Tacoma in August to hold the first meeting of the committee. But when other business kept him in Winona, he asked Long to go ahead without him, taking "action upon any particular matters which you think should have attention." As a matter of fact, Long had nothing on the agenda, and suggested that they simply postpone their meeting until late September or early October, at which time they would probably have business worth considering.

Long continued his preoccupation with the activities of the West Coast Lumbermen's Association and the National Lumber Manufacturers Association. After reminding Bell that they were contributing liberally to these organizations, Long observed that each of them needed the "best brains in the country to direct their affairs, in other words, they need our direction and supervision almost as much, if not more, than they need our liberal contributions." In addition, they should be paying attention to results. "Too often," he noted, "we follow the line of least resistance in many worthwhile endeavors, finding it easier to contribute money than to contribute high grade effort."

The start of Weyerhaeuser's Longview operation had preceded the association meetings by a matter of days, and Long soon was forced to deal with Greeley in an entirely different capacity. As West Coast Association secretary, Greeley was the watchdog for the curtailment program, such as it was, and his office was the depository of facts and figures. Long soon provided the production data for the new facility, which, he explained, was "built exclusively for double shift mills and has a rating capacity of 1,200,000 feet as two shifts." But, in the light of current circumstances, they had decided they "should do even more than our own part in the curtailment program for the remainder of this year and so have instructed our Manager at Longview to make no arrangements for night shifts . . . but in lieu of that . . . run one shift six days per week for the remainder of this year which will give us a production of 600,000 feet per day." Further, as everyone must understand, it would be several weeks before the new mills reached their capacity. And finally, such a program was in line with the six-day policy of the Long-Bell Company, and would thus "not give rise to two conflicting situations in our own bailiwick." Greeley had to wish that things were as they had been—without Weyerhaeuser at Longview—but all he could do was agree and offer congratulations.

Even without the new Longview mills, there was already more lumber than the market could absorb. Another general meeting was called, at which a joint announcement was signed by practically every prominent lumberman in the Northwest. The announcement noted the "unmistakable trends" pointing toward "a slump that might easily prove to be one of the most disastrous in the history of the industry." Every timber owner, logger, and lumber manufacturer was urged to attend a September 5 gathering in Tacoma to discuss "the welfare of the labor which we employ, and the interest of the communities where our properties and operations are located."

But a general admission of difficulties mattered only if a solution was at hand, and in this instance it wasn't. The problem of overproduction seemed beyond solution, for, as one lumberman observed, "If five operations the size of Long-Bell's should shut down entirely, it would not accomplish necessary results."

Moreover, the industry was witnessing a changing of the guard. Pacific Spruce's president C. D. Johnson summarized the situation: "Everett Griggs is not in good health and has resigned as a trustee in the West Coast Lumbermen's Association, and he is compelled to get away from his own and all other business for a period. George Long announced that he would be off the job in three months. John Tennant is not in first-class health and has been trying to get away from the business, and he announced that somebody would have to be selected to take his place for a time. Here are three very important men without whose assistance, hard work, advice and influence the lumber industry must face calamity unless their successors are as capable and as worthy of the confidence of the industry as were their predecessors."

Long, however, wasn't retired just yet. Rather, he was preparing to receive the newly constituted Executive Committee for its initial meeting in his own office. The results were unimpressive, reflecting the spirit of the times. As Long subsequently described it, "There is somewhat of a deadly atmosphere around the lumber industry." For example, they decided against any new developments because uncertainty concerning both the national economy and Weyerhaeuser management. The consensus was that they should "market that which we now have without attempting to take on any greater loads."

Long's annual eastern tour was delayed until late October, due to "a very severe cold which hardly warranted taking the risk of adding to it until I had it under control." His itinerary was more varied than usual, with a couple of weeks planned for the

Atlantic Coast, then an Executive Committee meeting in St. Paul on November 4 and 5, after which he would proceed to Missouri for a few days, then on to Arizona, finally arriving in Los Angeles for a meeting of the West Coast Lumber Trade Committee on November 12. On the final leg of the journey, he hoped to stop off in San Francisco "to attend a hearing of the Interstate Commerce Commission on the petition of the Great Northern and the Western Pacific to make certain extensions and connections," in the Klamath Falls district. Long boarded the train on October 23, the day before Wall Street's Black Thursday.

Stock prices had been sliding for more than a month, but few observers expected the frantic selloff that occurred on October 24, and that was only the beginning of the fall. Long made his tour, but clearly the old enthusiasm was spent. He did have the unexpected opportunity of attending a Washington, D.C., conference sponsored by the American Forestry Association. The McNary Bill was the subject under scrutiny, "more liberal appropriations" being the objective. President Hoover was among those in attendance. Naturally, Long studied Hoover's appearance and felt sympathy. There "was some little thought or somewhat of an atmosphere," he later reported, "that the President was being overworked more or less by just such conferences as we asked and that he had many other heavy duties on his hands which probably outclassed the issue which we called to his attention." The Great Depression would furnish a different perspective on everything and everybody.

But like most of his countrymen in the fall of 1929, Long wasn't thinking about throwing in the towel. Rather, he held to the belief that at heart the economy was sound and that "any healthy, well managed business operation" would survive. He preferred to note that they had "many reasons to be thankful," admitting that this "remark applies to a whole lot of other things right now besides logs and lumber." One lesson that had to be relearned "every once in a while, [was] that you cannot lift yourself very high by pulling on the straps of your boots—in other words—that you cannot get rich by mere speculation and in order to get money and have money, you have to make it by some kind of actual production instead of making it altogether with pen and ink, or to put it in modern language, dabbling in options on Wall Street."

THIRTY-ONE

Life's Western Slope

GEORGE S. LONG SPENT THE FIRST FEW DAYS OF 1930 cleaning up odds and ends prior to a leisurely drive southward with his daughters. Their destination was Phoenix, and how long they would stay was undecided. Long had rented a house for two months, so they might be away until the middle of March. Then again, George Long might tire of being a tourist.

He couldn't have timed his escape any better, for almost as soon as they left Tacoma, winter arrived with a vengeance. On January 21, Longview recorded a temperature of $-19°F$, the mills having been shut down several days previous, with twenty inches of snow covering the ground. In the woods, more than two feet had accumulated, so logging was also halted. The Everett mills closed when the city water supply froze. In his monthly report to F.E., Titcomb told of the "extreme difficulties" at Klamath Falls, where temperatures dropped as low as $-40°F$. But as soon as the weather moderated he expected Klamath "to be going like an old organization."

Faint word of the Pacific Northwest's winter reached the ears of George Long, enough to make him feel smug under the blue skies of Arizona. He worried briefly about ice damaging the booms at Longview, but had no advice for Titcomb. "Am too wise now to offer any suggestions as to how the situation should be handled," he wrote, adding, "The weather is very delightful here, bright and clear with temperature ranging from about 35 to 65 daily."

It was a real vacation, one of the few that Long had ever enjoyed, but there was still a bit of work to be done. At Greeley's urging, Long agreed to write a paper for the January

31 annual meeting of the West Coast Lumbermen's Association. The subject was of great concern: "The Immediate Future of Hemlock." Long barely completed his assignment in time, Greeley receiving a copy on January 28. The message was hardly new, at least to those who had spent any time in Long's company. Simply put, since hemlock was such a prominent species in the western forests, accounting for possibly one-third of the timber, they had to find a profitable use for it. Long's paper was his only contribution, for he skipped the meeting and stayed in Phoenix. He did take a break in early February to attend a confab in San Francisco, the subject of which was the old Hewes merger plan. Once again, nothing came of it.

The San Francisco trip proved something of an ordeal for Long. He picked up a nasty bug, which he found difficult to defeat. He was, however, helped by word of honors coming his way. Long had been "greatly missed" at the February meeting of the West Coast Lumbermen's Association trustees, as Greeley reported, but he wasn't forgotten. "By unanimous action," Long was elected an honorary trustee, Greeley observing, "This is but a very feeble and inadequate way of expressing the regard in which you are held universally by the West Coast lumbermen, as well as our desire to still have your valued counsel from time to time when it may be possible for you to meet with us."

The Washington Forest Fire Association also honored him at its annual meeting on March 11:

> RESOLVED, That the Members of the Washington Forest Fire Association received with deep and sincere regret the resignation of Mr. George S. Long as President and Trustee of the Association.
>
> The founding and direction of the Association is in the largest degree the result of Mr. Long's genius for organization and its effective and successful policies the result of his extraordinary abilities.
>
> The absence of his voice from our counsels is a serious loss, but it is hoped that as Honorary Trustee he will continue to join in our deliberations and continue to guide this organization as he has so ably done in the past.

While Long's leadership and wise counsel would indeed be missed in many forums, without question the place where he was to be missed most desperately was the Tacoma office of the Weyerhaeuser Timber Company. Even under the best circumstances, trying to fill the shoes of a living legend would have been difficult, but the circumstances in the lumber industry in early 1930 were far from the best. Though Rod Titcomb had ability, experience, and ambition, and tried his best to do as Long would have done, the results were not the same.

A big part of Titcomb's problem was his style. Long had always seemed comfortable and at ease; in comparison, Titcomb was a bundle of nerves. Long had assumed the best and offered encouragement, always finding something to compliment while making his point or offering his suggestion. "I've been thinking about our dilemma," he might write, "and am wondering if we might not benefit by trying this approach." Conversely, Titcomb went immediately to what he saw as the core of the problem, without any apparent awareness of an individual's feelings. Those with whom he worked could hardly be blamed for shaking their heads and remembering how things used to be.

The mere existence of an active Executive Committee was evidence that Titcomb wasn't to be Long's replacement in fact—that whatever he did was certain to be thoroughly scrutinized. A meeting of that Executive Committee lured Long from Phoenix in early March. The meeting convened in Los Angeles, continuing in session for four days. There was much to consider. Committee members Long, F. S. Bell, F. E. Weyerhaeuser, and Bill McCormick invited Gus Clapp to join in the discussions. Rod Titcomb and Minot Davis were also present for the first three days.

The old Pacific Northwest merger effort occupied most of their time. A delegation consisting of J. D. Tennant, S. M. Hauptman, R. A. Long, C. D. Johnson, and E. G. Griggs had continued to champion a plan for a new corporation aimed at acquiring and operating certain key plants and timber properties, prominent among which were those owned by Weyerhaeuser. Thus, the Timber Company Executive Committee once more discussed the proposal to convey selected tracts of timber at the present market value to the super corporation, accepting in exchange nonparticipating shares of stock redeemable at par, without interest, as the timber was removed. In the end, that proposal was summarily rejected. According to the minutes, "No satisfactory plan was arrived at, and conferences were adjourned, but the members of the Executive Committee expressed their willingness to give some further consideration to the subject." The truth was, however, that for Weyerhaeuser, the subject was dead.

Long must have felt like the proverbial old fire horse as annual meeting time approached. He knew he didn't have to answer the bell, but habit told him otherwise. And so he prepared, almost happy in the anticipation, even though the news was anything but encouraging. Actually, the 1929 figures were better than most expected, with a profit of nearly $5 million. This, however, was deceiving, a result of "several timber transactions showing profit over March 1, 1913, valuations . . . coupled with the rather satisfactory returns of all of our logging companies." For the most part, "the situation was very bad and drifting worse day by day, with no immediate prospects for an improvement." Further, any hopes that the curtailment program would be maintained were fast fading. Measured by capacity, only about 60 percent of producers had ever "made constructive and serious efforts" along those lines, and now it seemed likely that any reduced production would be the result of closures due to economic distress rather than curtailment policy.

Despite the disappointing business news, Long found himself eagerly awaiting the arrival of his friends for the annual meeting. He always enjoyed playing the host, and now, with Longview and Klamath Falls fully operational, there was more to see than ever before. Touring might brighten an otherwise gloomy picture, Long fearing that by the annual meeting time they would probably be "in the midst of some price levels that will indicate that there won't be any money in the business year 1930, even for the Weyerhaeuser Timber Company."

Fortunately, there were other subjects to consider. Mason & Stevens, the Portland-based forestry consulting firm, had recently published a report on sustained yield for the Clearwater Timber Company. Phil Weyerhaeuser enthusiastically accepted its recommendations and shared Clearwater's plans with others, including brother F.K. and uncle F.E. Both were converted; the report convinced them that programs similar to Clearwater's merited consideration by other companies, including, of course, the Weyerhaeuser Timber Company.

George Long wanted to believe in sustained yield, and he assured F.K. that Dave Mason would be invited to a session of the Executive Committee meeting, where he could "tell the story which so impressed you and others." Further, Long suggested that Mason's presentation include "the broader question of contributing efforts that will result in the government and the timber owners working harmoniously on some practical line, having to do with the idea of perpetual yield basis, which of course calls for cooperation of both the [federal] government, the state and personal ownership."

Long thought the general discussion might be more beneficial. He wondered whether the Weyerhaeusers, particularly F.E. and F.K., appreciated how different the circumstances of Pacific Northwest logging were from those of the Inland Empire. The forest densities, for example, varied greatly, with "four or five times as much growth on the [fir country] lands as exists in western pine." Then too, their methods were not the same. High-lead logging—dragging logs to a central staging area by means of cables running from a tall "spar" tree—was the most economical method in western Washington, "which necessitates the cutting down of practically all the timber in the area which is to be logged before the actual hauling commences." For if the logging didn't get the leftover trees, the slash fires would. Long reminded everyone that the Timber Company had "excellent personnel to look after this matter in our own ranks and they are all keyed up, not only to do the most effective job that is possible, but to glean additional information from time to time by actual experience." Long's observations were on the mark. After all, he had been studying these matters for what seemed a lifetime.

The annual meeting was convened on May 30, and for the second year, Rod Titcomb read excerpts from the report. This time, however, the words were his own, not George Long's. Titcomb's sentences were a good deal shorter than Long's, and his vocabulary less quaint, but his message contained few surprises. "As a whole our industry is not in an enviable position. It is faced with declining markets and the substitution of other materials. Closer utilization of our logs, and the resultant by-products, is going to be a big help in solving our problem. Again, better merchandizing is essential and this entails not alone the selling of the lumber, but also its preparation."

Perhaps the most telling event of that day was not the meeting itself, nor anything else occurring in the Tacoma offices of Weyerhaeuser. Rather it took place that evening at Long's Prospect Hill home, when John Tennant of the Long-Bell Company paid George Long a visit. This appearance underscored the fact that hard times were beginning to take their toll; Tennant requested that his company be allowed to skip its October 1 payment of $765,112, making it instead "the last payment under the contract." They would, he assured, pay the interest due on October 1. Long indicated that he was personally "agreeable to the idea, providing it could be done without disturbing any legal phase of the contract," and he promised to bring the question before the Executive Committee. The very next evening, with the committee en route to Klamath Falls, the matter was considered, "and without very much of any hesitation decided it was the proper thing for us to do."

The debate on selective logging, introduced by David Mason's presentation, continued as the directors traveled about. In the course of these discussions, mostly informal, it became increasingly clear that the so-called progressive position was anything but. Hard times meant lower prices, and lower prices affected choices in the woods. As Long put it: "What is the use of bringing any log out of the woods on which you cannot make any money at all?" Given the current conditions, choosing logs that could bring a profit was the only selective logging that made sense.

Applied to the Weyerhaeuser Timber Company, selective logging would likely "prevent us from cutting hemlock and certainly will keep all the cull logs in the woods." Long had major doubts about the long-range wisdom of this sort of approach. "At the present time," he noted, they were "looking through glasses that are rather distorted with the present condition of the lumber market." And there was an additional consideration. As he advised Peabody, "Years have been spent in an organization trying to save everything. That atmosphere is worth money to us and I would not throw it away even in this dark hour by not saving your edgings and not trying to revamp your lumber so as to sell it, even if it is done at a loss, because times may change and it is hard to be fickle in teaching your employees to be very saving and careful one day and the next to encourage them to be wasteful."

No sooner had his guests departed than Long prepared to go to St. Paul for another Executive Committee meeting. The first item on the agenda was of great importance: "Whether or not we will build a pulp mill at Longview." Because everyone was likely to be on hand, Long encouraged Carl Hamilton to call a meeting of the Executive Committee of Weyerhaeuser Forest Products at the same time. Assuming that would happen, he wrote to his three mill managers, Tip O'Neil at Snoqualmie Falls, Bill Peabody at Everett, and Al Raught at Longview. He wanted to know whether "in our 4-Square packages," they were "eliminating from our regular grades any considerable quantity of stock which ordinarily would be admissable in the grade as bundled and as we ship it to the outside trade."

The responses were nearly identical. Peabody indicated that they had tried to include everything on grade, but in practice, "there have been certain defects which from the very nature of the 4-Square idea has prohibited its inclusion." For example, they could not include pieces having "even a slight curve in the length of the piece, as a crooked piece in a bundle will not permit of the 4-Square cap being put on." Also, pieces slightly short in length "which would without question be accepted in our regular bundles have also had to be rejected, and you will note that these rejects on account of length amount to .406% of our rejects." In sum, and just as Long had predicted, 4-Square was making an unprofitable difference. Viewed in another context, Weyerhaeuser had met industry standards "if our West Coast grading inspections showed grades to be above 95% on grade," but 4-Square had eliminated the 5 percent tolerance.

The Executive Committee met in the St. Paul offices on Monday morning, July 30. All members were present plus numerous others. William H. Kenety, manager of the Northwest Paper Company of Cloquet, offered a report on a possible Longview pulp mill, and general discussion followed. That same afternoon, the committee authorized Titcomb "to have plans prepared and to proceed at once with the construction of a 150 ton bleached sulphite pulp mill at Longview, Washington."

The next morning, O. C. Schoenwerk, one of the architects competing for the job, made his presentation. He was subsequently awarded the contract to design and construct the pulp mill. The final item on the agenda involved a change in the capital structure, and, after considerable discussion, it was agreed to recommend to the board that the capital stock of the Weyerhaeuser Timber Company be split and converted into 750,000 shares of common stock without par value.

St. Paul's heat and humidity made for a most uncomfortable stay, in keeping with the ongoing debate relating to 4-Square. But such matters soon lessened in significance. Long was called to Indianapolis to attend the funeral of brother-in-law Charles L. Kellogg, and then George's sister, Eva Kellogg, died on July 8. As a result of these sad events, Long missed most of the St. Paul deliberations, including the chance to vote on a resolution favoring continuance of the 4-Square program. It was just as well.

William Carson subsequently provided the details: "You will see by the minutes that I was as active as any one in passing resolutions, believing that we have undertaken the 4-Square end of the game and that the only thing for us to do is to give it a fair chance." But even Carson allowed that perhaps they had gone a bit too far: "I am not backing up on anything I did at the meeting, for my wish was and is to promote 4-Square in every way possible, but we may go to excess; cost of various items such as C grade and many others may make our lumber unprofitable. Intensive salesmanship, in my opinion, is all right, but do you not think we are in danger of having the tail wag the dog?" Long surely agreed with that, and he would have gone further.

George Long gathered some additional 4-Square ammunition from old friend Horace Rand, president of the Burlington (Iowa) Lumber Company, ammunition that was no longer useful. Rand cited a recent conversation with William Carson, Burlington's vice-president, regarding 4-Square. In Rand's words, "We have five retail yards here and we have never had an inquiry for 4-Square lumber so far." He added, "Personally, I do not want to have it handled in our yards, as species and grade marking are very detrimental to us in handling lumber in competition with others." There was another reason: Rand suspected that any plant that produced 4-Square lumber inevitably hurt their regular grade stock, "and I would not let any of our yards buy regular grade from any mill that was taking out the best of their lumber to make this special stock." Others, he suggested, felt the same way.

Long appreciated the thoughts, but he also wanted to set the record straight. He assured Rand that he had been sensitive to the possibility of lowering regular grades, "and I have reasonable hope to believe that it will not occur to any marked or noticeable degree." Weyerhaeuser Timber Company needed midwestern customers, 4-Square or not.

But happier prospects were at hand. Long was in the midst of making plans for a short trip to Klamath Falls. He notified Jack Kimball of the details, that he would leave home in the company of his daughters on Monday, July 28, "and in a leisurely way drive down . . . stopping off at Crater Lake and probably reaching Klamath Falls on Wednesday or Thursday, at which time I will hope to spend at least a couple of days at Klamath Falls and vicinity and have the pleasure and benefit of an all around discussion of various matters."

A few items of unfinished business needed to be addressed before departing, one of which was somewhat sensitive. It involved what could be interpreted as direct disagreement between Long and F. K. Weyerhaeuser. F.K. had phoned the Tacoma office, reporting that those in the Sales Company office believed that in terms of fir-sales policy they should, as Long remembered it, "let down the bars and go after business and I presume the proper interpretation of this would mean regardless of price." For starters, Long wasn't convinced there was any market at all—"regardless of price." But there was another critical reason for avoiding such a unilateral approach. Just an hour prior to F.K.'s phone call, Long had been conferring with Bill Greeley, discussing the colonel's efforts to encourage firm prices throughout the industry. Long saw no purpose in this and respectfully declined to participate. Now, however, "if upon the heels of having declined to join in the movement of the firm price policy, that immediately thereafter we turned loose," then the logical conclusion "would be that we were trying to smash the market all to pieces and likewise to smash many individuals who were engaged in the lumber business." Accordingly, Long cautioned F.K. to go slowly.

This would be one of Long's last letters, and in a way it was fitting that it concerned Weyerhaeuser's responsibility within the industry. Perhaps it was also appropriate that death came to him at Klamath Falls, his beautiful and very special dream. Late in the afternoon of Saturday, August 2, George S. Long suffered a massive heart attack in his room at the Hotel Willard. A few minutes later, with Helen and Margaret holding his hands, he died. The train carrying his body arrived back at Tacoma's Union Station Sunday evening, met by sad friends, family, and the entire Tacoma Company office staff.

Others read the news on the front page of Monday's *Tacoma News Tribune.* "The lumber industry of the United States," the piece began, "is in mourning for George S. Long, for many years general manager of the Weyerhaeuser Timber Company and known from coast to coast as the dean of the American lumber industry." Letters of condolence and tributes deluged the family and the Tacoma office. F. E. Weyerhaeuser wired George, Jr., prior to heading west for the funeral, "He has been a tower of strength for all of us because of his wisdom, his strength of character, and his appealing personality which have won the admiration and confidence of the whole lumber fraternity, but we shall miss him most because of our love for him."

E. T. Allen lost "the best friend I ever had." That was the consensus—loss of a great business colleague but, more important, loss of a dear friend. The Logger's Information Association resolution observed, "He went to his long sleep, serene in the knowledge that his well-earned rest would never be disturbed by any dreams of unfairness to his fellowman." And the West Coast Lumbermen's Association expressed similar thoughts: "In nearly every critical juncture or important development affecting the forest industries of the Northwest during the last quarter century, Mr. Long's wisdom, breadth of vision, and understanding of public as well as industrial needs have been of the utmost benefit to the lumber industry and the nation. Mr. Long was much more than a lumberman of outstanding vision and ability. His friendly, personal interest in people and sympathetic understanding of human affairs made him a most valued friend and counselor of nearly every lumberman in the Northwest. With our profound respect for his ability and leadership, we will always think of him as a warm and genial friend whose rare sagacity and helpfulness were ever at the service of his fellows."

O. D. Fisher honored him in this way: "He will be remembered because he was George Long and not primarily because of any business affiliations." Mark Reed felt much the same: "Personally," he wrote to George, Jr., "I never knew a man that I have felt any closer to, and I am sure in my business associations that I have never been thrown with any one whose judgment I respected more than I did your father's. He was not only a very able man but a most honorable and lovable character. He will be greatly missed in the councils of the lumber industry of the Northwest; and personally I feel that I have lost one of my very best friends." Then he added a most insightful observation: "I have often marveled at the harmonious relationship that existed between the Weyerhaeuser Timber Company and the public. The Weyerhaeuser Timber Company is probably the most dominating industrial factor in the Northwest in that it is decidedly the important factor in the major industry of this territory, and yet it is seldom if ever that you hear a word of criticism or an unkind remark concerning the organization, from the public in general. That is indeed a splendid demonstration of wise, constructive management for which I feel that your large-minded father was mostly responsible."

On August 7, 1930, Bishop S. Arthur Huston of the Diocese of Olympia conducted services for George S. Long at the Buckley-King Funeral Home in Tacoma. Of the family's few requests, George, Jr., made one. He had read a poem by Christine Cronemiller, written in tribute to her father, Oregon forester George H. Cecil. George, Jr., thought it fit his own father as well, and Bishop Huston closed the service by reading, "He Chose One Tree."

> The Faller once again has passed this way,
> And on Life's Western slope, at close of day,
> He chose one tree.
>
> So straight it stood; so sound of heart; despite
> The Storms that buffeted, and sought by might
> To make it bend.

Long years it scattered on the needled sward
Its seeds of kindness. Pushing ever upward
To the light—and God.

And the friends of the forest, they sigh and sway,
Since the Faller of trees passed by this way.

There would be some—outsiders and latecomers—who might wonder that any lumberman could be counted among the friends of the forest. But those who had watched him labor through the years knew that George S. Long was, faithfully, just such a friend. The practice of forestry was admittedly still in its infancy, but that did not diminish the contribution of Weyerhaeuser's first general manager in preparing a fertile seedbed in which future foresters might work to good purpose. And in a few short years, on millions of acres, public and private, young trees in new forests reached skyward. The oft-predicted timber famine would have to wait.

Bibliographic Essay

AS WAS THE CASE WITH *PHIL WEYERHAEUSER: LUMBERMAN,*
the decision to forgo footnotes was a difficult one. The reason for
doing so, while doubtless insufficient to many, was similar: the great
majority of citations refer to a single source—the George S. Long
papers of the Weyerhaeuser Company Archives in Federal Way, Washington. This is a magnificent collection, and the essential value of the
book—in keeping with Professor Stratton's admonition as noted in
the Preface—is that it is the product of extensive research into original, archival sources.

Regarding those sources, Long's incoming correspondence is
arranged in the archives as one would expect, chronologically and
alphabetically. The outgoing correspondence, however, may surprise
some researchers. It is available in 136 bound volumes, each of approximately 1,000 pages. Better methods of record-keeping became available, but Long could be stubbornly old-fashioned on occasion, overly
committed to familiar practices. After all, if it worked, why change?
And so it was with his letters. Each was copied in the same old
manner: moisture and pressure combined to transfer just enough of
the ink from the original onto the thin rice paper. The copies haven't
become more legible with the passing years, but they are probably
now stabilized, mostly readable with a plain sheet placed underneath.
The work may strain the eyes, but it is nonetheless worth the effort. I
again assume, as with *Phil,* that serious students of the subject will
visit the Weyerhaeuser Archives, where there is available a lengthier,
fully footnoted manuscript of *George S. Long.*

I should also note a few important secondary sources. Anyone
who studies a subject touching on Weyerhaeuser will surely consult
Ralph W. Hidy et al., *Timber and Men: The Weyerhaeuser Story* (New

399

York: The Macmillan Company, 1963). Concerning the life of Frederick Weyerhaeuser, there is, amazingly, no single satisfactory source. Of value but not generally available is Frederick Edward Weyerhaeuser's tribute to his father, *The Record*. F.E.'s work was truly a labor of love and devotion, whose purpose was to record as completely as possible the accomplishments of father Frederick. The task proved far more demanding than F.E. had imagined, and it wasn't until November 21, 1934, the 100th anniversary of Frederick's birth, that F.E. mailed copies of *The Record* to his brothers and sisters. The letter which accompanied this distribution clearly stated his intentions: "Kindly remember, first, this is a record for our family and no one else." Although succeeding generations have become less protective, *The Record* remains something of a rare item.

Those desiring a description of Long's physical setting will be amply rewarded by reading Arthur R. Kruckeberg's recent study, *The Natural History of Puget Sound Country* (Seattle: University of Washington Press, 1991), coincidentally the first publication in the Weyerhaeuser Environmental Series. One of the most useful sources on all matters of forest history continues to be the *Encyclopedia of American Forest and Conservation History* (1983), edited by Richard C. Davis and published by the Forest History Society.

A special debt is owed to Robert E. Ficken for his two studies, *The Forested Land: A History of Lumber in Western Washington* (Seattle: University of Washington Press, 1987); and *Lumber and Politics: The Career of Mark E. Reed* (Seattle: University of Washington Press, 1979). There are several other books which relate to regional lumber activities, including Ellis Lucia's *The Big Woods* (Garden City, N.Y.: Doubleday and Company, 1975) and Donald McKay's *The Lumberjacks* (Toronto: McGraw-Hill Ryerson Limited, 1978).

Murray Morgan's writings give the reader a real feel for the region. Two of his contributions will provide a general background, *Puget's Sound: A Narrative of Early Tacoma and the Southern Sound* (Seattle: University of Washington Press, 1979); and *Skid Road: An Informal Portrait of Seattle* (New York: Viking Press, 1960; Seattle: University of Washington Press, 1982). More specific but also of interest is Morgan's *The Mill on the Boot: The Story of the St. Paul & Tacoma Lumber Company* (Seattle: University of Washington Press [published in cooperation with the Washington State Historical Society], 1982). For the best look at Seattle during the period, I recommend Richard C.

Berner's *Seattle 1900–1920: From Boomtown, Urban Turbulence, to Restoration* (Seattle: Charles Press, 1991).

Labor matters have received considerable attention over the years. The official 1919 *History of Spruce Production Division* is interesting but hard to find. See also Harold M. Hyman's *Soldiers and Spruce: Origin of the Loyal Legion of Loggers and Lumbermen* (Los Angeles: Institute of Industrial Relations, University of California, 1963). Vernon H. Jensen's *Lumber and Labor* (New York and Toronto: Farrar and Rinehart, 1945) has been a standard for years. The subject of labor has received such varied treatments as the following: Norman Clark's study of Everett, Washington, *Mill Town* (Seattle: University of Washington Press, 1970); Charlotte Todes, *Labor and Lumber* (New York: International Publishers, 1931); and the objective, well-written *Wobbly War: The Centralia Story* by John McClelland, Jr. (Tacoma: Washington State Historical Society, 1987).

John McClelland also contributed *R. A. Long's Planned City: The Story of Longview* (Longview, Washington: Longview Publishing Company, 1976), which I often consulted. This Long, invariably confused with George S., was the subject of a biography by Lenore K. Bradley, *Robert Alexander Long: A Lumberman of the Gilded Age* (Durham, North Carolina: Forest History Society, 1989). Other biographies which contain specific information include *James J. Hill & the Opening of the Northwest* by Albro Martin (St. Paul: Minnesota Historical Society Press, 1976); and Elmo Richardson's *David T. Mason: Forestry Advocate* (Santa Cruz, California: Forest History Society, 1983). And on the autobiographical side there is William B. Greeley's *Forests and Men* (Garden City, N.Y.: Doubleday and Company, 1951).

There are, of course, some in-house Weyerhaeuser publications which are occasionally of help. Included among these is *From Jamestown to Coffin Rock: A History of Weyerhaeuser Operations in Southwest Washington* (1974) by Alden H. Jones; and Harry J. Drew's study of the eastern Oregon region, *Weyerhaeuser Company: A History of People, Land and Growth* (1979); and *Men, Mills and Timber: Fifty Years of Progress in the Forest Industry* (1950).

Frequent reference has been made to *Phil Weyerhaeuser: Lumberman* (Seattle: University of Washington Press, 1985). I will only add here that when we conceived of this George S. Long project, it was envisioned as something of a companion piece to *Phil*. Weyerhaeuser Company is unusual insofar as it has had very few leaders over its nearly one hundred years. Together Long and Phil Weyerhaeuser provided that leadership over the company's first fifty-plus years. Although an ideal interaction of people and the environment has yet to be achieved, we are much closer to that end, thanks to the efforts of these two. We still need the products of the forest, even with all the substitutes that so worried George Long; and Long's organization continues to supply these needs through new forests on familiar lands. Nothing could have brought greater satisfaction to either George Long or Phil Weyerhaeuser.

Index